NEW LIGHT ON GALAXY EVOLUTION

INTERNATIONAL ASTRONOMICAL UNION

UNION ASTRONOMIQUE INTERNATIONALE

NEW LIGHT ON GALAXY EVOLUTION

PROCEEDINGS OF THE 171ST SYMPOSIUM OF THE
INTERNATIONAL ASTRONOMICAL UNION,
HELD IN HEIDELBERG, GERMANY,
JUNE 26–30, 1995

EDITED BY

RALF BENDER

*Universitäts-Sternwarte München,
Germany*

and

ROGER L. DAVIES

*University of Durham,
United Kingdom*

KLUWER ACADEMIC PUBLISHERS

DORDRECHT / BOSTON / LONDON

A C.I.P. Catalogue record for this book is available from the Library of Congress

ISBN-13: 978-0-7923-3976-2   e-ISBN-13: 978-94-009-0229-9
DOI: 10.1007/ 978-94-009-0229-9

*Published on behalf of
the International Astronomical Union
by
Kluwer Academic Publishers, P.O. Box 17, 3300 AA Dordrecht, The Netherlands.*

*Kluwer Academic Publishers incorporates
the publishing programmes of
D. Reidel, Martinus Nijhoff, Dr W. Junk and MTP Press.*

*Sold and distributed in the U.S.A. and Canada
by Kluwer Academic Publishers,
101 Philip Drive, Norwell, MA 02061, U.S.A.*

*In all other countries, sold and distributed
by Kluwer Academic Publishers Group,
P.O. Box 322, 3300 AH Dordrecht, The Netherlands.*

Cover picture:
From art. Dickinson and Steidel, page 295.

Printed on acid-free paper

Printed in the Netherlands

CONTENTS

Spiral Galaxies

Elliptical Galaxies

Small Stellar Systems and Galaxy Cores

The Interstellar Medium

Dark Matter Halos around Galaxies

Mergers in the Local Universe

High Redshift Galaxies

Theory of Galaxy Formation

Galaxies in Formation

POSTER PAPERS

x

PREFACE

The IAU Symposium 171 *New Light on Galaxy Evolution* was held at the Max Planck Haus in Heidelberg, Germany from 25th-30th June 1995. It was attended by 214 registered participants from 24 countries. The meeting was organised under the auspices of Sonderforschungsbereich 328 *The Evolution of Galaxies* and sponsored by Commisions 28 (*Galaxies*) & 33 *Structure and Dynamics of the Galactic System*) of the International Astronomical Union.

The seed of the idea for a wide ranging IAU Symposium on Galaxy Evolution germinated in late 1993. We had in mind a meeting that would provide a 1990s perspective on the topics covered at the influential meeting held in Santa Cruz in 1986 : *Nearly Normal Galaxies from the Planck time to the Present.* This agenda proved popular and we were fortunate to recruit an outstanding scientific organising committee for this project. You will find in these proceedings less emphasis than was given in Santa Cruz on the physics of the early universe and the connection to the large scale structure. This is not to deny the importance of these fields but simply a practical consequence of the limited time available, indeed both topics could be the subject of meetings in their own right. Conversely other topics, which have more recently come into prominence, have been emphasised, for example the controversy over the ages of elliptical galaxies and the impact of high resolution imaging with HST and spectroscopy with the Keck, on the study of intermediate redshift galaxies. An important development since the mid-80's is the role of high performance computing. This has made possible, for example, the inclusion of gas physics and stellar evolutionary effects in the cosmological simulations of galaxy evolution, as well as increasing the sophistication, and the ease of visualisation, of simulations of the dynamical interactions between galaxies. While starting from the broad topics covered in Santa Cruz you will find in these pages our own cocktail mixed from the currently topical ingredients of galaxy research.

The first two days of the programme concentrated on those clues to the principal processes occurring in galaxy evolution that emerge from the study of nearby galaxies. We began by considering the star formation history of spirals (starting with the Milky Way!). Panoramic near infrared imaging now offers the possibility of disentangling the effects of reddening from those of real stellar population changes in spirals. Discussion of stellar populations dominated the session on elliptical galaxies.

In the session on small stellar systems an attempt was made to account for the differences in the globular cluster populations of spiral and elliptical galaxies by forming ellipticals in mergers, turning on its head an argument regularly used to undermine the merger hypothesis. The HST data are close to forcing us to accept that black holes are common in the cores of giant galaxies, although the best case for a BH arises from the velocities of water masers in the core of NGC 4258. Studies of the interstellar medium in galaxies at X-ray wavelengths have now matured with the advent of the spectroscopic ASCA satellite and more surprises, such as the low iron abundance found in the IGM, can be expected. In the session on dark matter, while the halos around spirals were universally accepted, we heard that the challenge now is to detect halos around elliptical galaxies with the same confidence.

In the final session on nearby galaxies our physical understanding of the process and consequences of galaxy merging was dramatically illustrated. It appears that the global properties of galaxies can be accounted for and the effects of *galaxy harrassment* have been both simulated, and observed, in the Coma cluster.

In the last two days we moved to galaxies at high redshift. Discussions were dominated by the the spectacular HST images, both in pointed observations and the serendipitous Medium Deep Survey. Attempts to characterise the morphologies and global properties of galaxies in intermediate redshift clusters were reported.

Understanding the evolution of galaxies through cosmological simulations including gas physics, star formation and stellar evolution is just beginning and results from this work are likely to be highly influential for many years. Refreshingly, the development of analytical models of galaxy formation, will give us an alternative view of what-happens-when, as galaxies form. The final sessions covered new techniques to detect galaxies in formation and what we can infer about the physical conditions in the IGM at high redshift from quasars.

There is very little in astronomy that does not have some bearing upon the evolution of galaxies, after nine years our survey of the subject, extending from our local neighbourhood to early epochs, has revealed how new facilities and techniques, from large ground based telescopes to models of stellar atmospheres, have developed our understanding of galaxy formation. Looking at Jim Gunn's summary of the '86 conference he gives seven observational and and two theoretical tasks in his pet list. Of these we reckon good progress has been made on four or five of the former but perhaps only one of the latter remains relevant. How will we fare over the next decade? A two week conference will be essential by then.

We were fortunate indeed to be presented with a sunny and dry week in Heidelberg for IAU 171. The 21 reviews and 26 contributed papers were delivered in the auditorium of the Max Planck Haus, which was the scene of lively discussions both in the sessions and in the periods set aside to peruse the 137 posters. The week proved to be an intensive exploration of the subject of galaxy evolution with animated discussions being observed throughout the evenings by the casual Heidelberger strolling through the Old City. As a break from these rigors, the good weather enabled the particpants to enjoy a wonderful cruise up the Necker River and a hike to the castles at Neckarsteinach. Our thanks go to all that made this celebration of our subject so successful.

Roger Davies   Ralf Bender

ACKNOWLEDGEMENTS

The organization and success of this conference relied on contributions from a diverse band of cohorts, our thanks go to them all: the authors, the Scientific Organising Committee, the Local Organising Committee and the staff and students of the Landessternwarte and Max-Plank-Haus. Our special thanks go to Claus Möllenhoff who, as Chair of the LOC, carried the major burden of the work.

These proceedings were prepared at the University of Munich Observatory. We thank Jan Beuing, Dörte Mehlert, Sabine Grötsch, and Ilse Holzinger for their efficient help.

We gratefully acknowledge the financial support of the International Astronomical Union and the Sonderforschungsbereich 328 *The Evolution of Galaxies* of the University of Heidelberg, without which a meeting of this scale could not have been attempted. The conference was also sponsored by the Max-Planck-Gesellschaft and the Bezirkssparkasse, Heidelberg.

SCIENTIFIC ORGANISING COMMITTEE

R. L. Davies (Chairman), Durham University, U.K.
N. Arimoto, Institute of Astronomy, Tokyo, Japan
R. Bender, University of Munich Observatory, Germany
J. Binney, University of Oxford, U.K.
G. Bruzual, CIDA, Merida, Venezuela
R. Ellis, Institute of Astronomy, Cambridge, U.K.
S. Faber, UCO Lick Observatory, U.S.A.
M. Franx, Kapteyn Astronomical Institute, Netherlands
K. Freeman, Mount Stromlo Observatory, Australia
J. Frogel, Ohio State University, Columbus, Ohio, U.S.A.
O. Gerhard, University of Basel, Switzerland
R. Kennicutt, Steward Observatory, U.S.A.
J. Kormendy, University of Hawaii, U.S.A.
A. Renzini, University of Bologna, Italy
S. White, Max-Planck-Institute for Astrophysics, Germany

LOCAL ORGANISING COMMITTEE

C. Möllenhoff (Chairman), Landessternwarte Heidelberg
E. Bär, Landessternwarte Heidelberg
J. Heidt, Landessternwarte Heidelberg
G. Ramge, Landessternwarte Heidelberg
H.-J. Röser, Max-Planck-Institute for Astronomy, Heidelberg
C. Scorza, Landessternwarte Heidelberg
W. Tscharnuter, Institute for Theoretical Astrophysics, Heidelberg
S. von Linden, Landessternwarte Heidelberg

SPIRAL GALAXIES

THE EVOLUTIONARY HISTORY OF THE MILKY WAY

K.C. FREEMAN
Mount Stromlo and Siding Spring Observatories
The Australian National University
Canberra, AUSTRALIA

Abstract. The accretion of small satellite galaxies appears to have been important in the formation of the metal-poor halo of the Galaxy. The disrupting Sgr dwarf galaxy and the recent discovery of a young, metal-poor component of the halo indicate that this is a continuing process. The evolution of the galactic disk, and some consequences of the bar-like nature of the galactic bulge are briefly discussed.

1. Introduction

The Milky Way is a large disk galaxy. Its main components are the rapidly rotating thin and thick disks, the very slowly rotating metal-poor halo, the bulge, and the dark corona. For the early evolutionary history, we must look to the kinematics and chemical properties of the old components that are well represented near the sun: these are the metal-poor halo and the thick disk. The thin disk gives information about the later evolutionary history. The bulge and the dark corona are less well understood and do not yet provide much useful insight.

The metal-poor halo is supported by its velocity dispersion. Eggen *et. al.* (1962: ELS) proposed that it formed before the disk, through the rapid collapse of part of the protogalaxy. This was based on evidence that the orbital and chemical properties of halo stars are correlated, with the most metal-poor stars having the largest orbital eccentricities, the largest range of velocities perpendicular to the galactic plane, and the lowest systemic rotation. Their galaxy formation picture suggests a dissipative time sequence of increasing chemical abundance, decreasing random velocities and increasing ordered rotation; the stellar abundances and kinematics are

3

R. Bender and R. L. Davies (eds.), New Light on Galaxy Evolution, 3–10.
© *1996 IAU. Printed in the Netherlands.*

then measures of the time elapsed since the beginning of the collapse. The ELS picture was the standard galaxy formation picture during the 1960s and 1970s, but two aspects were contentious: (i) The timescale for the rapid collapse, required to produce the high observed orbital eccentricities of the halo stars, was given by ELS as 2×10^8 years. So any indication of Gyr age spreads among halo objects was taken as evidence against the ELS picture. (ii) the progressive chemical enrichment of the halo implies a correlation of kinematics and metallicity among the halo objects. As more data became available for halo stars and clusters, it became clear that this correlation is weak or absent.

During the 1970s, an alternative view of halo formation developed. To quote Toomre (1977): "It seems almost inconceivable that there wasn't a great deal of merging of sizeable bits and pieces (including quite a few lesser galaxies) early in the career of every major galaxy, no matter what it now looks like. The process would obviously have yielded halos from the stars already born, whereas any leftover gas would have settled quickly into new disks embedded within such piles of stars". This view fits well with the hierarchical galaxy formation picture that is now widely accepted. At about the same time, Searle and Zinn (1978: SZ) showed that the chemical and dynamical properties of the outer halo globular clusters are decoupled, and independently proposed that the halo formed by the accretion of small metal-poor fragments or satellites. If the halo formed from accreted fragments, then the kinematical and chemical properties of the halo objects depend on the orbital and chemical properties and structure of the infalling fragments. There would be no reason to expect a tight correlation of dynamical and chemical properties. After some controversy, this is the currently favored picture.

The galactic disk is believed to be easily heated by the accretion of small fragments. The Galaxy has a substantial thin disk, so most of these accretion events must have occurred early, while the disk was mainly gaseous and could resettle. However, recent unpublished work by several groups (Athanassoula, Barnes, Walker, White and their associates) on accretion of satellites by disk galaxies with live halos shows that disks are more robust than was previously believed.

2. The Metal-Poor Population: Halo and Thick Disk

This population is the source of the clues that led to the ELS and SZ pictures of galaxy formation. A comment: Metal-poor stars were long believed to be very old. In the SZ picture, age and abundance do not necessarily go together: *eg.* we see star-forming dwarf galaxies at the present time with [Fe/H] values as low as -2.

Beers and Sommer-Larsen (1995) compiled velocity and abundance data for 1936 metal-poor ([Fe/H] < −0.6) stars, selected by their chemical properties. Their analysis provides a useful summary of present views about the properties of the metal-poor halo and thick disk of our Galaxy.

- in the abundance range [Fe/H]= −0.6 to −1.0, we see the rapidly rotating thick disk population. Its mean rotation $V_{\rm rot} = 190$ km s^{-1} and its (U, V, W) velocity dispersion components (in the usual sense) are $\sigma = (63, 42, 38)$ km s^{-1}. Compare this with $V_{\rm rot} = 210$ km s^{-1} and $\sigma \approx (40, 30, 20)$ km s^{-1} for the old thin disk. These parameters for the thick disk agree well with other estimates from the solar neighborhood and *in situ*. From these parameters, the inferred scale length h_R and scale height h_z of the thick disk are 4.7 kpc (similar to the thin disk) and 1 kpc (300 pc for the thin disk) respectively.

- for [Fe/H] < −1.0, we see two components:
(1) the hot halo with $V_{\rm rot} \approx 0$ and $\sigma \approx (150, 100, 100)$ km s^{-1}, indicating radially elongated orbits. Its density is about 0.1 to 0.2 percent of the thin disk density near the sun. This halo shows no dependence of kinematics on abundance. The halo does show an apparent age gradient: the globular clusters and BHB stars (Zinn 1980, Preston *et. al.* 1991) indicate that the outer regions of the halo are younger by a few Gyr in the mean.
(2) the thick disk, persisting on to [Fe/H] $\approx$ −2. In the abundance range [Fe/H]= −1.0 to −1.5, about 60% of the stars near the sun belong kinematically to this metal-poor tail of the thick disk. (For [Fe/H] < −1.6, the fraction is about 30%). Is this metal-poor thick disk the heated remnant of early disk star formation ? Or is it the debris of accreted metal-poor satellites whose orbits have been more or less circularized by dynamical friction (*eg.* Quinn *et. al.* 1993) ? In any case the discovery of this metal-poor thick disk population is a significant development (Norris *et. al.* 1985; Morrison *et. al.* 1990), and it now seems clear that low metallicity does not necessarily identify a star as extremely old, nor as a member of the metal-poor halo.

Near the galactic plane, the mean rotation $V_{\rm rot} \approx 0$ for the metal-poor halo, but is probably retrograde at larger heights z from the plane. From proper motions of faint stars towards the NGP, Majewski (1992) finds retrograde rotation $V_{\rm rot} \approx -55 \pm 16$ km s^{-1} for a wide range of z. Beers and Sommer-Larsen (1995) come to a similar conclusion for $z > 2$ kpc, as do Carney *et. al.* (1995) for stars with maximum orbital heights $z_{\rm max} > 5$ kpc. The presence of this slowly rotating or retrograde halo population in an otherwise rapidly rotating Galaxy requires some explanation. It is usually interpreted now in terms of the accretion of small fragments during or after the dissipative settling of the disk.

The distribution of RR Lyrae stars (Hartwick 1987, Layden 1995) and the BHB stars (Kinman *et. al.* 1994) indicate that the metal-poor halo itself may have two components, one of which is quite flattened. For example, the BHB stars appear to show a spherical component with density distribution $\propto r^{-3.5}$ and local density of 6 stars kpc^{-3}, and a flatter component with a scale height of about 2 kpc and local density of 24 stars kpc^{-3}.

2.1. STELLAR ORBITS IN THE OUTER HALO

Sommer-Larsen *et. al.* (1994) compiled radial velocities for halo BHB stars with galactocentric distances R_G between 5 and 55 kpc, and estimated the run of the radial and tangential components $\sigma_r, \sigma_t(R_G)$ of the velocity dispersion. The velocity ellipsoid in the halo is radially anisotropic near the sun but becomes tangentially anisotropic for $R_G > 15$ kpc. This would not be expected in a monolithic collapse picture but is qualitatively consistent with a halo built up by accretion.

In the accretion picture, the shape of the halo and the shape of the velocity ellipsoid $\sigma(R_G)$ reflect the properties of the fragments and their trajectories. The dominant processes leading to the breakup of fragments are dynamical friction and tidal destruction, which depends on the mean density $\bar{\rho}$ of the fragment. For example, the mean galactic density ρ_G within R_G is $2\,[R_G(\text{kpc})]^{-2}\,M_\odot\,\text{pc}^{-3}$, so $\rho_G\,(10\text{ kpc}) \approx 0.02\,M_\odot\,\text{pc}^{-3}$. We would then expect a typical fragment like a dIrr or dSph galaxy with $\bar{\rho} \approx 0.02$ to $0.05\,M_\odot\,\text{pc}^{-3}$ to break up near the solar radius. More substantial lumps like the Dra and UMi dSph galaxies with $\bar{\rho} \sim 1\,M_\odot\,\text{pc}^{-3}$ would survive in to smaller R_G. In this way, the kinematical properties of the halo can provide useful constraints on the properties of the small scale fluctuations represented by the fragments. High resolution simulations of galaxy formation from fluctuation spectra would be very interesting, to see if the radial variation of the halo shape and the shape of the velocity ellipsoid can be understood.

3. Direct Evidence for Past and Ongoing Accretion

The positions and kinematics of various dwarf spheroidal galaxies and younger globular clusters define galactic streams (Lynden-Bell and Lynden-Bell 1995, Majewski 1994) which are candidates for the debris of disrupted fragments. Similarly, the stellar moving groups in the halo (*eg.* Eggen 1979, Majewski *et. al.* 1994) may come from disrupted fragments or globular clusters. The Sagittarius dwarf (Ibata *et. al.* 1994) is an example of a large dSph galaxy which is now tidally disrupting. This dwarf contains a significant intermediate age population, as well as an underlying old population.

The four globular clusters associated with the Sgr dwarf include both old and younger clusters (Da Costa and Armandroff 1995), so it will contribute both old and younger clusters to the galactic globular cluster system.

Rodgers et. al. (1981) and Lance (1988) discuss the young metal-rich main sequence A stars found up to 11 kpc from the galactic plane. This is an unusual population: at the SGP its velocity dispersion $\sigma_W = 62$ km s^{-1}, while similar stars near $l = \pm 90°, b = -45°$ have a line of sight velocity dispersion of 40 km s^{-1} and $V_{\rm rot} = 210$ km s^{-1}. They argue that the formation of these stars is associated with the accretion of somewhat metal-poor gas, perhaps from a dwarf galaxy (or a high velocity cloud).

Preston et. al. (1994) discovered a population of blue metal-poor (BMP) main sequence stars near the solar circle. These stars have [Fe/H] < -1 and are bluer (ie. younger) than the old turnoff stars of the halo. The kinematics of the BMP stars are particularly interesting. Their velocity dispersion is about 90 km s^{-1} and isotropic, and their $V_{\rm rot} = 128 \pm 30$ km s^{-1}. (For an old halo sample with similar abundances, we would expect a lower $V_{\rm rot} \approx 55$ km s^{-1}). These early-type metal-poor stars appear to be kinematically intermediate between the rapidly rotating disk and the slowly rotating halo. What are they ? Their ages are > 3 Gyr and their abundances have [Fe/H] < -1. Where would such stars form ? Preston et. al. note that nearby satellites like the Carina dSph galaxy have major intermediate-age metal-poor components, and suggest that the galactic BMP stars may come from similar accreted dwarf galaxies, with the BMP stars representing the blue tips of the accreted populations. They estimate that the accreted population contributes about 10% of the local halo density, consistent with a similar limit by Unavane et. al. (1995) on the density of accreted objects, with a total accreted mass of about $10^8 \, M_\odot$ (ie. several dSph galaxies).

4. The Disk of the Galaxy

Like the disks of most other spirals, the old disk of the Galaxy appears to have settled to a double exponential distribution: $\rho(R, z) \propto \exp(-R/h_R - z/h_z)$. The reason is not fully understood, but the radial structure may have to do with the dynamics of star-forming viscous disks which settle to an exponential in R if the star formation and viscous timescales are comparable but longer than the dynamical timescale (eg. Lin and Pringle 1987). For exponential disks in which the anisotropy σ_z/σ_R is constant with R, we would expect $\sigma_R \propto \exp(-R/2h_R)$. This is seen in our Galaxy (Lewis and Freeman 1989) and is common in others (eg. Bottema 1993). In our Galaxy, the velocity dispersion in the inner disk is almost as high as it is in the bulge (about 110 km s^{-1}); the source of the heating (spiral waves, bar heating, scattering by giant molecular clouds) is not well understood.

¿From the white dwarf luminosity function and nucleochronology, the age of the galactic disk near the sun is about 10×10^9 years. The disk shows dynamical evolution. This is well seen in the sample of F stars by Edvardsson *et. al.* (1993), with accurate ages, velocities and abundances (see also Freeman 1991). Important results from this study include:

- the run of stellar velocity against age shows the appearance of the old hotter thick disk at an age of about 12×10^9 years, and also the secular heating of the disk, particularly for stars younger than about 3 Gyr.
- the [Fe/H] − age plane shows a large spread of abundance among the disk stars of a given age. This indicates inhomogeneous enrichment; the amplitude of the scatter is comparable to that seen among the HII regions in the disks of other galaxies (*eg.* Zaritsky 1992).

5. More on the Thick Disk

For our Galaxy, the mass of the thick disk is about 10% of the thin disk mass. However, the thick disk is not an essential feature of galaxy formation and evolution: many disk galaxies, particularly those with weaker bulges, do not show thick disks. Gilmore *et. al.* (1995) studied galactic thick disk stars up to about 3 kpc from the galactic plane, and found no abundance gradient. This argues against dissipational settling as the formation process. Quinn *et. al.* (1993) made simulations of satellite accretion by disk galaxies, and showed how the disk is heated and the debris of the satellite is dispersed throughout the disk. The dynamical properties of the heated disk near the solar radius ($\sigma, V_{\rm rot}, h_R, h_z$) are very similar to those observed for the galactic thick disk. We note again that the heating must have occurred early in the life of the galactic disk, while the disk was still mainly gaseous and could resettle to a thin disk after the accretion events.

6. The Galactic Bulge

de Vaucouleurs (1964) pointed out that our Galaxy has an inner bar structure, and classified it as SAB(rs)bc. The evidence for an inner bar/bulge is now strong, and includes:

- the asymmetry in l of the DIRBE galactic light distribution and the brightness of bulge IRAS sources, Mira variables and clump giants,
- the l−velocity distribution of molecular gas, and
- the large optical depth (3×10^{-6}) to microlensing in bulge fields.

See Gerhard (1995) for references. The effects of this bar/bulge on the evolution of the Galaxy need to be fully considered, and include:

- bar-driven gas infall and chemical enrichment of the inner bulge (note that only the inner parts of the bulge are metal-rich),

- effects on the chemical evolution of the disk via bar-induced gas flows,
- the dynamical heating of the inner stellar disk by the bar.

Bars can form from disks via dynamical instabilities. This makes sense in the galactic context: we note that (i) the scale height of the DIRBE bar/bulge is about 300 pc, similar to that of the old disk, and (ii) the chemical abundance of the bulge at about 1 kpc from the plane is similar to that of the thick disk. The stars of the bar/bulge are mostly old, but this does not mean that the *structure* of the bar/bulge is so old: it may well be relatively recent. Bars can self-destruct by developing a central mass of only a few percent (*eg.* Norman *et. al.* 1995). This will naturally happen if gas is present. The central mass destroys the bar-forming orbits, and the bar rapidly dissolves to a more axisymmetric structure. So it is possible that our galactic bar/bulge may be relatively young and impermanent if current ideas on the dynamics of bars are correct.

References

Beers, T., Sommer-Larsen, J. 1995. Astrophys.J.Suppl., 96, 175.
Bottema, R. 1993. Astron.Astrophys., , 275, 16.
Carney, B. *et. al.* 1995. Personal communication.
Da Costa, G., Armandroff, T. 1995. Astron.J., 109, 2533.
de Vaucouleurs, G. 1964. In "The Galaxy and the Magellanic Clouds" (Austr. Acad. Sci),
      ed. F. Kerr and A. Rodgers, p 195.
Edvardsson, B. *et. al.* 1993. Astron.Astrophys., 275, 101.
Eggen, O.J., Lynden-Bell, D., Sandage, A. 1962. Astrophys.J., 136, 748.
Eggen, O.J. 1979. Astrophys.J., 229, 158.
Freeman, K. 1991. In "Dynamics of Disc Galaxies" (Göteborg), ed B. Sundelius, p 15.
Gerhard, O. 1995. In "Unsolved Problems of the Milky Way" (Kluwer), ed L. Blitz, in
      press.
Gilmore, G., Wyse, R., Jones, B. 1985. Astron.J., 109, 1095.
Hartwick, F.D.A. 1986. in "The Galaxy" (NATO), ed G. Gilmore, R. Carswell, p 281.
Ibata, R., Gilmore, G., Irwin, M. 1994. Nature, 370, 194.
Kinman, T, Suntzeff, N., Kraft, R. 1994. Preprint.
Lance, C. 1988. Astrophys.J., 334, 927.
Layden, A. 1995. Preprint.
Lewis, J., Freeman, K. 1989. Astron.J., 97, 139.
Lin, D., Pringle, J. 1987. Astrophys.J., 320, L87.
Lynden-Bell, D., Lynden-Bell, R. 1995. Mon.Not.R.Astron.Soc., 275, 429.
Majewski, S. 1992. Astrophys.J.Suppl., 78 87.
Majewski, S. 1994. Astrophys.J., 431, L17.
Majewski, S., Munn, J., Hawley, S. 1994. Astrophys.J., 427, L37.
Morrison, H.L., Flynn, C., Freeman, K.C. 1990. Astron.J., 100, 1191.
Norman, C., Sellwood, J., Hasan, H. 1995. Preprint.
Norris, J., Bessell, M., Pickles, A. 1985. Astrophys.J.Suppl., 58, 463.
Preston, G., Shectman, S., Beers, T. 1991. Astrophys.J., 375, 121.
Preston, G., Beers, T., Shectman, S. 1994. Astron.J., 108, 538.
Quinn, P., Hernquist, L., Fullagar, D. 1993. Astrophys.J., 403, 74.
Rodgers, A., Harding, P., Sadler, E. 1981. Astrophys.J., 244, 912.
Searle, L., Zinn, R. 1978. Astrophys.J., 225, 357.
Sommer-Larsen, J., Flynn, C., Christensen, P.R. 1994, preprint.

Toomre, A. 1977. In "The Evolution of Galaxies and Stellar Populations" (Yale University Observatory), ed B. Tinsley and R. Larson, p 420.

Unavane, M., Wyse, R., Gilmore, G. 1995. Preprint.

Zaritsky, D. 1992. Astrophys.J., 390, L73.

Zinn, R. 1980. Astrophys.J., 241, 602.

Discussion

Gerhard: In the dynamical simulations, the bending instability taking disk stars out of the plane often occurs shortly after the bar instability in the disk. If the latter is caused by gas infall and cooling of the disk, one would then expect significant star formation to occur shortly before the bulge.

Minniti: From all that stellar evidence, can you tell what kind of satellites were accreted to form the halo ?

Freeman: The total mass of all fragments accreted into the halo should not exceed about $1 \times 10^9 \ M_\odot$ (the mass of the metal-poor halo). There should be many of them (because the rotation of the halo is low) and their masses should be relatively low (because the chemical abundance of the halo is low). The presence of the BMP stars suggests that accretion of several dSph galaxies with intermediate-age populations has also occurred more recently. And we see the Sgr dSph now in the process of being accreted.

Rix: You mentioned repeatedly the "intermediate age" populations of dSph galaxies. Has there been any progress in understanding how the galaxies managed to hold (or re-acquire) gas for a second star formation episode ?

Freeman: It will be interesting to learn if the dwarf irregular galaxies in the Local Group have had similar distinct episodes of star formation.

Harris: The supposed age gradient in the galactic halo might really be a result of an increased *range* in ages among the outer halo objects. Some new HST photometry of the outermost halo globular clusters ($R_G = 100$ kpc) shows that one or two of them are just as old as the inner halo.

Kormendy: If you plot the density profile of the bulge and halo as a function of radius, is there a clearcut transition or break between them ? Do you think that the bulge and halo are physically related ? If there is a transition between bulge and halo, at what radius does it occur ?

Freeman: The structure of the inner halo is not yet well enough known to make the test with the density profile that you suggest. I think the bulge and halo are probably not physically related, because the kinematics of the bulge (rotation at about 100 km s^{-1}) and the halo (low rotation, even in the inner parts) are so different.

ASSEMBLING SPIRAL GALAXIES

R.C. KENNICUTT
Steward Observatory
University of Arizona
Tucson, AZ 85721 USA

1. Introduction

Nearby spiral galaxies offer vital clues to some of the most fundamental questions about galaxy formation and evolution: What is the star formation history of the universe, past and future? When did disks form, during the final stages of a single primeval collapse, or as a continuous or episodic process? What is the evolutionary nature of the Hubble sequence, and what are the physical mechanisms that dictate the present-day Hubble type of a galaxy? Was Hubble type imprinted at birth, or can it be deterred or at least modified by infall, mergers, or secular dynamical evolution within the galaxy? These issues are not specific to spirals, of course, and much of this conference will address just these questions in a broader context. However present-day spirals offer unique advantages for studying these problems; they exhibit a broad range of dynamical and evolutionary properties, and the dynamical fragility of disks makes them excellent seismometers of galaxy interaction and merger rates at recent epochs.

2. Stellar Birthrate Histories of Disks

The only system for which detailed information is available is the Galactic disk (solar neighborhood). Several stellar age indicators are available, including isochrone fitting of stars near the main sequence turnoff, chromospheric dating of late-type stars, and cooling ages of white dwarfs. Although some details remain uncertain, such as the detailed time dependence of the stellar birthrate, the role of starbursts, and the age of the oldest disk stars, the observations are consistent with a roughly constant star formation rate (SFR) over the history of the disk, with the long-term SFR not varying by more than factors of 2–3. A convenient way to parameterize this history is

11

R. Bender and R. L. Davies (eds.), New Light on Galaxy Evolution, 11–18.
© *1996 IAU. Printed in the Netherlands.*

in terms of the ratio of the current disk SFR to the average past rate, which I denote as b following Scalo (1986). The observations of the Galactic disk are then consistent with $b \simeq 1 \pm 0.5$ (Scalo 1986; Noh & Scalo 1990).

Several methods have been applied to measure b for large samples of spirals, including modelling of integrated colors in the visible and ultraviolet, and use of integrated Hα emission-line fluxes to measure the SFRs directly (see Kennicutt 1992 for references). Kennicutt *et al.* (1994) have used Hα and broadband photometry of 210 Sa–Irr galaxies to estimate b for the spiral disks, corrected for the effects of bulge contamination. The star formation history shows a strong systematic dependence on type, with $b \sim 1$ in the disks of late-type galaxies (Sbc and later), decreasing to $b < 0.05$ in Sa disks. This means that the pronounced changes in the integrated spectra and stellar populations of spirals along the Hubble sequence are due overwhelmingly to changes in the *disk* star formation histories, with changes to the bulge/disk ratio being much less important. For example between types Sa and Sc the fractional contribution of young stars to the integrated disk population changes by at least a factor of 20, while the disk luminosity fraction changes by only a factor of two (Kennicutt *et al.* 1994). Bulges may play a pivotal role in causing these changes to disk populations, but even if the bulges were invisible there would be a strong trend in integrated spectra with spiral type.

The radial dependence of the star formation history within disks can provide important clues to the physical origin of the type-dependent trends. For example most formation models predict a pronounced "inside-out" evolution due to the higher densities and shorter dynamical times at small radii (e.g., Larson 1972). The advent of wide-field CCD and infrared array cameras now makes it possible to address this problem quantitatively. De Jong (1995) has used $BVRIHK$ imaging of 86 face-on spirals to study the radial color distributions. The spirals show a trend toward bluer colors with later type as expected, but most of the disks also show significant color gradients, with bluer colors at larger radii. Similar color gradients have been observed by Terndrop *et al.* (1994) and Peletier *et al.* (1995).

Several effects can produce the color gradients, including a radial change in the star formation history, a reddening gradient, or an abundance gradient. Models by de Jong suggest that all three effects are important, and this makes it difficult to accurately measure the magnitude of the variation in b. However an independent study by Ryder & Dopita (1994) offers strong evidence for radial birthrate gradients, based on a comparison of Hα and I-band scale lengths for 30 southern spirals. They find that the young disks (Hα) are considerably more extended, typically by $\sim$50% as compared to I. Taken together these results suggest that b increases by factors of up to a few over the radial extent of the star forming disks.

3. The Conventional Model: Closed-Box Evolution

The birthrate histories implied by these data are usually interpreted in the context of a passive closed-box model, in which the stellar birthrate is assumed to decline slowly (usually exponentially) with time. This implicitly assumes that most disks formed at early epochs, and that the differences in subsequent evolution along the Hubble sequence are due to physical differences in the properties of the proto-disks (e.g., Sandage 1986).

The simplest version of this picture treats disks as systems with identical ages and IMFs, varying only in the e-folding time of the SFR. Such models have been used to synthesize the colors, Hα emission, and integrated spectra of nearby galaxies. The model is remarkably successful at reproducing the spectra of galaxies over a wavelength range extending from the ionizing UV to the near-infrared (e.g., Kennicutt 1983; Buat *et al.* 1987; Bruzual & Charlot 1993). The same model predicts a relatively modest evolution in the photometric properties of most L^* spirals with lookback time, which appears consistent with faint galaxy surveys and Mg II absorber statistics.

Given these successes why should one be dissatisfied with the closed-box picture at all? Perhaps the most disconcerting problem is the lack of any physical basis for why the disk star formation histories should vary so strongly with parent galaxy type. One normally thinks of the SFR being driven by the local gas surface density (e.g., Kennicutt 1989), but if density controlled the evolutionary time scale we would expect to observe very strong radial trends in b and only weak trends with galaxy type, whereas the opposite trends are observed. Closed-box models also have well-known difficulties in reproducing the observed metal abundance distributions and radial abundance gradients in disks (e.g., Zaritsky 1995).

Hence we are forced to conclude that either the closed-box assumption is invalid (see next section), or that a physical mechanism other than the Schmidt gas density law is responsible for the strong changes in star formation timescale along the Hubble sequence. Early-type spirals possess large central bulges, and the bulges might influence the disk evolution in several ways: by increasing the angular rotation speeds in the inner disks and stimulating star formation; by raising the threshold density for star formation at later epochs; or by inducing strong spiral density waves at early times. Alternatively the evolutionary differences may be connected to the systematic trend toward higher galaxy mass toward early types. Or it is possible that some spirals evolve dynamically from late to early types, for instance by forming bulges by disk heating after the formation of a massive bar (Pfenniger & Norman 1990; Friedli & Martinet 1993). Some combination of these mechanisms might account for the evolutionary changes of disks along the Hubble sequence without having to abandon the closed-system assumption.

4. Revisionist Interpretations: Infall, Captures, Mergers

In a hierarchical model of galaxy formation one would not expect galaxies to evolve as closed systems at all, and indeed it is conceivable that the present-day Hubble type of a spiral could be largely dictated by its disk accretion or merger history. Here we discuss models for accretion-driven spiral galaxy evolution, including infall models, in which the accretion of gas is continuous, and merger models, in which disks grow by the capture of discrete clumps of gas and stars.

4.1. INFALL-DOMINATED EVOLUTION?

Infall can solve several of the problems alluded to earlier, extending the star formation timescales of late-type galaxies and more readily explaining the abundance distributions and gradients in disks. However recent observations appear to limit the amount of infall to rates which are well below the present-day SFRs in massive spirals. HI measurements reveal no large reservoirs of cold gas (Bothun 1985), and the inflow rate from high-velocity clouds in the Galaxy are several times smaller than that needed to sustain the current SFR (Mirabel & Morris 1990). Perhaps the strongest limit comes from the soft X-ray luminosities of spirals (White & Frenk 1991). *ROSAT* observations of nearby spirals yield diffuse X-ray luminosities that are several times smaller than the values expected from a cooling flow if the infall rate were equal to the gas depletion rate from star formation. Hence an extreme spherical infall model appears to be ruled out. However some fraction of the star formation could be fueled by infall or by a radial inflow of HI from the outer disk.

4.2. MERGER-DOMINATED EVOLUTION?

These constraints can be avoided if the accretion of gas takes place in discrete events, such as by the merger of a spiral with a companion or satellite galaxy. Indeed such hierarchical growth is a fundamental consequence of many galaxy formation models (e.g., White & Frenk 1991; Kauffmann & White 1993). A serious problem with this picture however is the observed thinness and dynamical coldness of disks. Numerical simulations show that much of the dynamical friction loss in a merger is converted into thermal energy of the disk (Quinn *et al.* 1993). Tóth & Ostriker (1992) have estimated that no more than 4% of the mass of the disk within the solar radius could have been accreted in the last 5 Gyr. These results appear to rule out recent mergers of *stellar* systems (e.g., satellite galaxies) as the predominant mechanism for assembling disks, but these limits may not apply to the accretion of largely gaseous (dissipative) material.

4.3. OBSERVATIONAL EVIDENCE FOR ACCRETION EVENTS

A very exciting development in this field has been the discovery of numerous examples of disks with distinct kinematic subsystems in stars and/or gas, which must represent the products of at least two discrete accretion events. These observations offer direct evidence that captures of galaxies or gas clouds play some role in disk evolution, though as discussed later the role of such mergers at recent epochs is probably a minor one.

Polar Ring Galaxies: Polar ring galaxies represent the best known and perhaps the best studied examples of multiple kinematic subsystems in spirals. Whitmore *et al.* (1990) catalog the ∼70 known systems and provide a bibliography to the extensive literature on these objects. They used these data to estimate that roughly 5% of S0 galaxies are estimated to have had a polar ring over their lifetime.

Star–Star Counterrotation: Multiple kinematic components in a coplanar disk are more difficult to detect, but two spectacular instances of such disks are now known. The disk of the edge-on S0 galaxy NGC 4550 is comprised of two counterrotating components with nearly identical masses and scale lengths (Rubin *et al.* 1992; Rix *et al.* 1992). Both disks are cold with dispersions of 45 and 60 km s^{-1}. In the normal Sb galaxy NGC 7217 20–30% of the disk stars are in a retrograde cold disk (Merrifield & Kuijken 1994). Counterrotating stellar subsystems are seen frequently in elliptical galaxies, and are readily explained as resulting from stellar mergers or gas infall. NGC 4550 and NGC 7217 show that it is possible to form a stable counterrotating disk without violating the disk heating constraint (Sellwood & Merritt 1993). Such systems may form from two distinct generations of disk accretion and star formation, or by an extended period of infall in which the angular momentum of the accreted material undergoes a rapid change (Quinn & Binney 1992; Merrifield & Kuijken 1994).

Star–Gas Counterrotation: The most frequently observed instances of counterrotation involve gas disks counterrotating with respect to the stars. A compilation of S0 galaxies by Bertola *et al.* (1992) lists 9 clear examples of gas-stellar counterrotation. In most instances the gas is confined to the nuclear regions, and the mass involved is small, of order 10^6 $M_\odot$ or less. However in at least two galaxies, NGC 4526 (Bettoni *et al.* 1991) and NGC 3626 (Cirri *et al.* 1995) the counterrotation extends over the entire disk and the gas masses are of order $10^8 - 10^9$ $M_\odot$. These systems most likely originate from the capture of a gas-rich companion in a retrograde orbit.

Gas–Gas Counterrotation: NGC 4826 (M64) represents a rare example of a galaxy in which the rotation of the gas disk reverses itself, at a radius of ∼1 kpc, coincident with its prominent dust lane (Braun *et al.* 1992). The inner system, probably the remnant of the original gas disk, corotates with the stars. Outside of 1 kpc the gas is in retrograde rotation (Rubin 1994; Rix *et al.* 1995). The gas motions in the transition region are complex, with much of the gas moving at less than the circular velocity. However the stellar disk exhibits a normal exponential profile, a dynamically cold disk, and no discernable counterrotation (Walterbos *et al.* 1994; Rix *et al.* 1995). As with the systems described above NGC 4826 is most likely the result of the capture of a gas-rich companion. The absence of retrograde stars and the cold stellar disk appear to rule out a significant mass in stars in the captured object, but the presence of strong [OIII], [NII], and [SII] forbidden lines in the outer disk suggest a significant degree of chemical enrichment (and hence star formation) in the captured object.

Infalling Gas Clouds Evidence for infalling gas in nearby spirals comes from a variety of sources. High-velocity HI clouds with masses of up to 2×10^8 $M_\odot$ have been detected in M101, NGC 628, and NGC 6946 (Kamphuis 1993). High-resolution HI observations also reveal large-scale kinematic structures in the form of massive arms and arcs of gas that may still be settling into disks, as in IC 10 (Hodge *et al.* 1994). See Sancisi's review in this volume for a more extensive discussion of this subject. Infalling ionized gas has been observed in the early-type spirals NGC 4258 (Rubin & Graham 1990) and NGC 4826 (Rubin 1994). Dynamical modelling of mergers suggests that a steady infall of gas can persist for billions of years after a merger (Hibbard & Mihos 1995).

4.4. IMPLICATIONS FOR DISK FORMATION AND EVOLUTION

The many examples described above indicate that episodic accretion does play a significant role in disk evolution, and that significant accretion can take place without disrupting or overheating the disks. However these types of subsystems probably comprise no more than a few percent of the present-day mass of disks. Rix and Kuijken (private communications) have completed surveys for stellar counterrotation of the type seen in NGC 4550 for ~30–40 S0 and spiral galaxies and have failed to detect counterrotation in any, implying upper limits of a few percent on the fraction of retrograde stars. Similar fractions were derived from polar rings by Whitmore *et al.* (1990). Counterrotating gas disks may be more common, but the masses involved are rarely more than a few percent of the total disk mass. These numbers are reminiscent of the limits based on disk heating derived for the Galactic disk by Tóth & Ostriker (1992).

This result may have important implications for current models of galaxy formation. In an $\Omega = 1$ CDM model most L^* galaxies experience major halo mergers over the past few Gyr (Kauffmann & White 1993), and even if only a fraction of these events lead to stellar/gas disk mergers the predictions would appear to be in conflict with the observations described above. On the other hand in an $\Omega = 0.2$ model most of the galaxy assembling occurs at early cosmological epochs, and the fraction of recent mergers, of order a few tens of percent or less, is in reasonable accord with the observations of disks. While hardly conclusive this example illustrates the importance of disk galaxies in constraining the merger rate and the nature of the galaxy formation paradigm.

I am very grateful to several colleagues for stimulating discussions and for sharing new results prior to discussion, especially Marc Balcells, Robert Braun, Roelof de Jong, John Hibbard, Konrad Kuijken, Reynier Peletier, Hans-Walter Rix, Renzo Sancisi, Peter Tamblyn, Chris Taylor, Thijs van der Hulst, Jacqueline van Gorkom, Rene Walterbos, Simon White, and Eric Wilcots. My work was supported in part by the U.S. National Science Foundation through grants AST-9019150 and AST-9421145.

References

Bertola, F., Buson, L. M., & Zeilinger, W. W. 1992, ApJ, 401, L79
Bettoni, D., Galletta, G., & Oosterloo, T. 1991, MNRAS, 248, 544
Bothun, G. D. 1985, AJ, 90, 1982
Braun, R., Walterbos, R. A. M., & Kennicutt, R. C. 1992, Nature, 360, 442
Bruzual, G., & Charlot, S. 1993, ApJ, 405, 538
Buat, V., Donas, J., & Deharveng, J. M. 1987, A&A, 185, 33
Cirri, R., Bettoni, D., & Galletta, G. 1995, Nature, in press
de Jong, R. 1995, Ph.D. thesis, Univ. of Groningen
Friedli, D., & Martinet, L. 1993, A&A, 277, 27
Hibbard, J. E., & Mihos, J. C. 1995, AJ, 110, 140
Hodge, P., Wilcots, E., & Miller, B. 1994, BAAS, 26, 1436
Kamphuis, J. 1993, Ph.D. thesis, Univ. of Groningen
Kauffmann, G., & White, S. D. M. 1993, MNRAS, 261, 921
Kennicutt, R. C. 1983, ApJ, 272, 54
Kennicutt, R. C. 1989, ApJ, 344, 685
Kennicutt, R. C. 1992, in Star Formation in Stellar Systems, eds. G. Tenorio-Tagle, M. Prieto, & F. Sánchez (Cambridge: Cambridge Univ. Press), p191
Kennicutt, R. C., Tamblyn, P., & Congdon, C. W. 1994, ApJ, 435, 22 (KTC)
Larson, R. B. 1972, Nature, 236, 21
Merrifield, M. R., & Kuijken, K. 1994, ApJ, 432, 575
Mirabel, F., & Morris, M. 1990, ApJ, 356, 130
Noh, H.-R., & Scalo, J. M. 1990, ApJ, 352, 605
Peletier, R. F., Valentijn, E. A., Moorwood, A. F. M., Freudling, W., Knapen, J. H., Beckman, J. E. 1995, A&A, in press
Pfenniger, D., & Norman, C. 1990, ApJ, 363, 391
Quinn, P. J., Hernquist, L, & Fullagar, D. P. 1993, ApJ, 403, 74
Quinn, T., & Binney, J. 1992, MNRAS, 255, 729
Rix, H.-W., Franx, M., Fisher, D., & Illingworth, G. 1992, ApJ, 400, L5
Rix, H.-W., Kennicutt, R. C., Braun, R., & Walterbos, R. A. M. 1995, Ap.ʹ
Rubin, V. C. 1994, AJ, 107, 173
Rubin, V. C., & Graham, J. A. 1990, ApJ, L5

Rubin, V., Graham, J., & Kenney, J. 1992, ApJ, 394, L9
Ryder, S. D., & Dopita, M. A. 1994, ApJ, 430, 142
Sandage, A. 1986, A&A, 161, 89
Scalo, J. M. 1986, Fund. Cos. Phys., 5, 287
Sellwood, J. A., & Merritt, D. 1994, ApJ, 425, 530
Terndrop, D. M., Davies, R. L., Frogel, J. A., DePoy, D. L., & Wells, L. A. 1994, ApJ, 432, 518
Tóth, G., & Ostriker, J. P. 1992, ApJ, 389, 5
Walterbos, R. A. M., Braun, R., & Kennicutt, R. C. 1994, AJ, 107, 184
White, S. D. M., & Frenk, C. S. 1991, ApJ, 379, 52
Whitmore, B. C., Lucas, R. A., McElroy, D. B., Steiman-Cameron, T. Y., Sackett, P. D., & Olling, R. P. 1990, AJ, 100, 1489
Zaritsky, D. 1995, ApJ, 448, L17

Discussion

McCall: The Hubble sequence is essentially a disk sequence because in the Revised Hubble System disk properties are favoured over bulge/disk ratio. At a given Hubble stage, the bulge/disk ratio can vary systematically. Is there any evidence for a variation of the star formation rate (characterized by b) with the bulge/disk ratio at a given Hubble stage?

Kennicutt: This is difficult to test, because of the scarcity of good bulge/disk photometry for a large enough sample. I have looked a correlations of SFR with Yerkes type, which classifies on the basis of bulge/disk ratio, and the scatter is similar to that seen when the SFR is correlated with RSA or RC2 type.

Minniti: What fraction of these galaxies with counter-rotating disks are barred?

Kennicutt: Counterrotation is seen in barred and non-barred galaxies, but the number of known systems is too small to test whether bars are more or less common in counterrotating systems. For nearly edge-on systems such as NGC 4550 it is difficult to tell if a bar is present.

Athanassoula: It is possible to produce a retrograde component without invoking a merger. Namely, if a bar decayed and turned into an axisymmetric component (e.g., because of an increase in the central mass concentration) then a fraction of the orbits which were ergodic before the decay will become retrograde.

Freeman: I heard secondhand that Hernquist et al. have modelled accretion events on disk galaxies with live halos and this makes the disks more robust. Does anyone know about this?

Athanassoula: The simulations of Quinn et al. have a rigid halo. I have run a series of simulations in which the halo, as well as the disk and companion, are live. In some of these simulations the companion loses a substantial fraction of its mass while it is still in the halo. This depends on how dense the companion is.

THE STELLAR POPULATION OF THE GALACTIC BULGE

R. MICHAEL RICH
Dept. of Astronomy, Columbia University
538 W. 120th St., NY NY 10027

How old is the bulge? Are bulges in general as old as halos? Do bulges form rapidly or slowly? Are they formed from disks via dynamical instabilities, or perhaps by starbursts? In general, do luminous spheroids form at the same time as the oldest stars?

For populations older than 10 Gyr, it is hard even to constrain relative ages (Renzini, 1992). And while there is much current excitement about the prospect of direct lookback studies, imaging of galaxies at a particular redshift give us no information about the evolutionary path of a particular galaxy. These studies tell us about the general state of galaxies at a given epoch and environment, and as the distances become greater, a sample more extreme in properties (activity, intrinsic luminosity) is likely to be selected. While some phases of galaxy formation may be luminous and easily visible (formation of a $2 \times 10^{10} M_\odot$ bulge in 10^8 yr) important evolutionary phases might be less readily observed (formation of the first halo stars, or the thick disk). The relative age dating of the oldest stars is still an important project, even with successful direct imaging of high redshift galaxies.

1. The Galactic Bulge

The bulge of the Galaxy is that stellar population which dominates the inner kpc, with approximately Solar metallicity and age > 5 Gyr. The baryonic mass of the bulge is $\approx$ 10 times that of the stellar halo and $\approx$ 30% that of the disk. If proto-galaxies and an era of galaxy formation is discovered, we might expect the formation of bulges such as that of our galaxy, at 10-100 $M_\odot yr^{-1}$, to be among the more easily observable phases of galaxy formation.

The age of the bulge has been the subject of much recent debate. Observationally, there are a number of reasons to be skeptical that the bulge is as old as the extreme halo. Rich (1992) has emphasized that even if

19

R. Bender and R. L. Davies (eds.), New Light on Galaxy Evolution, 19–22.
© 1996 IAU. *Printed in the Netherlands.*

one assumes that all the stars > 1 mag brighter than the globular cluster giant branch tips in Frogel & Whitford (1987) can be igored, the large numbers of long period Miras cannot be overlooked. Since Feast's (1963) classic study, it has been known that Miras with $P > 300$ days in the Solar neighborhood have disk kinematics, whereas shorter period Miras look like halo members. Whitelock *et al.* (1991) report a substantial number of Miras with $P > 300$ days in bulge fields 1 kpc from the nucleus, and Glass *et al.* (1995) find approximate half the Miras in the low-latitude Sagittarius field have long periods. Metallicity differences cannot explain this, since the mean metallicity of the bulge is approximately that of the Solar neighborhood (McWilliam & Rich, 1984). Finally, Holtzmann's (1993) *HST* main sequence photometry favors an age < 10 Gyr for the Galactic bulge.

There are other reasons to suspect that the Galactic bulge and perhaps bulges in general may form late or undergo extended periods of star formation. The inner bulge is a bar (Blitz &Spergel, 1991) and has clearly formed as a result of highly dissipative processes. Local group galaxies all have extended giant branches in their spheroids (e.g. Rich & Mighell, 1995). A rapid, violent formation history should leave a clear chemical imprint on the bulge (enhancing the alpha-capture elements, especially oxygen), but no clear evidence of this is found. While McWilliam & Rich (1994) find Mg and Ti enhanced in bulge giants, Ca and Si are found in their Solar abundance ratios. And for bulge giants covering the entire range of Fe metallicity, s-process elements are found in Solar abundance ratios. Since s-process elements are made on the AGB, it is a mystery how an early, violent starburst could leave such a signature. Massive stars are expected to have made large amounts of oxygen. In a careful re-analysis of their data, McWilliam, Tomaney, & Rich (1995) rule out a very high oxygen abundance for the bulge. The chemical evidence (with the current small sample sizes) favors a slower, more extended formation scenario.

1.1. BULGE FORMATION SCENARIOS

A number of plausible scenarios for the formation of the Galactic bulge have been discussed.

Classical: The bulge collapsed on a free-fall timescale. Since $t_{ff} \sim \rho^{-1/2}$, the densest part of the spheroid would be expected to collapse first. Renzini (1992) has emphatically pointed out the logic of this hypothesis and Lee (1992) argues that RR Lyraes in the Galactic bulge are the oldest stars in the Galaxy.

Bulge From Disk: A massive starburst, happening at any time during the life of a galaxy, might have formed a massive inner disk. This disk is unstable to bar modes, which can vertically thicken into a bulge (Pfenniger

& Norman, 1990; Merrit & Sellwood, 1994). This idea explains the high density of the bulge and the peanut/bar morphology observed for many galaxies. The longevity of the bar may be a problem.

Bulge from Halo: The metal poor halo could have formed early on, perhaps in an early era of "frustrated" star formation. The hot 10^7K gas could cool and dissipate to the center, resulting in either an extended or starburst formation scenario (Larson, 1975; Wyse & Gilmore, 1992).

Bulge from Mergers: The Sgr I dwarf is observed to be in an encounter with the Milky Way (Ibata *et al.* 1995); were there others? If so, such galaxies must not have been rich in carbon stars, since virtually no thermally pulsing carbon stars are known in the bulge (Blanco, 1988).

In principle, measurements of age and age dispersion can distinguish among these scenarios, but differences between a 10 and 14 Gyr old metal rich population are subtle.

2. Ages of Metal Rich Globular Clusters and the Galactic Bulge

Direct measurement of the age of the bulge from turnoff photometry is complicated by reddening, dispersions in age and metallicity, and the depth of the bar, problems that affect both ground and space based photometry. An approach which addresses this problem (and in which I am involved) is to determine secure ages for the metal rich Galactic center globular clusters and to estimate their ages by the parameter ΔV_{TO}^{HB} (Ortolani *et al.* 1995). The bulge can be compared with the clusters by tying both luminosity functions at the red clump, removing reddening and distance uncertainties. The metal rich clusters have the same ΔV_{TO}^{HB} as the old halo clusters (Figure 1). Luminosity functions of the bulge field Baade's Window (500 pc S of the nucleus) and the clusters agree almost perfectly near the turnoff, admitting virtually no age spread in the bulge field population. It is important to repeat this experiment at various locations in the bulge, particularly at lower latitudes—note that Catchpole *et al.* (1990) see a concentration of luminous AGB stars toward the nucleus, with a clear excess in the inner 140 pc.

An absolute age measurement of the bulge is within sight. At the time of this writing, deep *HST* imaging of the old globular cluster 47 Tuc has been obtained (GO6114, A. Renzini PI). Reaching deep enough to solidly measure the white dwarfs, the distance modulus will be obtained by comparison with the white dwarf cooling sequence rather than with the main sequence. Since Ortolani *et al.* (1995) show that the loci of the metal rich clusters overlay those of 47 Tuc nearly exactly, this approach will indirectly give an absolute age for the Galactic bulge. If the initial findings of Ortolani *et al.* (1995) are confirmed and the bulge is found to be old, then it will

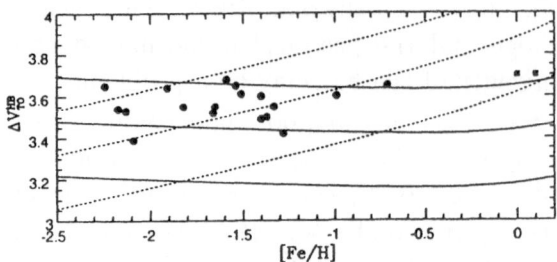

Figure 1.   Metal rich globular clusters are probably as old as the oldest halo globular clusters (Ortolani *et al.* 1995). The luminosity difference between the HB and turnoff ΔV_{TO}^{HB} is plotted for a representative set of halo globular clusters (filled circles) and for the metal rich clusters NGC 6528 and NGC 6553 (filled squares). The lines refer to ages of 18, 15, and 12 Gyr (upper, middle, and lower lines) for HB luminosity-metallicity relationship of Sandage &Cacciari, 1990 (solid lines) and Walker, 1992 (dotted lines); see Ortolani *et al.* (1995) for additional discussion.

mean that the Mira period distribution and absolute luminosities will be of little use in the relative age dating of stellar populations, a fact that will complicate our extension of an "age ladder" into the Local Group spheroids and beyond.

References

Blanco, V.M. 1988, AJ, **95**, 1400
Blitz, L., &Spergel, D.N. 1991, ApJ, **379**, 631
Catchpole, R.M., Whitelock, P.A., & Glass, I.S. (1990): MNRAS **247**, 479
Feast, M.W. (1963): MRAS **125**, 367
Frogel, J.A., & Whitford, A.E. (1987): ApJ **320**, 199
Glass, I.S., Whitelock, P.A., Catchpole, R.M., & Feast, M.W. 1995, MNRAS, **273**, 383
Holtzmann, J.A., Light, R.M., & WFPC-IDT (1993): AJ **106**, 1826
Ibata, R.A., Gilmore, G., & Irwin, M.J. 1994, Nat, **370**, 194
Larson, R.B. 1975, MNRAS **173**, 671
Lee, Y.-W. (1992): AJ **104**, 1780
McWilliam, A., & Rich, R.M. (1994): ApJS **91**, 749
McWilliam, A., Tomaney, A., & Rich, R.M. (1995): in preparation
Merritt, D., & Sellwood, J.A. (1994): ApJ **425** 551
Ortolani, S., Renzini, A., Gilmozzi, R., Marconi, G., Barbuy, B., Bica, E., & Rich, R.M. (1995): Nat (in press)
Pfenniger, D., & Norman, C. (1990): ApJ **363**, 391
Renzini, A. (1992): *Stellar Populations* (eds.) B. Barbuy· & A. Renzini (Kluwer, Dordrecht), 325
Renzini, A. (1994): A&A **285**, L5
Rich, R.M. (1992): *The Center, Bulge, and Disk of the Milky Way* (ed.) L. Blitz (Kluwer, Dordrecht) 47
Rich, R.M., Mould, J.R., & Graham, J.R. (1993): AJ **106**, 2252
Rich, R.M., & Mighell, K.J. (1995): ApJ **439**, l45
Sandage, A.R., &Cacciari, C. (1990): ApJ **350**, 645
Walker, A.R. (1992): ApJ **390**, L81-L84
Whitelock, P., Feast, M. & Catchpole, R. (1991): MNRAS **248**, 276
Wyse, R.F.G., & Gilmore, G. 1992, AJ, **104**, 144

CHEMO-DYNAMICAL MODELS AND THE STAR FORMATION HISTORY OF GALAXIES

M. SAMLAND, G. HENSLER
Institut für Astronomie und Astrophysik
Universität Kiel, D-24098 Kiel, Germany

1. Introduction

Sandage (1986) showed what the star formation rate (SFR) of galaxies of different Hubble type might look like. His curves are based on the study of Gallagher et al. (1984), who determined the SFR at three different epochs of galactic evolution. Sandage's figure establishes a connection of SFR and Hubble type but, as was already mentioned by Sandage, it "contains no physics". In order to explore the background of this connection, however, it is necessary either to improve the observations or to model the evolution of galaxies self-consistently. However, the results of modelling the SFR are not reliable, if simplified models are used, which only describe some structural properties of galaxies. It is necessary to apply self-consistent models which take into consideration all relevant interaction processes between the gaseous and stellar components of a galaxy. Such models can be checked by comparison with observations like density and abundance distributions, star-gas content, velocities, velocity dispersions, mass-luminosity relations and age distributions of stars. A detailed model will show, whether the initial conditions, the feedback mechanisms during the evolution or the environment determine the evolution of a galaxy.

2. Chemo-dynamical Models: A self-consistent way of discribing the evolution of galaxies

Galaxies consist of complex structures and processes like heating and cooling, dissipation and supernova (SN) are important. In addition, the dynamics of stars and gas influence the evolution. We have developed a 2-d code which describes the dynamics of the stellar and gaseous components, and includes interaction processes like heating and cooling, evaporation and condensation, SNI and SNII, planetary nebulae, dissipation due to cloud-cloud

R. Bender and R. L. Davies (eds.), New Light on Galaxy Evolution, 23–27.
© 1996 *IAU. Printed in the Netherlands.*

collisions, drag, star formation (SF), element synthesis and gravitation, all of which influence the evolution of galaxies. The program named CoDEx (**Chemo-Dynamical Evolution of Galaxies**) is described in detail in Samland (1994), Samland et al. (1995) and Samland & Hensler (1995).

As a first application for the chemo-dynamical approach we modelled the evolution of a galaxy, the mass (M) and angular momentum (L) of which is comparable to the Milky Way. If we take into account the observational error margins, our model explains the abundance distributions of nitrogen, oxygen and iron for the stars as well as for the interstellar medium in the different parts of the galaxy. Furthermore, the velocity dispersions of bulge, disk and halo stars can be reproduced. Since the spatial resolution in the model amounts to only 200pc, it is not possible to distinguish between a thick and a thin disk component. Average properties of the disk stars, like the anisotropy of the velocity dispersion, are, however, in agreement with the observations. Detailed results of that model are published in Samland (1994), Samland et al. (1995) and Samland & Hensler (1995).

3. The Star Formation History of Galaxies

Our calculations set in at the time when a gaseous protogalaxy starts cooling and clouds of 10^5 to 10^7 M$_\odot$ are forming. Due to dissipation of kinetic energy the protogalaxy collapses. As feedback mechanisms like SN heating dominate, the collapse time is about a factor 10 longer than the dynamical timescale. The multi-phase character of the ISM also prolongs the collapse time, because hot gas, heated by SNe is driven out of the star forming regions into the halo.

In our models the halo forms early during a period of $2 - 3 \cdot 10^9$ years, while bulges are on average $2 \cdot 10^9$ years younger with a formation time of about $5 \cdot 10^9$ years. The disk is the youngest component of a rotating galaxy. While the inner disk forms at the same time as the bulge, the outer disk formation extends over more than 10^{10} years. The details of the formation process, in particular the evolution of densities, velocity dispersions, abundances and SFRs are described in Samland (1994) and Samland et al. (1995). It is likely that total mass (Tully et al. 1982), angular momentum (Sandage et al. 1970) and the initial density distribution (Gott & Thuan 1976) determine the SF history of a galaxy. Because of the complex character of galaxy formation it is, however, not possible to estimate a priori the influence of these quantities on the SFR. Only chemo-dynamical models enable ones to simulate the evolution of different types of galaxies and thus to investigate what determines the global SFR. A number of different models with masses of $0.2 - 5 \cdot 10^{11}$M$_\odot$ and different L have been calculated. For rotating galaxies, independent of their specific angular momentum, we derive the relation

$$\frac{SFR_{max}}{M_\odot yr^{-1}} = 16.6 \pm 5.7 \cdot \left(\frac{M}{10^{11}M_\odot}\right)^{1.5}$$

For the non-rotating galaxies the dependence remains the same, except that the coefficient 16.6 has to be doubled. Fig. 1 shows the total SFR of four models differing by the total M, total L and specific angular momentum L/M given in the table.

The SF histories of non-rotating galaxies differ from other galaxies because they consume most of their gas during a collapse phase of $2 - 4 \cdot 10^9$ years, so that their later gas content is provided by the stellar mass return. If galaxies are rotating, the collapse is less violent, leading to a lower peak SF (SFR_{max}) and an increasing gas consumption time. Rotation, however, changes the consumption time and the SFR_{max} only by a factor 2, which is a minor effect compared to the mass dependency of SFR_{max}. We found no correlation between the Hubble type of a disk galaxy and the total L or the specific angular momentum. This is confirmed by Miyamoto et al. (1980), who determined the L and M of disk and irregular galaxies from observational data. In general the late-type galaxies tend to have lower M than early-type galaxies (Roberts & Haynes 1994). The question has to be addressed which additional parameters determine the galaxy type, because galaxies with equal M but of different type are observed. As a further example, in Fig.1 we also plotted the SFR of a galaxy (straight line) with $2.2 \cdot 10^{11}M_\odot$ and low specific angular momentum. Contrary to the other galaxies, this model retains a nearly constant SFR for a long time. Evidently, the initial density of this galaxy had been lower and in addition the density distribution was flat.

At the moment, the number of simulations is still too small to conclude how the initial density distribution affects the evolution of galaxies, but we are sure that the precursors of early-type galaxies must have had higher densities in the centre than the ones of the late types. This is consistent with the model of Sofue and Habe (1992) and the observations by Dressler (1979) according to which the environment influences the formation of these different types of protogalaxies and therefore also the galaxies. M of a galaxy also determines the SF history. In Fig.2 we show, how the SFR and the M of two different galaxies evolve in comparison with observational data. The plot clearly reveals the different phases of the galactic evolution. At the beginning, M and SFRs of both galaxies increase due to the collapse. The collapse phase ends when the kinetic energy input of the SNII is of the order of the dissipation rate and when the inflow of cloudy medium is balanced by the outflow of hot gas. Only 10% of this hot gas is ejected by the SNII, while the rest is contributed by evaporating clouds. The further

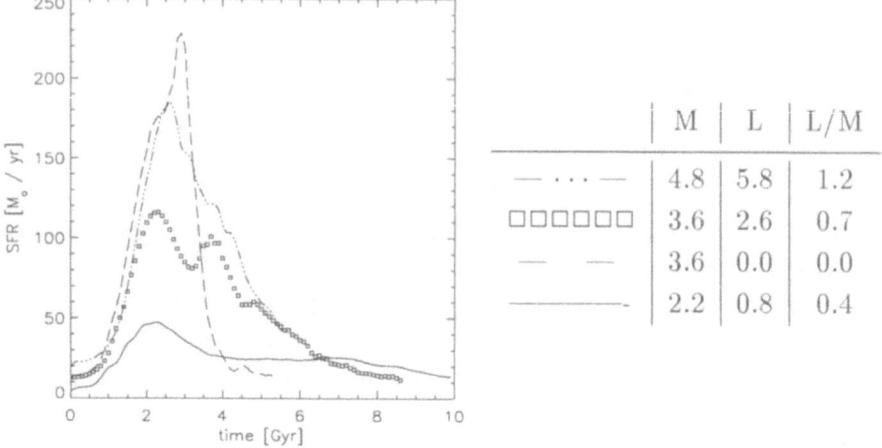

Figure 1. Fig.1: Evolution of the SFR for 4 different galaxies. Mass $[10^{11}M_{\odot}]$, angular momentum $[10^{11}M_{\odot}pc^2yr^{-1}]$ and specific angular momentum $[pc^2yr^{-1}]$ of the models are shown by the table.

evolution depends on the total M. The more massive a galaxy is, the better it can keep the hot gas bound. Therefore the SFR drops with constant M because of gas consumption. The less violent the collapse of a galaxy is, the longer the galaxy needs to convert the clouds into stars. Low-mass galaxies $(M < 10^{11}M_{\odot})$, on the other hand, expel some amount of hot gas by a SN-driven galactic wind. In Fig.2 it can be recognized that M of such a low-mass galaxy decreases during the evolution. It is remarkable that after attaining the maximum the SFR of the low-mass galaxy decreases by one order of magnitude within $5 \cdot 10^9$ years, while the more massive galaxy requires $12 \cdot 10^9$ years, although the gas fraction of the low-mass galaxy is higher. In order to compare the calculated evolution tracks with the observations, one has to take into account that the errors of the determined SFR and M for individual galaxies are still high and that the data (Hunter & Gallagher, 1986) are biased towards high SFRs (Some of their objects are classified as starburst galaxies.). Keeping this in mind, we can conclude that the results of the chemo-dynamical models are in good agreement with the observations.

4. Conclusion

One of the most important results of the chemo-dynamical approach is that most galaxies evolve as self-regulated systems. We find that the evolution of a galaxy depends strongly on the initial M and the density distribution of the protogalaxy. Compared to the M, L plays a minor rôle for the evolution. The simulations predict that early-type galaxies convert most of their

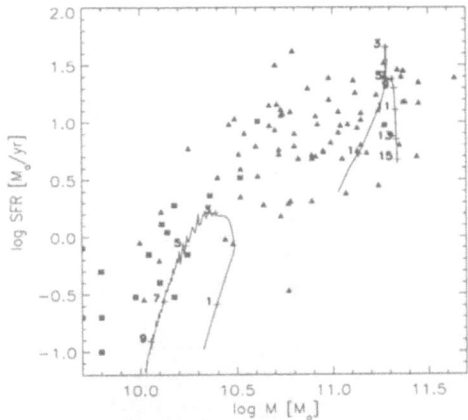

Figure 2. Fig.2: M and SFR of two models (lines) in comparison with observational data (squares and triangles). The numbers in the plot indicate the age of the galaxies in units of 10^9 years. The squares are data from Hunter & Gallagher (1986) and the triangles are galaxies from the IRAS bright galaxies catalogue (Condon et al. 1990). Their SFRs are calculated using the formula of Thronson & Telesco (1986) and their total M are taken from Tully (1988).

gas during an initial collapse phase, which is of the order of $3 \cdot 10^9$ years long, and that the late type galaxies convert more gas after the collapse phase. The initial density distribution proposed by Gott & Thuan (1976) determines the evolution. The precursors of early-type galaxies have higher densities in the centre than the ones of the late types. The evolution of low-mass galaxies is different because they lose mass by galactic winds, what affects the SF history. The chemo-dynamical models show that the SFR of galaxies cannot be described by simple formulae, but in general the results confirm Sandage's ideas concerning the SF history of galaxies.

References

Condon, J.J., Helou, G., Sanders, D.B., Soifer, B.T., 1990, ApJS 73, 359
Dressler, A., 1979, ApJ 231,659
Gallagher, J.S., Hunter, D.A., Tutukov, A.V., 1984, ApJ 284, 544
Gott, J.R., Thuan, T.X., 1976, ApJ 204, 649
Hunter, D.A., Gallagher, J.S., 1986, PASP 98, 5
Miyamoto, M., Satoh, C., Ohashi, M., 1980, Astrophys. Space Sci. 67, 147
Roberts, M.S., Haynes, M.P., 1994, ARAA 32,115
Samland, M., 1994, Dissertation, University of Kiel, Germany
Samland, M., Hensler, G., 1995, IAU-Symp. 169, in press
Samland, M., Hensler, G., Theis, C., 1995, ApJ submitted
Sandage, A., Freeman, K., Stokes, N., 1970, ApJ 160, 831
Sandage, A., 1986, A&A 161, 89
Sofue, Y., Habe, A., 1992, PASJ 44,325
Thronson, H.A.J., Telesco, C.M., 1986, ApJ 311,98
Tully, R.B., Mould, J.R., Aaronson, M., 1982, ApJ 257, 527
Tully, R.B., 1988, *Nearby Galaxies Catalog*, Cambridge University Press

Roland Wielen and Ken Freeman

AGES OF GALAXY BULGES AND DISKS FROM OPTICAL AND NEAR-INFRARED COLOURS

R. F. PELETIER
Kapteyn Astronomical Institute
Postbus 800, 9700 AV Groningen, Netherlands
and
Instituto de Astrofísica de Canarias
Via Lactea s/n, 38200 La Laguna, Tenerife, Spain

AND

M. BALCELLS
Kapteyn Astronomical Institute
Postbus 800, 9700 AV Groningen, Netherlands

Abstract. For a sample of bright nearby early-type galaxies we have obtained surface photometry in bands ranging from U to K. Since the galaxies have inclinations larger than 50° it is easy to separate bulges and disks. By measuring the colours in special regions, we minimize the effects of extinction, and by looking at $B - K$ colour gradients we can show that for these type of spirals the colours mainly give information about stellar populations, and not extinction. We find that the differences between bulges and disks in all colours is very small, and using simple population models we can show that on the average the age difference between the bulge and the disk at 2 scale length is smaller than 30%, and much smaller if part of the difference is caused by a gradient in metallicity.

1. Introduction

Age differences between bulges and disks can provide important basic checks on theories of the assembly of disk galaxies. What do galaxy colours tell us about the age of formation of bulges and disks? According to the classical concept of Population I and II disks should be Population I and young,

R. Bender and R. L. Davies (eds.), New Light on Galaxy Evolution, 29–33.
© 1996 *IAU. Printed in the Netherlands.*

while bulges are named Population II and old. Galaxy photographs give the same picture. To address this question, external galaxies might give better clues than our Galaxy, due to the large obscuration towards the Bulge. The drawback is still that one has to analyze integrated colours or spectra.

Here we discuss optical and near-infrared colours. As has been shown by e.g. Frogel (1985) age and metallicity can be disentangled by using a blue optical colour index as a measure of recent star formation, and a near-infrared colour index as a measure of metallicity. Bothun & Gregg (1990) applied this method to S0 galaxies. They found that for a given $J - K$ disks are generally bluer by ~0.4 mag in $B - H$ than bulges, corresponding to an age difference of more than 5 Gyr. Their infrared measurements however are from off-nuclear aperture photometry, which is extremely complicated, and can produce large errors. Here we use near-infrared images, combined with optical CCD-data, allowing us to obtain high-quality measurements. We present some colour-colour relations, and concentrate on the determination of ages. More details are given in Balcells & Peletier (1994, **BP**) and Peletier & Balcells (1995).

2. The sample, observations and the determination of colours

The sample discussed here is a diameter-limited sample of inclined galaxies (i $\geq$ 50°) of type S0 - Sbc. Apart from 2 objects it is the same as described in BP, where the optical data for the bulges are presented. The sample was observed at K at UKIRT with IRCAM3, equipped with a 256 × 256 InSb array. 20 of the 30 galaxies were also observed at J. Mosaics were made with a total size of 100" × 100", a pixel size of 0.291" and an effective seeing between 0.8" and 1".

Colour profiles were obtained in wedges centered on the center of the galaxy in K. For the bulges these wedges were centered on the minor axis, with a width of 22.5°. For the disks, at 10° wide wedge was taken 15° away from the major axis. In both cases, colours were obtained on the non-dusty side, this way minimizing the effects of extinction. Given the orientation of these galaxies, it happens that the bulge profiles, except sometimes very close to the center, are really dustfree, and only suffer from disk light behind the bulge. Given the fact that the surface brightness of bulges is generally much larger than in disks, this does not affect the colours. Some disk-profiles, even on the non-dusty side, are still rather dusty. However, scale length ratios between B and K in most of our galaxies are close to 1, and only the ones in some Sb's and Sbc's are similar to the ratios in the galaxies of Peletier *et al.* (1995), so that we can still presume that in general the effects of extinction are unimportant.

3. Colours of bulges and disks

Galaxy disks and bulges have negative colour gradients, i.e. colours become bluer radially outward (de Jong 1995; BP) , but the gradients are small enough that we could assign representative values for the colour of each component. For bulges, we have taken the colour at $0.5 \times r_{eff}$ or at 5 arcsec, whichever is larger. For disks we use the colours at 2 major axis scale lengths. We find that disk colours, while somewhat bluer, are very similar to bulge colours for all the galaxies (see Fig. 1). Here the diagonal line indicates the locus where both colours are equal. The average differences between disk and bulge and bulge colours are 0.126 ± 0.165 for $U - R$, 0.045 ± 0.097 for $B - R$, 0.078 ± 0.165 for $R - K$ and 0.016 ± 0.087 for $J - K$. Unless there is a conspiracy between metallicity and age, this diagram implies that age differences between bulges and disks are small. This information however is only useful if we also know the ages themselves. We can obtain this information from Fig. 2, where we plot $R - K$ of the bulges vs. $U - R$, analogous to Bothun & Gregg (1990). Also plotted are single age - single metallicity models by Vazdekis et al. (1995) for 17, 12, 8, 4 and 1 Gyr. We estimate the systematic error in the model in the colours to be ~ 0.2 mag. Comparison of the data with the models shows that there is a large range in ages between the bulges. Note also the scale. The range in colours between the galaxies is much larger than the average colour difference between bulge and disk of the same galaxy.

Using the same models, we now quantify the colour differences between bulges and disks in terms of age differences (in case of constant metallicity), or metallicity differences (in case of constant age). Since it is likely, from e.g. abundance measurements in HII regions, that there are metallicity gradients, the real situation will probably be somewhere in between. For an average galaxy, we find that the difference in log Z will be ~ 0.12, or ~ 0.11 in log(t). The numbers imply that a bulge is at most 30% older than its disk, or much less, if metallicity gradients are taken into account.

4. Discussion

The data of this paper put the strong constraint that disks of early-type spirals must have been together with, or just after the bulge. It does not endanger the traditional model of Eggen, Lynden-Bell & Sandage (1963) of galaxy formation, but on the other hand implies that it is very unlikely that there were three discrete events of massive star formation: for the bulge, the disk, and the thick disk. The properties of bulge and inner disk seem to merge smoothly in terms of for example kinematical properties and kind of stars. The continuous infall model of Gunn (1982), where the age of the stars is determined by the free-fall collapse time of the infalling gas, would

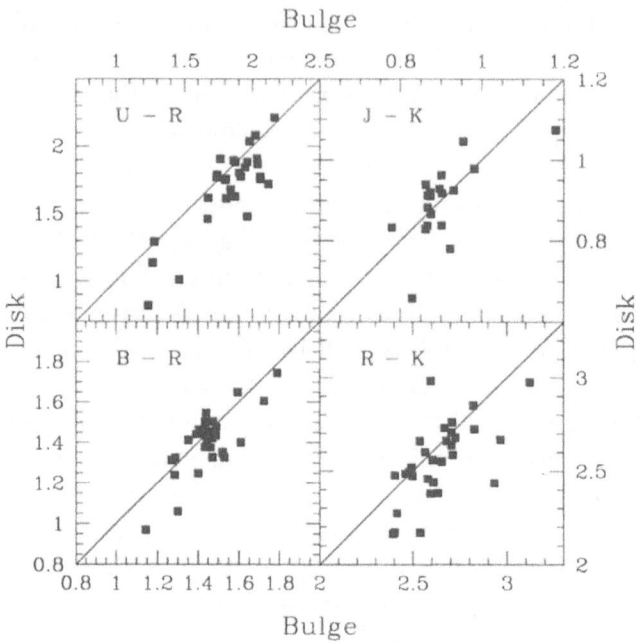

Figure 1.   Disk colours as a function of bulge colours

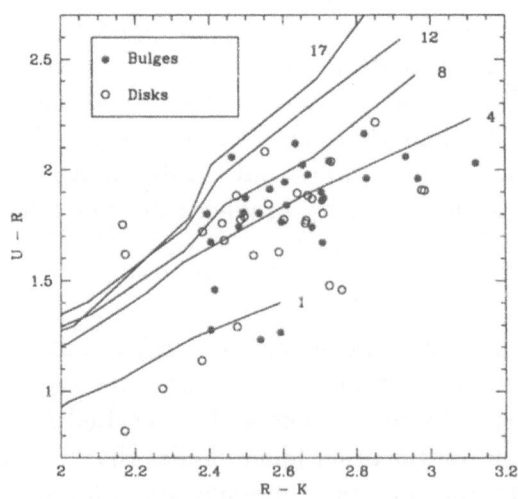

Figure 2. $U - R$ vs. $R - K$ relations for bulge colours. Also included are single age, single metallicity models of Vazdekis *et al.* (1995)

produce such a system, since these time scales will be similar for bulge and inner disk.

An alternative way of forming bulges has been proposed by Combes *et al.* (1990) and others. They form a bulge through the formation, and later desctruction of a bar due to dynamical instabillities. This method does not predict any age differences, and is in good agreement with our results here. However, galaxies with large bulges like the Sombrero cannot be formed in this way, and also one would not expect a continuous change in colours, colour gradient and surface brightness profile shape from large bulges of S0's to small Sb bulges (Andredakis *et al.* 1995).

To summarise, we have found that colour differences in many optical and near-infrared colours between bulges and disks at 2 scale lengths are very small, implying very small age differences. Our result is in disagreement with Bothun & Gregg's (1990) result for S0 galaxies.

References

Andredakis, Y.C., Peletier, R.F. & Balcells, M., 1995, MNRAS 275, 874
Balcells, M. & Peletier, R.F., 1994, AJ 107, 135 **(BP)**
Bothun, G.D. & Gregg, M., 1990, ApJ 350, 73
Combes, F., Debbasch, F., Friedli, D. & Pfenniger, D., 1990, A&A 233, 82.
de Jong, R.S., 1995, Ph.D. Thesis, Univ. of Groningen
Eggen, O., ...
Frogel, J. A. 1985, ApJ 298, 528
Gunn, J.E., 1982, in *Astrophysical Cosmology*, ed. H.A. Brück, G.V. Coyne, and M.S. Longair; Vatican City: Pontifica Academia Scientiarum, p. 233
Peletier, R.F. & Balcells, 1995, in preparation
Peletier, R.F., Valentijn, E.A., Moorwood, A. & Freudling, W., Knapen, J.H. & Beckman, J.E., 1995, A&A 300, L1
Vazdekis, A., Casuso, E., Peletier, R.F. & Beckman, J.E., 1995, in preparation

Dante Minnitti and Mike Rich

ELLIPTICAL GALAXIES

THE STAR FORMATION HISTORY OF ELLIPTICAL GALAXIES

ROGER L. DAVIES
Department of Physics, University of Durham,
South Rd., Durham, U.K., DH1 3LE.

Abstract. There is a growing body of evidence indicating young ages, 8 ± 3 Gyrs, for elliptical galaxies and significant age gradients with a younger population residing at the centre. The data appear to be consistent with a scenario where elliptical galaxies are assembled hierarchically with low luminosity galaxies forming first. Late star formation, associated with the last merging event and usually involving only a small fraction of the galaxy mass, could then account for the low age estimates of some luminous galaxies.

1. Introduction

Conventionally elliptical galaxies are thought to be are old and coeval, having experienced an initial burst of star formation about 15 Gyrs ago, with little or no star formation since. An increase in mean metallicity with luminosity then accounts for the observed systematic variation in galaxy properties with luminosity, for example the colour magnitude relation. The principle empirical problem in testing this scenario is that the observational effects of a decrease in metallicity and a decrease in age are degenerate.

In contemporary cosmologies where dark matter halos evolve through a sequence of hierarchical mergers, low luminosity ellipticals form before their brighter brethren. Luminous elliptical galaxies form late (at say $z \sim 0.5$) by the merging of old, mostly *stellar* subsystems. Some observations, such as the existence of dynamically decoupled cores and the first results of cosmological simulations including gas physics, suggest that this merging process can be accompanied by star formation. Bender, Burstein & Faber, 1992, have suggested that many dynamical and stellar population characteristics of ellipticals can be accounted for if gas dissipative processes are more important in the evolution of low luminosity systems and stellar, dissipa-

R. Bender and R. L. Davies (eds.), New Light on Galaxy Evolution, 37–45.

tionless process dominate in luminous ellipticals. In order to understand the dominant physical processes in galaxy evolution we need to date the star formation episodes in elliptical galaxies and estimate the mass involved at each stage of star formation activity.

From this written record of the review I have omitted the discussion of the important work on M32 which I have discussed elsewhere (Davies, 1996).

2. The Uniformity of Elliptical Galaxies

2.1. STRUCTURE & DYNAMICS

Elliptical galaxies are remarkably uniform, they have smooth luminosity profiles with no azimuthal structure (Peletier *et al*, 1990). They are well fit by an $R^{\frac{1}{4}}$ law with residuals of typically less than 0.1 magnitudes over $R < 4R_e$. Van Albada, 1982, has used N-body simulations to show that such an $R^{\frac{1}{4}}$ profile arises from the dissipationless collapse and violent relaxation of an initially clumpy distribution of stars. The most significant residuals from an $R^{\frac{1}{4}}$ law (typically less than a few % of the light of any isophote) are the boxy and disky distortions that show up most clearly in the analysis of the isophote shapes.

The tight relationship between the central velocity dispersion, σ, effective radius, R_e & effective surface brightness, I_e, of elliptical galaxies, the "Fundamental Plane", discovered by Djorgovski & Davis (1987) & Dressler *et al* (1987) sets constraints on mass-to-light ratio changes amongst the population of ellipticals and therefore constrains variations in their star formation history. The typical scatter is $\sim 20\%$ in R_e, for galaxies in the Coma cluster Jorgensen, Franx & Kjaergaard (1993) report that the scatter is 11%. Renzini & Ciotti (1993) have explored the thickness and tilt of the FP, they conclude that at any location in the FP the thickness of the plane limits the dispersion in M/L to 12% and that the tilt of the plane implies a variation in M/L amongst ellipticals to less than a factor of three. The star formation history of elliptical galaxies gives rise to remarkable degree of structural and dynamical homogeneity.

2.2. SPECTROPHOTOMETRIC PROPERTIES

Sandage & Visvanathan (1978) demonstrated the systematic reddening of elliptical galaxies with increasing luminosity. Elliptical galaxies also become redder and have stronger metal absorption lines with decreasing radius (Faber, 1977, Cohen, 1979). These observations led to the suggestion that *both* effects were driven by increases in metallicity. The tightness of the colour magnitude relations for galaxies in clusters has been emphasised by

Bower et al (1992) and Renzini (1995) as putting tight constraints in the possible spread in age amongst cluster ellipticals. Recently near infrared surface photometry, has shown that the colours of disks and bulges are similar (Terndrup *et al* 1994, de Jong 1995). This may indicate that the *old* stellar population in ellipticals, bulges & disks have similar ages and metallicities.

There is a remarkably tight relationship between the central Mg_2 line strength and the central velocity dispersion in ellipticals:

$$Mg_2 = 0.2 \text{Log}\sigma - 0.166$$

The scatter corresponds only 20-30% in metallicity. This tight relationship may arise from special processes that occur in the centres of galaxies, however the tightness of the relationship between a dynamical parameter and one that arises entirely through the physics of stellar evolution suggests a remarkably close connection between the chemical and dynamical evolution of ellipticals.

3. Late Star Formation in Elliptical Galaxies?

There have been hints for many years that some elliptical galaxies have experienced star formation much less than a Hubble time ago. Larson, Tinsley and Caldwell (1980), showed that while cluster galaxies exhibit a tight colour magnitude diagram, those in the field show more scatter, which they attributed to late star formation in galaxies in low density environments. Similarly Schweizer *et al*, 1990, showed that galaxies with well developed symptoms of past interactions, jets, shells, etc. have stronger $H\beta$ and weaker metal lines than those that appear to have had a calmer history. They interpreted this in terms of late star formation in the disturbed objects. At high redshift, direct evidence that star formation in cluster environments continued until relatively recently is provided by the observation that the fraction of blue galaxies in clusters of redshift $\sim \frac{1}{2}$ is much higher than it is at zero redshift (Butcher & Oemler 1978).

In recent years a number of in depth studies have attempted to combine high metallicity isochrones and models of stellar atmospheres, with spectral libraries of high abundance stars and models of post main sequence evolution, to produce predictions of the spectral energy distribution (SED) of the integrated light of elliptical galaxies. These models span a range of philosophies, some provide model SEDs for populations of a given metallicity and age such as Peletier (1989), Buzzoni (1989,1995), Bruzual & Charlot (1993) and Worthey (1994), others take into account the large spread in metallicity expected in ellipticals eg. Arimoto & Yoshii 1987; the most ambitious also include the effects of galaxy evolution eg. Bressan, Chiosi & Fagotto

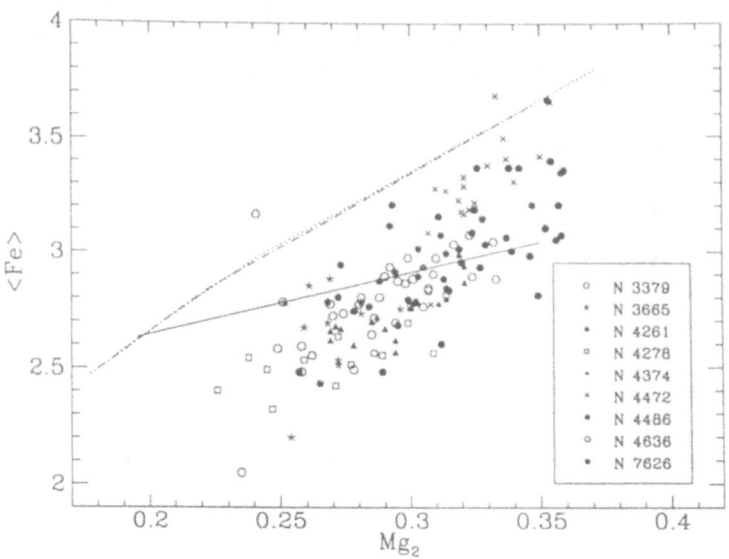

Figure 1. The variation of the mean of two iron features, $< Fe >$, plotted against Mg_2 for points within 9 elliptical galaxies. The solid line is the fit to the cores of elliptical galaxies taken from Burstein *et al 1984.* Two models of single age stellar populations (with ages 12 & 20 Gyrs) from Peletier (1989) are plotted as dash-dot and dotted lines, each spanning a range of metallicity from 0.25 to 2.5 times solar.

(1994). However it remains difficult to unambiguously separate the effects of younger age from those of decreased metallicity, especially in luminous galaxies where the sensitivity of the measurements is reduced by the high velocity dispersions which blend and broaden absorption features. Some results are however emerging.

3.1. NON-SOLAR ABUNDANCE RATIOS

The simplest interpretation of the increase in line strengths with decreasing radius and increasing luminosity (or σ) is an increase in [Fe/H], the single parameter metallicity. Spectral features would then be expected to strengthen in "lockstep" as metallicity increased. It has become clear that no models with solar abundance ratios can account simultaneously for the variation of Mg_2 to Fe within and between elliptical galaxies (Peletier 1989, Gorgas, Efstathiou & Salamanca, 1990, Worthey, Faber & González, WFG, 1992, Davies, Sadler & Peletier, (DSP), 1993). This is illustrated in Figure 1, taken from DSP, which shows the variation of Fe against Mg_2.

The measured points do not correspond to models with solar ratios of Fe to Mg, but are shifted to higher values of Mg at fixed Fe. WFG draw the same conclusiopn using their models. The discrepancy cannot be accounted for by reducing the age as the effects of decreasing age and lowering the metallicity are the same in this diagram. It appears that either magnesium is enhanced or iron depleted compared to solar values throughout the giant galaxies. Similarly it is clear that the trend from galaxy to galaxy exhibits a shallower slope than that found within galaxies and on this basis we would expect the low metallicity galaxies (with $Mg_2 \sim 0.2$) to have roughly solar ratios which WFG find to be the case. For an explanation of these anomalies we look to the processes that produce iron & magnesium. Mg is produced predominantly in massive (20-40 M_{solar}) Type II supernovae (Woosley & Weaver 1986), their lifetimes are typically 10 million years, whereas Fe is produced in both Type II and in Type Ia supernovae. The low mass stars that give rise to Type Ia SN have lifetimes typically a few hundred times longer than the Type II progenitors and these provide the dominant source of Fe in old populations. There are several ways in which the galaxies with strong absorption lines might have an excess ratio of Mg/Fe compared to solar values:

1. The stars that dominate the V-band light output were formed in a short burst after which star formation ceased. This would produce a population enriched in Mg from rapid, Type II supernovae. Stars (with solar abundance ratios) would not be formed from the Fe enriched gas generated by Type Ia supernovae.

2. The Initial Mass Function was skewed to more massive stars in comparison to the solar neighbourhood, this would require the IMF to vary continuously from the giants to low luminosity ellipticals which do not exhibit the abundance anomaly.

3. Fe-rich gas produced by Type Ia SN is lost through the onset of a galactic wind that is established after Mg production but before significant Fe production. This would imply an *over*-abundance of Fe in the hot IGM. The status of measurements of the abundance of Fe in the IGM, infered from ASCA spectra, is discussed by Arimoto elsewhere in this volume.

4. The star formation processes in ellipticals produced fewer binary stars, this is unlikely given the observed nova rates and LMXB populations.

5. Metal rich SN could produce higher ratios of [Mg/Fe], (Peterson, 1976).

It is not clear how a short, early burst of star formation fits into hierarchical galaxy formation models where the stars that are assembled into giant galaxies formed in a range of localities. Nevertheless, wherever they formed, if the process was rapid and there was no subsequent star formation (perhaps because the system was disrupted) the abundance anomaly would be established.

3.2. THE AGES OF ELLIPTICAL GALAXIES

To separate the effects of age and metallicity we need to identify colours or absorption line features which change differentially with metallicity changes at constant age, compared to age changes at fixed metallicity. Worthey (1992), for example, showed that while the weak Fe features in the Faber-Burstein indices are the most sensitive to metallicity they are difficult to measure accurately leaving the widely used Mg_2 as a practical indicator. He found that $H\beta$ is relatively insensitive to metallicity and can be used as an age indicator, the largest drawback being the correction for emission observed in many galaxies. Buzzoni (1989, 1995) and Buzzoni, Gariboldi & Mantegazza (1992, 1994) have generated models that produce almost the same effects in Mg_2 and $H\beta$. In his thesis González (1993), reports line strength gradient measurements in 41 ellipticals. He applied Worthey's models using a mean MgFe index (to reduce the effect of the magnesium overabundance) and $H\beta$ to estimate ages/metallicities. He made an empirical correction for the $H\beta$ emission based on the emission in O[III]. He shows that the mean stellar age of the bulk of ellipticals is 8 ± 3 Gyrs. He also found that the inner regions of elliptical galaxies are younger by about 3 Gyrs than the outer parts. These results are reported and amplified in WFG and Faber, Trager, Gonzalez & Worthey, 1995. These age estimates are uncomfortably low for galaxies that were thought to form 15 Gyrs ago in a single burst of star formation. There are two possible systematic effects which could account for the high values of $H\beta$ absorption which drive these authors to low age estimates. The first is the possibility that the correction for $H\beta$ emission is too large, Carrasco (1996) shows that $H\beta$ emission varies with radius in ellipticals and rarely corresponds to Gonzalez prescription. The second possibility is that the high $H\beta$ values arise not from stars at the tip of the main sequence, but from blue horizontal branch stars and therefore $H\beta$ is not a good age indicator. This would seem to be unlikely in a metal rich population although it may be that a considerable range of metallicity exists in elliptical galaxies and therefore BHB stars may contribute.

Elsewhere in this volume (Halliday *et al*) we report preliminary measurements of line strength gradients in low luminosity ellipticals. The data for NGC 4478 are shown in Figure 2. The same age gradient, with a younger population in the centre, is apparent as is found in the luminous galaxies. We have made preliminary estimates of the ages and metallicities of the LLEs relative to the giant ellipticals. We find metallicities down by factors of 4-6 and ages, for the bulk of the galaxy, to be 20-40% older (or the last star formation occurred that much longer ago) than in the corresponding regions of the luminous galaxies. This appears to support a hierarchical

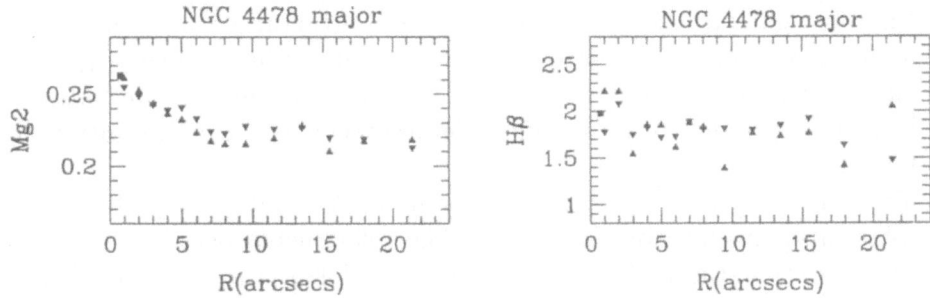

Figure 2.   **Grad**ients in the Mg_2 and $H\beta$ indices along the major axis of NGC 4478. These are **preliminary measureme**nts on the Faber-Burstein scale but not corrected for velocity dispersion effects.

picture where the low luminosity systems experienced their last merger at an earlier stage in their evolution and, associating this merger event with the last opportunity for star formation, we expect their distribution of ages to be skewed to larger values, as we observe.

Gas accreted in mergers would be likely to have high angular momentum (from the relative orbit) and so we might expect any associated star formation to be distributed in a disk. (Franx & Illingworth (1989) made this suggestion to account for the peculiar core kinematics of IC1459). With de Jong, I have tested this hypothesis (reported elsewhere in these proceedings) by examining the isophote shapes for galaxies with measured $H\beta$ strengths. As shown in the figure in our poster we find that there is a tendency for galaxies with stronger $H\beta$ to have disky isophotes. This result is at least consistent with the $H\beta$ arising from stars formed in gas accreted in a merger.

So how old are the stars in *luminous* elliptical galaxies? Gonzalez' results indicate a large spread of ages for galaxies with essentially the same metallicity. Hierarchical models of galaxy formation assemble the most massive galaxies latest, therefore if we hypothesise that star formation is sometimes associated with a merger event, then we may conclude that the luminous ellipticals with stronger $H\beta$ lines experienced a burst of star formation associated with their last merger and those with weaker $H\beta$ lines did not.

The age of the *oldest* stars in ellipticals, (this is certainly what we mean when we estimate the "age" of the Milk Way even though the oldest population, the globular clusters, provide an insignificant fraction of the total luminosity), remains uncertain.

4. Conclusions & Questions

1. Elliptical galaxies exhibit a high degree of structural and dynamical uniformity, some of this arises from identified physical process eg. violent relaxation. If ellipticals experience a turbulent star formation history associated with a merging hierarchy how is this uniformity retained?
2. The light elements and Fe peak elements appear not to occur in their solar ratios in luminous ellipticals. The relationship between Fe and Mg line strengths *within* galaxies does not follow the same trend as is found *between* galaxies suggesting a different physical origin. Rapid star formation will produce Mg quickly in Type II supernovae. If star formation is truncated, no significant population will form from gas enriched in Fe, and a Mg overabundance will result. This also implies that the V-band light of elliptical galaxies is dominated by a stellar population that was formed in a single burst. What physical processes control the Fe/Mg ratio in ellipticals?
3. The large range of ages for ellipticals of a given metal line strength, the age gradients, the older age estimates of LLEs and the tendency for galaxies with high $H\beta$ to have disky isophotes, all support a scenario where galaxy formation proceeds in a merging hierarchy in which late star formation, involving only a small fraction of the galaxy mass, is sometimes associated with the mergers.

References

Arimoto N. & Yoshii Y., 1987, A&A, 173, 23.
Bender R., Burstein D. & Faber S.M., 1992, ApJ, 399, 462.
Bower, R. G., Lucey, J. R., & Ellis, R. S., 1992, MNRAS, 254, 601.
Bressan A., Chiosi C. & Fagotto F., 1994, ApJ Supp, 94, 63.
Bruzual G.A. & Charlot S., 1993, ApJ, 405, 538.
Burstein D., Faber S.M., Gaskell C.M. & Krumm N., 1984, ApJ, 287, 586.
Butcher H. & Oemler A., 1978, ApJ, 219, 18.
Buzzoni A., 1989, ApJ Supp, 71, 817.
Buzzoni A., 1995, ApJ Supp, 98, 69.
Buzzoni A., Gariboldi G. & Mantegazza L., 1992, AJ, 103, 1814.
Buzzoni A., Gariboldi G. & Mantegazza L., 1994, AJ, 107, 513.
Carrasco M., 1996, in Proceedings of "Fresh Views of Elliptical Galaxies", P. 235, ASP Conf. Series No.86 (San Francisco, ASP), eds. A. Buzzoni, A. Renzini and A. Serrano.
Cohen J.G., 1979, ApJ, 228, 405.
de Jong R.S., 1995, PhD Thesis, University of Groningen.
Davies R.L., Sadler E.M. & Peletier R.F., 1993, MNRAS, 262, 650.
Davies R. L., 1996, in Proceedings of "Fresh Views of Elliptical Galaxies", P. 175, ASP Conf. Series No.86 (San Francisco, ASP), eds. A. Buzzoni, A. Renzini and A. Serrano.
Dressler A., Lynden-Bell D., Burstein D., Davies R.L., Faber S.M.,Terlevich R.J. & Wegner G., 1987, ApJ, 313, 42.
Djorgovski S. & Davis M., 1987, ApJ, 313, 59.
Elston R. & Silva D., 1992, AJ, 104, 1360.

Faber S.M., 1977, in *"The Evolution of Galaxies and Stellar Populations"*, Yale University Press, p157, ed. Tinsley B.T. & Larson R.B.

Faber S. M., Trager S. C., Gonzalez J.J. & Worthey G., 1995, in *"Stellar Populations"*, IAU Symp. 164, ed. G. Gilmore & P. van der Kruit, (Kluwer).

Franx M. & Illingworth G.D., 1990, ApJ Lett, 359, L41.

González J.J., 1993, PhD Thesis, University of Santa Cruz.

Gorgas J., Efstathiou G. & Aragón Salamanca A., 1990, MNRAS, 245, 217.

Jørgensen I., Franx, M., & Kjærgaard, P., 1993, , 411, 34.

Larson R.B., Tinsley B.M. & Caldwell C.N., 1980, ApJ, 237, 692.

Peletier, 1989, PhD Thesis, University of Groningen.

Peletier R.F., Davies R.L., Illingworth G.D., Davis L. E. & Cawson, 1990, AJ, 100, 1091.

Peterson, R., 1976, ApJ, 210, L123.

Renzini A., & Ciotti, L., 1993, ApJ Lett, 416, L49.

Renzini A., 1995, in *Stellar Populations*, Proc. of IAU Symp 164, P. 325, ed. Gilmore G. & van der Kruit P., (Kluwer).

Sandage A.R. & Visvanathan N., 1978, ApJ, 225, 742.

Schweizer F., Seitzer P., Faber S.M., Burstein D., Dalle Ore C. & González J.J., 1990, ApJ Lett, 364, L33.

Terndrup D.M., Davies R.L., Frogel J.A., Depoy D.L., & Wells L.A., 1994, ApJ, 432, 518.

van Albada T. S., 1982, MNRAS, 201, 939.

Woosley S.E. & Weaver T.A., 1986, Ann. Rev. A&A, 24, 205.

Worthey G., Faber S.M. & González J.J. (WFG), 1992, ApJ, 398, 69.

Worthey G., 1994, ApJ Supp, 95, 107.

Discussion

Tim de Zeeuw: What would the luminosity weighted age of our own galaxy be?

Roger Davies: The star formation rate in our own galaxy's disk has been roughly constant over its age, so it would be roughly 5Gyrs, younger if luminosity weighted in the *blue*.

Marshall McCall: Both the ages and metallicities are luminosity weighted. Is it possible that this weighting varies with metallicity? Is it possible that Mg is weighted differently from Fe? What stars are setting the metallicity we "see" in ellipticals?

Roger Davies: In general we would expect the luminosity weighting to be a function of metallicity. It is unlikely that Fe and Mg will be weighted differently as they arise in the atmospheres of stars of rather similar temperatures namely G, & K-giants, these are the stars that dominate the metallicity determination.

Hans-Walter Rix: You have shown how degenerate the SED is to age/metallicity trade-offs. Is there any hope of breaking this degeneracy by considering the colour dependence of surface brightness fluctuations?

Roger Davies: In principal yes, calculations of the SBFs were made by Worthey in his thesis.

Josh Barnes and Tim de Zeeuw

DYNAMICAL CONSTRAINTS ON THE FORMATION OF ELLIPTICAL GALAXIES

P.T. DE ZEEUW AND C.M. CAROLLO

Sterrewacht Leiden

Postbus 9513, 2300 RA Leiden, The Netherlands

Abstract. Recent work on the construction of spherical, axisymmetric and triaxial dynamical models for elliptical galaxies is reviewed briefly, including their role in providing evidence for dark halos and central black holes. The different orbital structures and shapes of low-mass and giant elliptical galaxies provide essential constraints on scenarios of galaxy formation.

1. Introduction

The aim of stellar dynamics is to reconcile the observed morphology and kinematics of galaxies, in order to determine their internal orbital structure. This provides information on the presence and importance of dark halos and massive central black holes, and on the connection between kinematics and stellar populations, i.e., between the motions and the physical properties of the stars.

Elliptical galaxies, as a class, are triaxial stellar systems with stationary or slowly tumbling figures (Binney 1976; Franx, Illingworth & de Zeeuw 1991). Their orbital structure depends on: (i) the rate of tumbling of their figure, (ii) their degree of triaxiality, and (iii) the presence and strength of a central concentration of mass (such as a stellar cusp or a black hole). Observations with the Hubble Space Telescope (HST) show that their luminosity distributions approach a power-law form $\rho(r) \propto r^{-\gamma}$ at small radii r, with γ ranging between 0 and 2.5 (Crane et al. 1993; Ferrarese et al. 1994; Forbes, Franx & Illingworth 1995; Lauer et al. 1995; Carollo et al. 1995b). Thus, some galaxies have nearly-flat cores ($\gamma \sim 0$), while others have steep

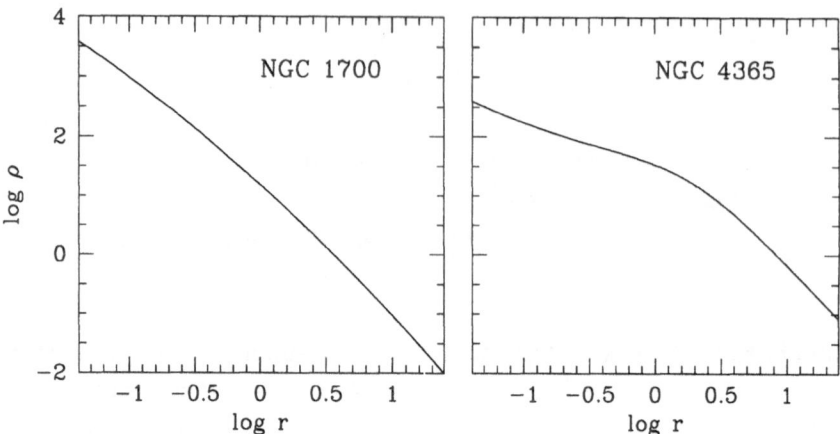

Figure 1. Two types of deprojected luminosity-density profiles (in $L_\odot/pc^3$) for elliptical galaxies. The left panel shows the profile of the intermediate luminosity elliptical NGC 1700, which has a logarithmic slope $\gamma \approx 2$ down to the innermost measurable radii (given in arcsec). This galaxy has a steep cusp. The right panel shows the profile of the giant elliptical NGC 4365. It has a clear bend at $r_b \approx 2$ arcsec, inside which the profile is very shallow. Based on HST data from Carollo et al. (1995b). See also Forbes et al. (1995).

Theoretical dynamical modeling must answer the following key questions: (i) what is the full range of shapes, density profiles, and tumbling rates covered by equilibrium models, and (ii) what kind of internal velocity distributions do these have? The comparison of dynamical models with observations must subsequently establish which subset of all equilibrium models is occupied by elliptical galaxies. It is very likely that physical processes acting at formation, and/or subsequent dynamical evolution, have banished galaxies from certain parts of parameter space otherwise allowed by the simple requirement of stability. Delineating the region of permitted equilibria will provide essential information about scenarios of galaxy formation and mechanisms of galaxy evolution.

2. Spheres

An equilibrium model of a galaxy is fully specified by its phase-space distribution function (DF) $f(\vec{x}, \vec{v}) \geq 0$, which gives the density of stars at each position $\vec{x}$ and velocity $\vec{v}$. Jeans' theorem states that f depends on $\vec{x}$ and $\vec{v}$ through the isolating integrals of motion admitted by the gravitational potential of the system. These integrals label the individual stellar orbits. Jeans' theorem therefore states that a dynamical model is defined once the number of stars that populate each orbit has been specified.

Jeans' theorem is most effective for spherical systems, in which the energy E and the angular momentum $\vec{L}$ are conserved. Most models that have

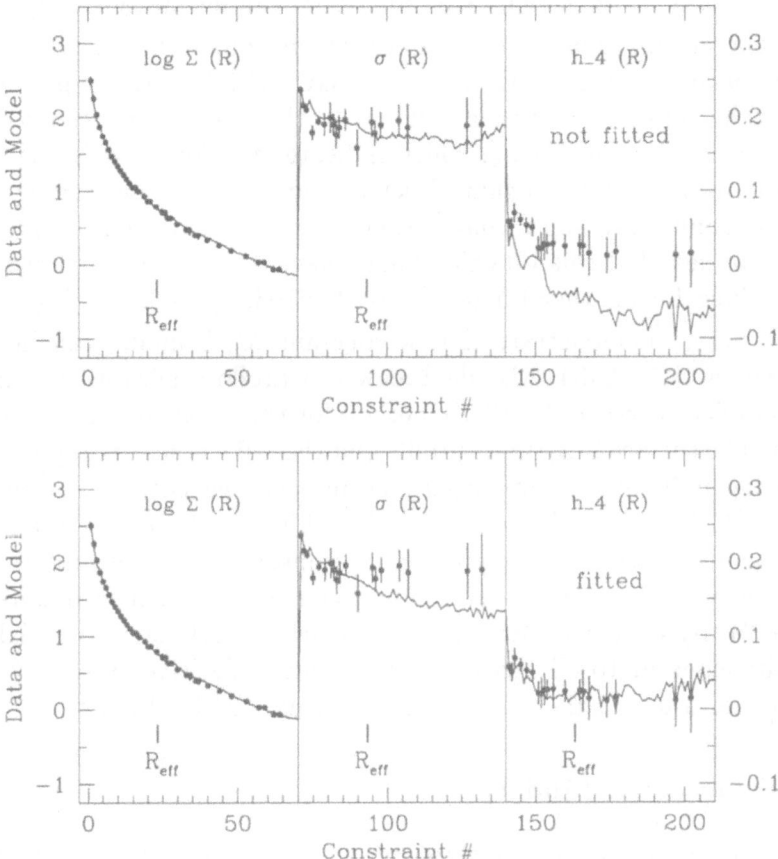

Figure 2. Models without dark halo for the E0 galaxy NGC 2434, compared with available photometry and kinematics. The upper panel shows the best fit to the radial profiles of the surface brightness Σ and the velocity dispersion σ. The predicted values for the profile of the VP shape-parameter h_4 are shown as well, and are clearly inconsistent with the observed profile. The lower panel shows the best simultaneous fit to Σ, σ and h_4. This again is unable to match the data, and in particular cannot reproduce the constant value of σ out to large radii. NGC 2434 must therefore have a dark halo (Rix, de Zeeuw & Carollo 1995).

been constructed have DFs of the form $f(E, L^2)$, so that the velocity distribution is anisotropic, but does not have a preferred tangential direction. These models are particularly useful in studies of E0 galaxies. Important applications include the searches for dynamical evidence for massive dark halos and central black holes, i.e., for variations of the mass-to-light ratio M/L with radius, in such systems. These studies require measurements of the entire line-of-sight velocity distribution — the velocity profile VP — as a function of radius, in order to distinguish variations in velocity anisotropy from variations in M/L (e.g., Gerhard 1993; van der Marel 1994).

Carollo et al. (1995a) measured VP shapes out to about two half-light radii R_e in the normal elliptical galaxy NGC 2434. The radial profile of the line-of-sight velocity dispersion σ is flat to the outermost measured point. The VP shape parameter h_4, which indicates the lowest-order symmetric deviation from a Gaussian-shaped velocity profile, rules out strong tangential anisotropy. Based on a comparison with simple dynamical models, the authors argued that this E0 galaxy must have a dark halo. A similar result was obtained for another E0 galaxy, NGC 6703, by Jeske et al. (1995).

In order to demonstrate that a constant M/L model can indeed be ruled out in NGC 2434, Rix, de Zeeuw & Carollo (1995) used a modified version of Schwarzschild's (1979) method for the construction of dynamical models. The surface brightness profile and the VP shapes are reproduced by populating individual orbits calculated in a chosen potential. Application to the NGC 2434 data showed that although the stellar orbits in a constant M/L spherical model can reproduce the observed surface brightness and velocity dispersion profile, no such model can also fit the measured h_4-values (Figure 2). By contrast, inclusion of a plausible dark halo potential allows a simultaneous fit to all measurements. This is decisive stellar dynamical evidence for a dark halo surrounding a normal elliptical galaxy.

3. Axisymmetric Models

Despite our ignorance on the third integral I_3 in general axisymmetric potentials, knowledge of the two classical ones E and L_z has allowed investigation of axisymmetric dynamical models with DF $f(E, L_z)$. The actual calculation of such DFs has long been impeded by certain technical difficulties. Hunter & Qian (1993, hereafter HQ) developed a contour integral method that circumvents these difficulties, and allows calculation of $f(E, L_z)$ for a wide variety of finite and infinite mass models. When applied to the density $\rho(R, z)$ in a potential $\Psi(R, z)$, the method gives the unique $f_e(E, L_z)$ that is *even* in L_z and that generates ρ. When applied to $R\rho\langle v_\phi\rangle(R, z)$, it gives the unique $f_o(E, L_z)$ that is *odd* in L_z and generates the mean azimuthal streaming motions $\langle v_\phi\rangle$. It is not easy to obtain good observational data on the full two-dimensional mean line-of-sight velocity $\langle v_{\text{los}}\rangle$ on the plane of the sky, from which the intrinsic azimuthal mean streaming field $\langle v_\phi\rangle(R, z)$ must be found. Therefore, a popular approach is to take the odd part f_o as a product of the (positive) even part f_e and a prescribed function such that $f = f_e + f_o$ is physical (non-negative).

Qian et al. (1995) applied the HQ method to high–quality kinematic data for M32 (van der Marel et al. 1994). No self-consistent $f(E, L_z)$ model can fit either the observed central peak in the line-of-sight velocity dispersion σ_{los}, or the steep central $\langle v_{\text{los}} \rangle$ gradient. However, an $f(E, L_z)$ model with a central dark mass of $1.8 \times 10^6 M_\odot$ provides an astoundingly good fit to all available photometric and kinematic measurements (see Figure 11 in Qian et al. 1995; and Dehnen 1995). This does not prove that M32 has a large central black hole, since it remains to be demonstrated that a three–integral model without a central dark mass can be ruled out. However, the measured VP shapes show that in order to avoid a central dark mass, any model must combine a tangentially anisotropic velocity distribution outside 3 pc with a strongly radially anisotropic velocity distribution inside 3 pc. Even if a model with these properties and a DF $f \geq 0$ exists, it is not clear whether it would be stable, or, in fact, plausible. Cycle 5 HST/FOS observations have been collected to investigate this issue further. If, for the time being, we assume that the good fit of the simplest axisymmetric model with a central point mass is not a mere coincidence, then it is legitimate to ask: is the DF of M32 the end-product of stellar dynamical processes which have operated after its formation, possibly caused by the presence of a central dark mass? A positive answer would imply that such processes must have been effective in removing any dependence of the DF on a third integral of motion in less than the Hubble time.

Other recent applications of the HQ method include the construction of models with dark halos used to fit the extended kinematic absorption line measurements of elliptical galaxies (Carollo et al. 1995a), an analysis of the observable properties of small stellar disks embedded in the nuclei of ellipticals (van den Bosch & de Zeeuw 1995), and construction of models for the Galactic Bulge (Kuijken 1995; Hoogerwerf & Arnold 1995). In addition, following the HQ paper, a variety of other methods for generating $f(E, L_z)$–models have been developed by Dehnen & Gerhard (1994), Dehnen (1995), Kuijken (1995) and Magorrian (1995). It is safe to conclude that $f(E, L_z)$ axisymmetric models have now largely replaced spherical models as the standard theoretical template for a zeroth-order comparison with kinematic observations of flattened galaxies.

4. Triaxial Models

Triaxial potentials generally admit only one exact isolating integral, the orbital energy E. Although there are three planes of reflection symmetry, there are no symmetry axes, and no component of the angular momentum vector is conserved. Thus, two of the three integrals of motion are unknown. As a result, realistic models must be constructed by numerical means, such as Schwarzschild's method.

Since the orbital structure in a triaxial system depends on its central mass concentration, the internal dynamical structure of galaxies with flat

cores differs from that of galaxies with steep cusps. Triaxial stellar systems
with cores have a rich internal dynamical structure (Schwarzschild 1979;
de Zeeuw 1985). There are four major orbit families. Two families circulate
around the longest axis, and a third circulates around the shortest axis —
these are called *tube* orbits; they carry all the angular momentum of the
model. The fourth family is formed by the *box* orbits which have no net circulation and penetrate to the center of the model. Orbits of all four families
must be occupied in any dynamical model, but many different orbit combinations, i.e., many different DFs, can reproduce the same triaxial density.
Since there can be net streaming around both the long and the short axis,
the total angular momentum vector (the 'rotation axis') can lie anywhere
in the plane containing these two axes. The resulting projected kinematics
can display a complex structure (Statler 1991, 1994; Arnold, de Zeeuw &
Hunter 1994). Such complexity is indeed observed in the kinematics of giant
elliptical galaxies (e.g., de Zeeuw & Franx 1991).

The three families of tube orbits are also present in models with cusps.
Their mean streaming fields are therefore very similar to those of the models with cores. However, the few numerical constructions carried out to date
indicate that whereas equilibrium models with cusp slopes $\gamma \lesssim 0.5$ probably can be built for all triaxial shapes, those with steeper inner profiles
may exist only for near-axisymmetric shapes (Kuijken 1993; Schwarzschild
1993; Merritt & Fridman 1995). The reason is that in models with steep
cusps or massive central black holes, the box orbits are replaced by minor orbit families and irregular orbits (Gerhard & Binney 1985; Merritt &
Fridman 1995). These are associated with stable and unstable higher-order
resonances between the oscillation frequencies along and perpendicular to
the principal axes, and have been christened boxlets (Miralda–Escudé &
Schwarzschild 1989). Boxlets display a large variety of shapes, but in substantially flattened triaxial models they cannot reproduce the characteristics of very elongated box orbits that remain close to the long axis. The
latter are needed in any equilibrium model, owing to the fact that all tube
orbits are elongated opposite to the figure of the model. These results are
based on a small number of experiments, and need to be extended.

The fact that elliptical galaxies might have slowly tumbling figures has
significant consequences for their orbital structure, since the Coriolis force
distinguishes between direct and retrograde motion. Very little work has
been done on the construction of tumbling triaxial galaxy models. The
presence of a central density cusp or black hole is expected to influence the
orbital structure less strongly than in a stationary triaxial system, because
the Coriolis force causes box orbits to acquire net angular momentum and to
become centrophobic (Vietri & Schwarzschild 1983). It is unknown whether
a tumbling triaxial system can support a strong cusp.

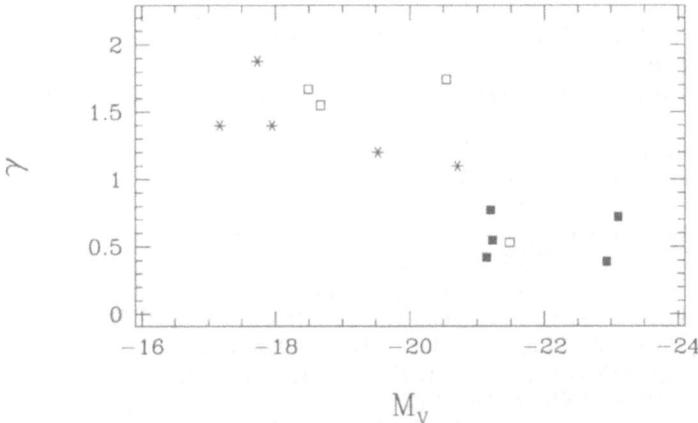

Figure 3. The inner logarithmic cusp slope of the deprojected density as a function of absolute V magnitude, for a sample of elliptical galaxies. Different symbols indicate the quality of the deprojection (solid square = highest; open square = intermediate; stars = lowest; see Carollo et al. 1995b for details). Based on HST data from Lauer et al. (1995).

5. Implications and Conclusions

It is by now well established that the global properties of elliptical galaxies correlate with their mass. Large elliptical galaxies are red, have a high metal content, and have boxy isophotes. They rotate slowly, and are supported by anisotropic stellar velocity distributions. By contrast, low-luminosity ellipticals are bluer, are possibly less metal-rich, have disky isophotes, and rotate relatively fast (e.g., de Zeeuw & Franx 1991). Recent work with the HST has shown that also the nuclear properties, in particular the steepness of the central light profile, correlate with total luminosity (Figure 3).

   When combined with the results on dynamical modeling of triaxial systems with cusps discussed in §4, these correlations suggest that (i) the giant ellipticals are indeed triaxial; (ii) the low-luminosity ellipticals are likely to have near-axisymmetric shapes. This difference in shape needs to be confirmed, and its origin needs to be investigated. It might have been imposed by the galaxy formation process, or might be the result of subsequent dynamical evolution, or both.

It is a pleasure to thank Hans-Walter Rix, Garth Illingworth, Marijn Franx, and Duncan Forbes, for permission to quote results prior to publication, and to thank Roeland van der Marel and Martin Schwarzschild for a careful reading of the manuscript. Part of this paper was written at the Institute for Advanced Study, with support from NSF Grant PHY 92–45317. CMC was supported by HCM grant ERBCHBICT940967 of the European Community.

References

Arnold R.A., de Zeeuw P.T., Hunter C., 1994, MNRAS, 271, 924
Binney J.J., 1976, MNRAS, 177, 19
Carollo C.M., de Zeeuw P.T., van der Marel R.P., Danziger I.J., Qian E.E., 1995a, ApJL, 441, L25
Carollo C.M., Franx, M., Forbes D.A., Illingworth G.D., 1995b, AJ, submitted
Crane P., et al. 1993, AJ, 106, 1371
Dehnen W., Gerhard O.E., 1994, MNRAS, 268, 1019
Dehnen W., 1995, MNRAS, 274, 919
de Zeeuw P.T., 1985, MNRAS, 216, 273
de Zeeuw P.T., Franx M., 1991, ARAA, 29, 239
Ferrarese L., van den Bosch F.C., Jaffe W., Ford H.C., O'Connell R., 1994. AJ, 108, 1598
Forbes D.A., Franx M., Illingworth G.D., 1995, AJ, 109, 1988
Franx M., Illingworth G.D., de Zeeuw P.T., 1991, ApJ, 383, 112
Gerhard O.E., 1993, MNRAS, 265, 213
Gerhard O.E., Binney J.J., 1985, MNRAS, 216, 467
Hoogerwerf R., Arnold R.A., 1995, preprint
Hunter C., Qian E.E., 1993, MNRAS, 262, 401 (HQ)
Jeske G., Gerhard O.E., Saglia R.P., Bender R., 1995, preprint.
Kuijken K., 1993, ApJ, 409, 68
Kuijken K., 1995, ApJ, in press
Lauer T., et al., 1995, preprint
Magorrian J., 1995, MNRAS, in press
Merritt D.R., Fridman T., 1995, preprint.
Miralda-Escudé J., Schwarzschild M., 1989, ApJ. 339, 752
Qian E.E., de Zeeuw P.T., van der Marel R.P., Hunter C., 1995, MNRAS, 274, 602
Rix H.-W., de Zeeuw P.T., Carollo C.M., 1995, in preparation.
Schwarzschild M., 1979, ApJ, 232, 236
Schwarzschild M., 1993, ApJ, 409, 563
Statler T.S., 1991, ApJ, 382, L11
Statler T.S., 1994, ApJ, 425, 458
van den Bosch F.C., de Zeeuw P.T., 1995, in preparation
van der Marel R.P., 1994, MNRAS, 270, 271
van der Marel R.P., Rix H.W., Carter D., Franx M., White S.D.M., de Zeeuw P.T., 1994. MNRAS, 268, 521
Vietri M., Schwarzschild M., 1983. ApJ, 269, 487

Discussion

KOO: Given that luminous ellipticals seem to show the best evidence for massive black holes that can destroy the box orbits associated with tri-axial shapes, how do you reconcile this with your conclusion that massive ellipticals are triaxial?

DE ZEEUW: I assume that the evidence you allude to is the fact that powerful radio sources are associated with giant ellipticals. Yet, as far as I am aware, we simply don't know whether all ellipticals have (quiescent) black holes, and in particular, whether the low-luminosity ones have them. The case of M32 is very interesting in this respect. As a result, it is hard to estimate the dynamical importance of these central point masses. Their effects may well be swamped by those induced by the central cusp.

DISKY ELLIPTICALS IN THE HUBBLE SEQUENCE

CECILIA SCORZA
Landessternwarte
Königstuhl, D-69117 Heidelberg, Germany

AND

RALF BENDER
Universitätssternwarte
Scheinerstr. 1, D-81679 München, Germany

1. Introduction

About one third of the elliptical galaxies show disky isophotes (Bender et al. 1989). These are low luminosity objects, are rotationally flattened (Carter 1987, Bender 1988b, Nieto 1988), are much weaker radio emitters than ellipticals in average and show no strong X-ray emission in excess of their discrete sources contribution. They have pointed isophotes which are thought to reflect the presence of embedded stellar disks. A large number of these disks could be hidden due to low inclinations (Rix and White 1990). Because of these properties, which makes them strongly resemble S0 galaxies, it has been suggested that disky ellipticals form together with S0s and spirals a continuous transition in D/B ratio in the Hubble sequence (Bender 1988b, Nieto 1988). However, until the present, this hypothesis had not been investigated throughout.

The aim of the present study is to place the E-S0 link on a more quantitative base. As a first step, we investigate whether the observed photometric and kinematic properties of these galaxies can be understood within the frame of work of bulge+disk models. Then we quantify the link between disky Es and S0 galaxies in terms of their D/B ratios. An important question to be addressed in this context is to what extend the observed kinematics of disky Es are influenced by the embedded disks. Once the kinematic contribution of the disks is separated, are the bulges of these objects rotationally flattened or are they anisotropic? In doing so, we hope to derive constraints that help to answer the following question: Is the dichotomy in

55

R. Bender and R. L. Davies (eds.), New Light on Galaxy Evolution, 55–59.

in the elliptical family indicating two different origins for disky and boxy ellipticals? or in other words, did all ellipticals form from mergers?

2. The Internal Structure of Disky Elliptical Galaxies

In order to investigate the internal structure of the galaxies and see whether their properties can be understood within disk+bulge models, a photometric and kinematic decomposition was applied to a sample of representative objects (for the description of the methods and results see Scorza and Bender 1990,1995). The photometric decomposition allowed to determine the disk parameters, namely the central surface brightness SB_0, the scale length r_0 and inclination i. The kinematic decomposition relies on the fact that the velocity line profiles (VLP) of disky ellipticals are asymmetric (Bender 1990, Bender, Saglia and Gerhard 1994). We find that the radii at which the asymmetries are largest corresponds to the radii at which the disks constribute with most of their light (see Fig 5 in Scorza and Bender 1995). In view of this, a double-gaussian fit to the VLPs was carried out under the constraint that the flux ratio of the two Gaussians is similar to the D/B derived from the photometry. The results do not change if we allow for a restricted range of intrinsic asymmetries in the VLPs of the bulges. We find that the VLPs of the galaxies can be very nicely modelled and that the kinematic contribution of the disk and bulge components can be, in first approximation, separated.

3. The continuity of physical properties between E-S0 Galaxies

3.1. D/B RATIO-SEQUENCE BETWEEN E AND S0 GALAXIES

Fig 1.a shows a histogram of the D/B ratios determined by Kent (1985) for S0 and spiral galaxies together with the values derived here from the photometric decomposition of disky ellipticals. The dependence of the D/B ratios on the Hubble type is clear. To better visualize the distribution of disky ellipticals, a similar histogram is shown in Fig 1.b but covering only the range $D/B < 1$. It can be seen that the D/B ratios of disky ellipticals are confined to < 0.20.

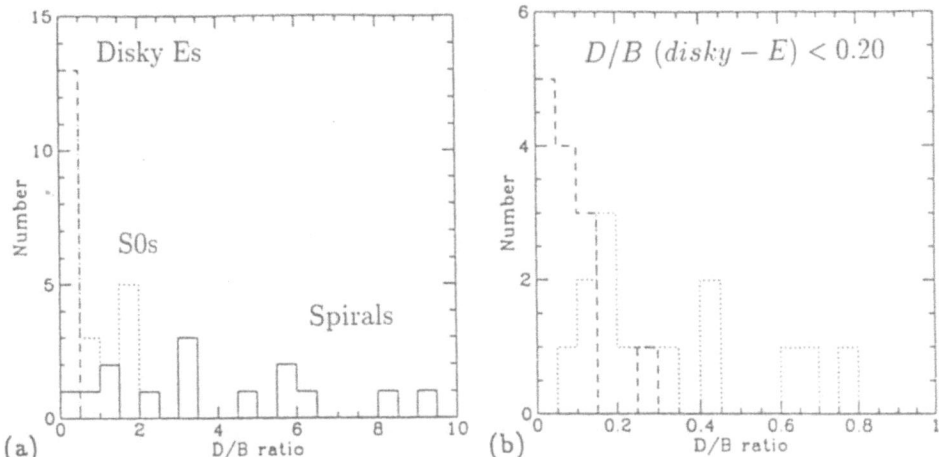

Figure 1. (a) The D/B ratios estimated by Kent (1985) for spiral (solid line) and S0 galaxies (pointed lines) together with the values determined here for disky ellipticals (dashed lines). (b) the same as (a) but confined to D/B ratios < 1

3.2. BULGE AND DISK LUMINOSITIES ALONG THE HUBBLE SEQUENCE

The behaviour of the bulge and disk luminosities along the Hubble sequence was also investigated. Figures 2.a and 2.b show a plot of the absolute magnitudes of the bulges and disks as a function of the D/B ratios. It was found that while the bulge luminosities increase from spirals towards S0s and reach a constant value among S0 and disky Es, the disk luminosities decrease along the entire sequence. This suggests that disks in ellipticals could have formed in a similar way as disks in S0s (we refer here to intrinsic formation theories) with the difference that in the disky E case, only small amounts of gas (10% of the whole mass) led to the formation of the disks.

4. Conclusions and Implications for Formation Scenarios of Disky Elliptical Galaxies

The present study indicates that there is a continuous sequence in D/B ratio between E-S0 galaxies. It was found that the angular momentum vectors of disks and bulges are aligned in nearly all objects, which excludes random accretion events as the origin of the disks and therefore supports the connection to S0 galaxies. After disk subtraction, the bulges are still rotating. A correlation was found indicating that more compact bulges are more rotationally flattened than less compact ones (see Fig. 18 in Scorza and Bender 1995). We find that bulges of galaxies showing merger signatures deviate from this correlation and conclude that those objects following the correlation have most likely an undirsturbed history.

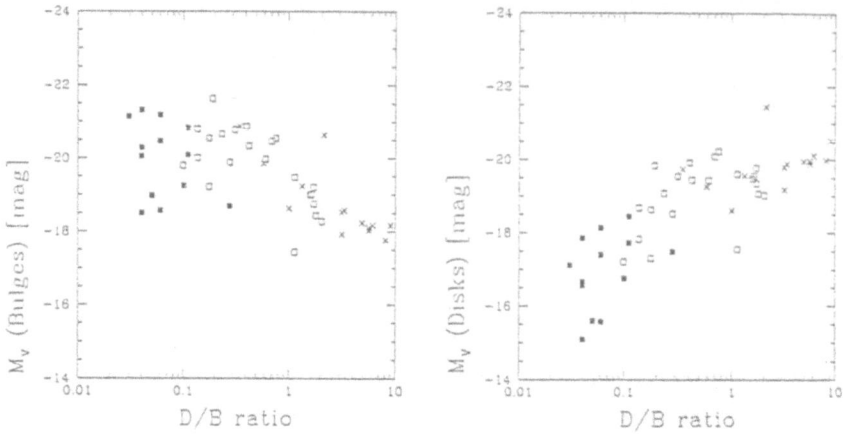

Figure 2. (a) Absolute magnitudes of the bulges and (b) of the disks as a function of the D/B ratios for spirals (crosses), S0 galaxies (open squares) and disky ellipticals (filled squares)

4.1. DID DISKY ELLIPTICALS FORM FROM MERGERS?

Collisionless mergers: this scenario cannot explain the high phase-space densities (Carlberg 1987) and unresolved cores of disky ellipticals (Kormendy 1987, Kormendy et al. 1995). Moreover, Steinmetz and Buchner (1995) have recently investigated the kinematic properties of merger remnants in dissipationless simulations. They find that all remnants are anisotropic and most of them boxy, and conclude that disky ellipticals could not have formed in collisionless merger events.

Mergers "con gas": this type of mergers give account for the formation of kinematic sub-components in elliptical galaxies (counter-rotating cores and central disks) as shown by Hernquist and Barnes (1991) and Barnes (1993). The question whether it could also be possible to form the rotationally flattened bulges of disky ellipticals in this way, is still open. Barnes (1993) has shown how small amounts of gas can change the orbit structure of galaxies during mergers; namely the box and X-tube orbits are transformed into Z-tube orbits and the remnant becomes a more oblate structure. A clearer answer will be given by measuring the V/σ ratio of the bulges of these remnants and by comparing them with the observed values.

References

Barnes, J.E., 1993, *The Formation of Galaxies*, proceedings of the V Canary Islands Winter School of Astrophysics, ed. Munoz-Tunon, C., Cambridge University Press
Bender, R., 1988b, *Astr. Ap.*, **202**, L5
Bender, R., Surma, P., Döbereiner, S., Möllenhoff, C., Madejsky, R., 1989, *Astr. Ap.*, **217**, 35

Bender, R., 1990, *Astr. Ap.*, **229**, 441.

Bender, R., Saglia, R.P, Gerhard, O.E., 1994, *MNRAS*, **269**, 785

Carlberg, R.G., 1986, *Ap. J.*, **310**, 593

Carter, 1987, *Ap. J.*, **312**, 514

Hernquist, L., Barnes., J.E., 1991, *Letters to Nature* 354, 210

Kormendy, J., 1987, in the IAU Symposium 127 *Structure and Dynamics of Elliptical Galaxies*, eds. T. de Zeeuw

Kormendy, J., et al. 1995, this proceedings

Kent, S., 1985, *Ap. J. Suppl. Series*, **59**, 115

Nieto, J-L., 1988, 2da Reunion de Astronomia Extragalactica, Academia Nacional de Ciencias de Cordoba, Cordoba Argentina, p. 239

Rix, H.-W., White, S., 1990, *Ap. J.*, **362**, 52

Scorza, C., Bender, R., 1990, *Astr. Ap.*, **235**, 49

Scorza, C., Bender, R., 1995, *Astr. Ap.*, **293**, 20

Steinmetz, M., Buchner, S., 1995, in preparation

DISCUSSION

Rich: How do bulges of ellipticals compare with spiral bulges?

Scorza: From the kinematic point of view, most bulges of spirals, S0s and disky ellipticals are rotationally flatenned. There exists also a smooth trend in absolute luminosities, the bulges of spirals being less luminous than those of S0 and ellipticals (see Fig. 2.a in this contribution).

Zepf: Is there evidence for an environmental dependence in the D/B ratio for early-type galaxies?

Scorza: An answer to this question will be provided after an analyses of a larger sample. Segregration studies are on the way in the Sternwarte-München, under the supervision of Ralf Bender.

Forbes: Is there any evidence for an age difference between the disk and the bulge in an individual elliptical galaxy, say from colors?

Scorza: We haven't look systematically into the colors of the galaxies. There are some evidences for the disks being younger as the bulges in some objects, as indicated by the higher H_β absorbtion along the major axis (where the disks lie). de Jong et al. (see this conference) have found also a correlation between H_β absorbtion and diskyness, suggesting the same result.

Huizinga: Do the exponential disks extend all the way to the centre, or do some show central holes?

Scorza: We find both types of behaviour. However most of all objects have disk profiles which extend all the way to the center but deviate sometimes from the exponential law. They seem to follow more an $r^{1/4}$ profile. Note however, that seeing makes conclusions concerning the central SB_0 quite uncertain.

Cecilia Scorza and Ortwin Gerhard

POPULATION SYNTHESIS MODELS

GUSTAVO BRUZUAL A.
C.I.D.A.
Apartado 264, Mérida 5101-A, Venezuela

1. Introduction

Increasingly complex population synthesis models have been developed during the last few years. Among the different authors working in the field, these groups have been particularly active: Arimoto & Yoshii (1987), Guiderdoni & Rocca-Volmerange (1987), Buzzoni (1989), Fritze-v. Alvensleben & Gerhard (1994), Bressan, Chiosi, & Fagotto (1994), Bruzual & Charlot (1993, 1995). The basic astrophysical ingredients used in these models are the stellar evolutionary tracks and the stellar spectral libraries (either empirical or theoretical). The computational algorithms are different in each set of models, but, one way or the other, these models depend on the same adjustable parametric functions: *(1)* the stellar initial mass function (IMF); *(2)* the star formation rate (SFR); and *(3)* the rate of chemical enrichment (in some models $Z = Z_\odot = constant$). Most of the recent codes use the isochrone synthesis algorithm, which allows to compute the evolution of a simple stellar population (SSP), and from it, by a convolution integral, the properties of more complex composite stellar populations (CSP).

From the population synthesis models one obtains basically the time evolution of the galaxy spectral energy distribution (SED), integrated colors, and spectral indices. From the model SEDs one can derive a *spectral age* for a given problem galaxy SED (i.e. the age of the model that reproduces best the problem SED). The spectral age depends, in general, on the IMF, the SFR, the assumed metallicity, and the particular set of models used. From the selected best fitting model and assuming a cosmological model, one can relate galaxy magnitude and colors with redshift (z), or galaxy number counts with galaxy apparent magnitude, etc. All these topics will be discussed in detail during this meeting.

R. Bender and R. L. Davies (eds.), New Light on Galaxy Evolution, 61–69.
© *1996 IAU. Printed in the Netherlands.*

Are population synthesis models consistent? A desirable property of population synthesis models is that all the models produce the same answer for the same set of input parameters. This is in fact not always the case. Some of the discrepancies among the models may be expected from the differences in the ingredients used to build them, but one should understand these differences and their origin. To investigate the origin of the remaining differences, one can fix a subset of the input parameters and vary the remaining part.

2. Comparison of 3 Sets of Models

A detailed comparison of the differences in the predictions from the models by Bruzual & Charlot (1995, hereafter BC95), Worthey (1994, hereafter W94), and Bertelli, Bressan, Chiosi, Fagotto, & Nasi (1994, hereafter BBCFN) has been performed by Charlot, Worthey & Bressan (1996, hereafter CWB96). I will comment here on the main conclusions of this work.

The BC95 models constitute a revised, updated, and debugged version of the Bruzual & Charlot (1993, hereafter BC93) models. Evolutionary tracks from the Geneva (Schaller et al. 1992, Charbonnel et al. 1995) and Padova (Bressan et al. 1993) groups for solar metallicity have been used to construct two parallel sets of models. The spectral library is mostly empirical, based on the Gunn & Stryker (1983) atlas, and improves the coverage of the coolest giants in the optical range with the observations by Fluks et al. (1994). The color vs. T_{eff} and bolometric correction (BC) vs. T_{eff} relationships used to transform the theoretical HR diagram to the observational plane have been revised to follow as close as possible the standard calibrations.

The W94 models are based on the evolutionary tracks by VandenBerg (1985) and the revised Yale isochrones (Green, Demarque, & King 1987). The stellar SEDs are obtained from the Kurucz (1992) set of model atmospheres for $T_{eff} > 3750$ K, and Bessel et al. (1991, and references therein) for cooler stars (supplemented by optical SEDs from the Gunn & Stryker atlas). These models cover a wide range of metallicities.

The BBCFN models are based on evolutionary tracks computed by Alongi et al. (1993), Bressan et al. (1993), and Fagotto et al. (1994). The Kurucz (1992) set of model atmospheres is used for stars of $T_{eff} > 3500$ K, and observed spectra for cooler stars. These models as well cover a wide range of metallicities.

The BC95, W94, and BBCFN sets of models sample and combine a good fraction of the theoretical and empirical data available in the literature. By computing the same galaxy model with the 3 codes one can explore the sensitivity of the population synthesis models to changes in the input in-

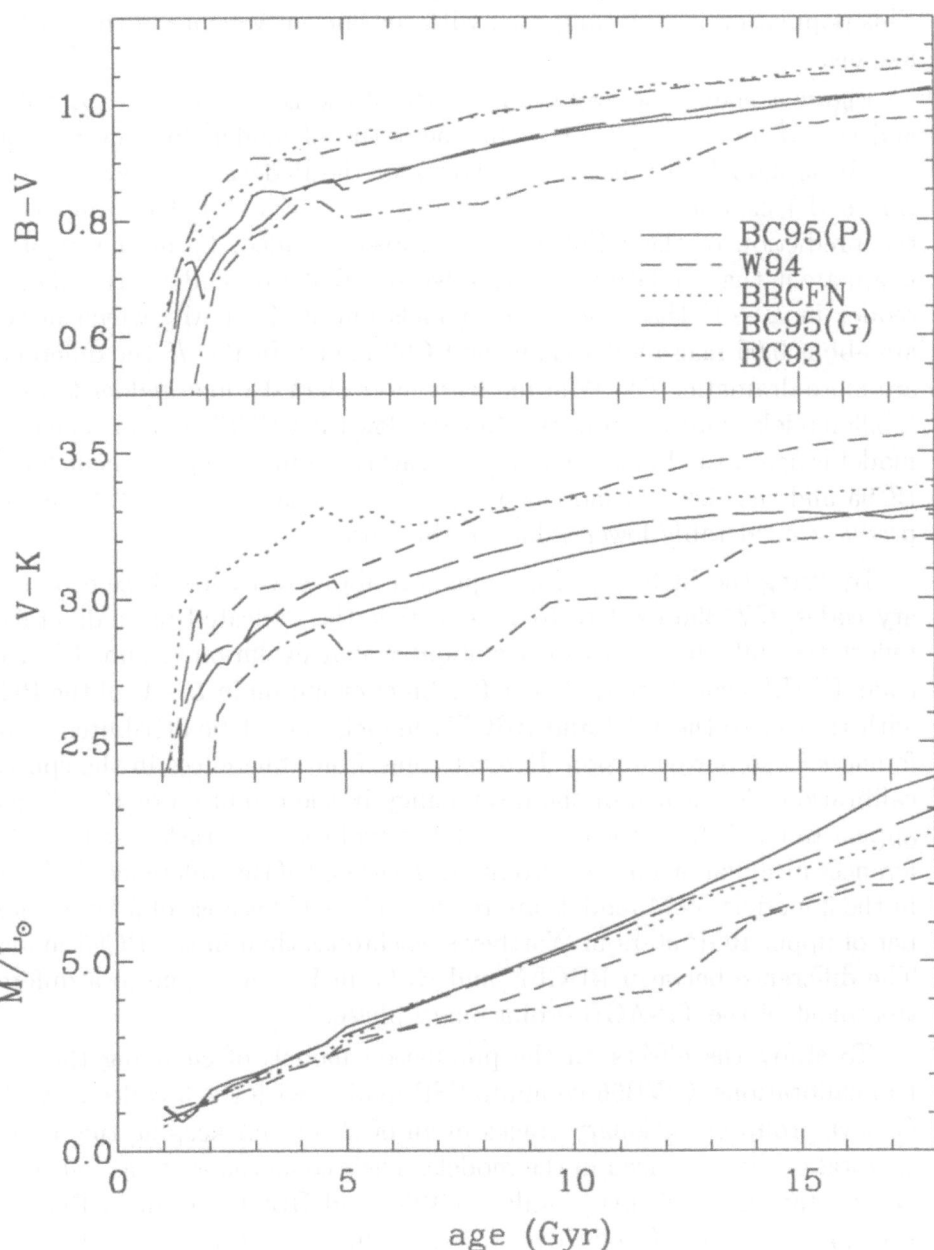

Figure 1. Evolution in time of the $B - V$, and $V - K$ colors, and the M/L_V ratio for the 3 sets of models discussed in the text.

gredients. CWB96 chose to compute a solar metallicity SSP for the Salpeter (1955) IMF ($m_L = 0.15, m_U = 2 \ M_\odot$) in the age range from 1 to 17 Gyr. This population can be computed with the current versions of the 3 sets of models.

Figure 1 shows the evolution in time of the $B - V$ and $V - K$ colors, and the M/L_V ratio predicted by the 3 sets of models for this SSP. For the BC95 models the results are shown for the Padova group evolutionary tracks (P) as well as for the Geneva group tracks (G). For comparison, the prediction of the BC93 models is also included. From this figure it is apparent that at late ages the W94 and BBCFN models are 0.05 mag redder in $B - V$ than the BC95 models (either C or M), which in turn are about 0.06 mag redder than the BC93 model. In $V - K$ the differences are more dramatic. The W94 model is more than 0.2 mag redder than the BC95 models, which are only 0.04 mag below the BBCFN model. The BC93 model is again the bluest. The M/L_V ratios are in good agreement for the BC95 and the BBCFN models up to 15 Gyr. The W94 and BC93 models predict considerably lower values of this ratio.

By using the BC95 empirical spectral library with the three evolutionary codes, CWB96 explore to what extent the indicated color differences reflect the different prescriptions about stellar evolution assumed in each code. CWB96 conclude that: (1) The bluer evolution in $B - V$ of the BC95 with respect to the W94 and BBCFN models cannot be attributed to differences in stellar evolution. It must come from differences in the spectral calibrations. (2) Much of the discrepancy in the evolution of $V - K$ does appear to result from the different stellar evolutionary tracks used. (3) Differences in stellar evolution account for nearly all of the differences in M/L_V in the 3 models. W94 models are redder in $V - K$ because of a larger number of upper RGB stars in Worthey's isochrones than in the BC95 models. The difference between BBCFN and BC95 in $V - K$ is due to a different treatment of the TP-AGB evolutionary phase.

To study the effects on the population models of changing the spectral calibrations, CWB96 compute SSP models with the 3 codes using the Geneva group evolutionary tracks in all of them, but keeping the original spectral calibration used in the models. Their conclusions are as follow: (1) Dwarf stars are 0.05 mag redder in W94 and BBCFN than in BC95 for $\log T_{eff} > 3.6$. (2) Cooler dwarfs are about 0.1 mag redder in $B - V$ in W94 and BC95 than in BBCFN. These cool dwarfs contribute very little to the total SED. (3) For giants with $\log T_{eff} > 3.65$, $B - V$ is 0.06 mag redder in W94 and BBCFN than in BC95 or the Johnson (1966) calibration. CWB96 think that this difference is introduced by the Kurucz (1992) model atmospheres, which are redder than the observed SEDs. (4) BC95 use the $V - K$ vs. T_{eff} calibration of Ridgway et al. (1980). W94 stars are 1

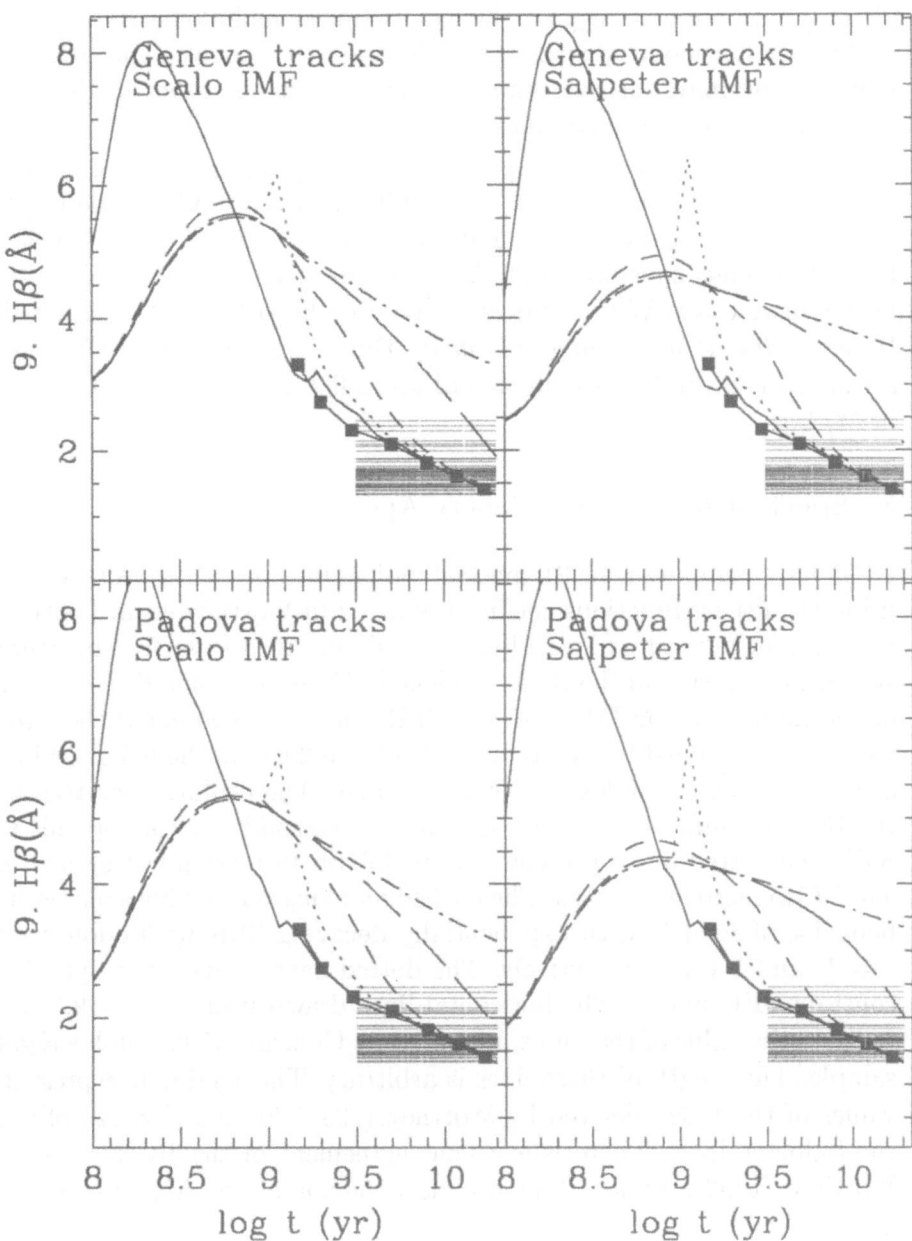

Figure 2. Evolution in time of the Hβ spectral index. The meaning of the different lines and symbols is explained in the text.

mag redder than this calibration cooler than $\log T_{eff} = 3.51$. BBCFN stars are in turn 1 mag bluer than the Ridgway et al. calibration. *(5)* Cool giants (3300 K) are about 1 mag brighter at V (for same bolometric luminosity) in the BBCFN model than in the BC95 and W94 models. The BC used by BBCFN are more positive than the ones used in the other models, which follow more closely the standard calibration.

CWB96 conclude that despite the large differences in the photometric calibrations, color evolution is very similar in the 3 models. The large differences in the adopted BC's and T_{eff} scale for giants in the BC95 and BBCFN models conspire to produce an almost identical $V - K$ evolution in these two cases. W94 giants are not as bright in V and $V - K$ is redder. However, *this is misleading agreement*. Differences in stellar evolution prescriptions can produce differences of up to 0.2 in $V - K$ and of 15 to 20% in M/L_V.

3. Spectral Indices and Galaxy Ages

BC95 have computed the 21 spectral indices introduced by Worthey (1992) using the fitting functions coefficients listed in Worthey et al. (1994). The pseudo-continuum flux levels have been determined from the spectrum assigned to each star in the BC95 models. BC95 have explored the sensitivity of the indices to the IMF and the SFR, and have compared their results with those obtained by Worthey et al. Figure 2 shows the behavior in time of the Hβ spectral index for several models. The evolutionary tracks and the IMF are indicated inside each panel. The solid line corresponds to an SSP. The dotted line represents a burst of star formation lasting for 1 Gyr. The SFR is zero afterwards. The dashed and long-dashed lines represent the behavior of models with exponentially decaying SFR with e-folding time $\tau = 1$, and 5 Gyr, respectively. The dotted-dashed lines corresponds to a constant SFR model. The horizontal lines drawn from $\log t = 9.5$ to 10.3 indicate the value of the index measured by González (1993) in his E galaxy sample. The length of these lines is arbitrary. The solid dots represent the values of the index derived by Worthey (1992) for the $Z = Z_\odot$ SSP with the Salpeter IMF. There is excellent agreement of the BC95 values with Worthey (1992). Figure 3 shows the behavior of the Mgb index for the same models.

Figures 2 and 3 show clearly how the spectral indices depend on the SFR, and to a lesser amount on the IMF and stellar physics. Even though some of the E galaxies in González sample have values of the indices reached at by an SSP at ages around 3 Gyr, the same values are reached at later ages by mostly old populations in which star formation has been going on over a longer time span. Galaxy ages > 10 Gyr do not seem to be ruled out

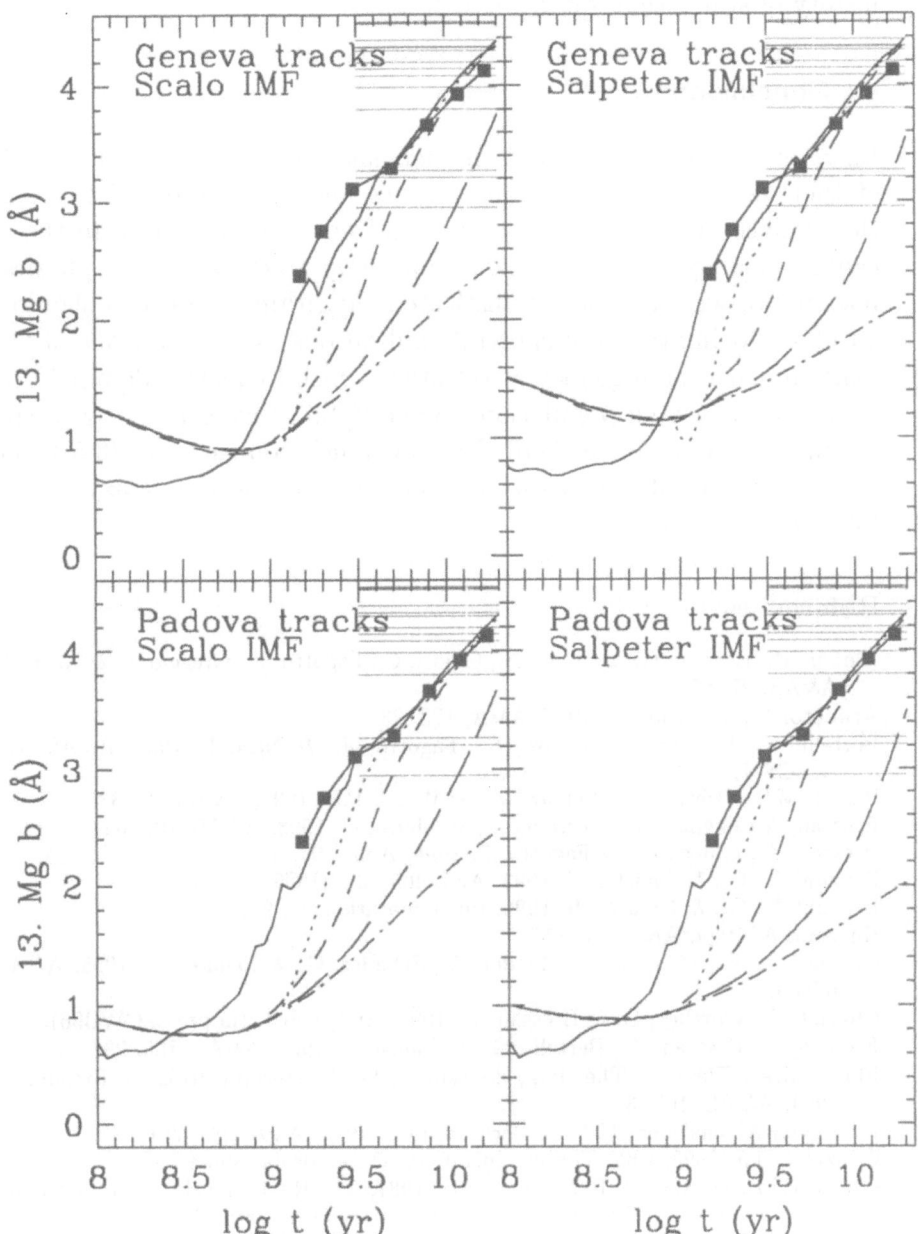

Figure 3. Evolution in time of the Mgb spectral index. The meaning of the different lines and symbols is explained in the text.

by the observations. Young galaxy ages based on fits to the values of the indices predicted for SSP's are uncertain as much as our knowledge of the history of star formation in these systems.

4. Conclusions

Users of population synthesis models should be aware of the sensitivity of their favorite models to the physics and assumptions that went into the models. We, model builders, should pay more attention to the stellar calibrations, spectral libraries. evolutionary tracks and their physics, and not only to writing more sophisticated algorithms. Observers should make an effort to improve the spectral calibrations in those regions of the HR diagram with the highest uncertainties, which frequently dominate the behavior of integrated populations. Some E/S0 galaxies may have a spectral (or index) age as low as 8-10 Gyr, but much younger ages (3-5 Gyr) may not be required if the history of star formation in the galaxy is properly taken into account.

References

Alongi, M., Bertelli, G., Bressan, A., Chiosi, C., Fagotto, F., Greggio, L. & Nasi, E. 1993, A&AS, 97, 851
Arimoto, N., & Yoshii, Y. 1987, A&A, 173, 23
Bertelli, G., Bressan, A., Chiosi, C., Fagotto, F., & Nasi, E. 1994, A&AS, 106, 275 (BBCFN)
Bessel, M.S., Brett, J.M., Scholz, M., & Wood, P.R. 1991, A&AS, 89, 335
Bressan, A., Fagotto, F., Bertelli, G., & Chiosi, C. 1993, A&AS, 100, 647
Bressan, A., Chiosi, C., & Fagotto, F. 1994, ApJS, 94, 63
Bruzual A., G., & Charlot, S. 1993, ApJ, 405, 538 (BC93)
Bruzual A., G., & Charlot, S. 1995, (in preparation, BC95)
Buzzoni, A. 1989, ApJS, 71, 817
Charbonnel, C., Meynet, G., Maeder, A., Schaller, G., & Schärer, D. 1995, A&AS, submitted
Charlot, S., Worthey, G. & Bressan, A. 1996, ApJ, Feb 1, (in press, CWB96).
Fagotto, F., Bressan, A., Bertelli, G., & Chiosi, C. 1994, A&AS, 105, 29
Fluks, M.A., Plez, B., Thé, P.S., de Winter, D., Westerlund, B.E., & Steenman, H.C. 1994, A&AS, 105, 311
Fritze-von Alvensleben, U.A. & Gerhard, O.E. 1994, A&A, 285, 751
González, J.J. 1993, PhD Thesis, University of California, Santa Cruz
Green, E.M., Demarque, P. & King, C.R. 1987, The Revised Yale Isochrones and Luminosity Functions (New Haven: Yale University Observatory)
Guiderdoni, B. & Rocca-Volmerange, B. 1987, A&A, 186, 1
Gunn, J.E. & Stryker, L.L. 1983, ApJS, 52, 121
Johnson, H.L. 1966, ARAA, 4, 193
Kurucz, R.L. 1992, in The Stellar Populations of Galaxies, IAU Symp. 149, eds. B. Barbuy & A. Renzini (Dordrecht: Kluwer), 225
Ridgway, S.T., Joyce, R.R., White, N.M., & Wing, R.F. 1980, ApJ, 235, 126
Salpeter, E.E. 1955, ApJ, 121, 161
Schaller, G., Schärer, Meynet, G., & Maeder, A. 1992, A&AS, 96, 269

VandenBerg, D.A. 1985, ApJS, 58, 711
Worthey, G. 1992, PhD Thesis, University of California, Santa Cruz
Worthey, G. 1994, ApJS, 95, 107 (W94),
Worthey, G., Faber, S.M., González, J.J., & Burstein, D. 1994, ApJS, 94, 687

DISCUSSION:

Peletier: The discrepancy you showed between your Mgb index and Worthey's might well be related to the way you have measured the indices. The way indices are measured on the Lick system is very hard to reproduce, since spectra of individual stars have to be transformed to a certain resolution and continuum shape. It might well be worth recalibrating the indices on flux calibrated spectra for all the stars of the Worthey et al. paper.

Bruzual: Your comment is well taken. However, since the oral version of this talk, G. Worthey pointed out to me that the fitting function coefficients in the copy of his thesis that I have are in error, and that I should use the ones in Worthey et al. (1994). After doing this, the discrepancies that I mentioned become less significant. But I still think that it may be useful to recalibrate the indices.

Bender: You showed that colors and line strengths can differ considerably for different stellar population models. What about the slopes? Can, e.g., relative ages be determined more accurately than absolute ages from measurements of Mgb or Hβ?

Bruzual: The slopes are considerably more stable than the intercepts. From the figures presented here you can see that it is possible to guess how much younger (or older) a stellar population is than a fiducial age. Even for different SFR's and IMF's the slope of the indices vs. age lines (at large ages) are not far from one another.

Worthey: Are the short timescale bumps and wiggles seen in all recent population synthesis models real, or can we smooth them away?

Bruzual: I think they are produced by numerical noise and that they can be smoothed out.

Renzini: Those little wiggles are just due to the numerics, i.e. to the specific numerical algorithm used in the synthesis. They are not real.

Renzini: All the comparisons you have shown refer to solar metallicity. Yet, the situation may be even worse for $Z > Z_\odot$, as theoretical model atmospheres become more and more unreliable, and empirical spectral libraries more and more incomplete. Do you agree with this statement?

Bruzual: I could not agree more with your comment. In fact these uncertainties have prevented S. Charlot and myself from extending our population synthesis models to metallicities above solar. The below solar regime is less uncertain and population synthesis for $Z \leq Z_\odot$ is more reliable than for $Z > Z_\odot$.

Gustavo Bruzual, Javier Gorgas, Nicolas Cardiel

FINDING AGES FOR OLD STELLAR POPULATIONS

GUY WORTHEY
Department of Astronomy
University of Michigan
Ann Arbor, MI 48109-1090

We are far from being able to populate a histogram of star formation versus time for an elliptical galaxy based solely on observations of its spectrum, but the path toward such a dream is becoming more clear. Still, we are denied the easiest paths, and most of what I have been thinking about in the last six months are the *obstacles* rather than the *opportunities* in age determinations for old stellar populations, and Es in particular. The following list should illustrate what I mean.

1. 3/2 degeneracy: we need abundance to get age

Age effects and metallicity effects have similar effects on broad-band colors and spectral features. Aaronson et al. (1978) found that this degeneracy followed the (approximate) rule $\delta \log \text{Age}/\delta \log Z \approx 3/2$ and this has been confirmed by subsequent studies. Thus a population 15 Gyr old but a little less than solar abundance mimics almost perfectly the colors and line strengths of a 5 Gyr population that is twice as metal-rich. The immediate implication is that one needs to know the abundance to 7% in order to get a 10%-accuracy age estimate.

2. 3/2 degeneracy: colors alone are not adequate

Since the usual case is that one knows neither the age nor the abundance at first, it is needful to find spectral indices that are dependent on only one parameter. Broad-band colors simply do not fit this description, and neither do most prominent absorption features. Worthey (1994) points out some indices which are relatively more age-sensitive than the average (Balmer lines) and some indices which are relatively more Z-sensitive. Taken in maximally contrasting pairs, the indices can be used to find a simultaneous "mean age" and "mean Z," both of which must be regarded as heuristic at

71

R. Bender and R. L. Davies (eds.), New Light on Galaxy Evolution, 71–74.
© 1996 *IAU. Printed in the Netherlands.*

this point in time. Work on finding more and better age- and Z-sensitive indices should command a high priority. To date, we have not explored much of what the spectrum can offer us.

3. model age uncertainties are ±35%

Charlot, Worthey, & Bressan (1996) have recently compared different synthesis models to find that the author-to-author spread in predictions correspond to an age error of ±35%. This spread comes from models with the same input metallicity, age, and IMF, and so is independent of any astrophysical uncertainties. Uncertainties in both the input stellar evolution and the assumed stellar fluxes contribute to this error. The 35% uncertainty extends also to Balmer-metal diagrams. The error can be reduced over time, but the process is not straightforward due to its complexity.

4. abundance ratios in real galaxies are not solar

O'Connell (1976) noticed that solar-neighborhood stars could not, in any combination, match the absorption features of giant ellipticals. Features due to Mg, Na, and CN are too strong compared to Fe blends (Worthey 1994). Since the deviations do not mimic age-Z degeneracy, variations in element-to-element abundance ratios are unambiguously implicated; giant Es have an abundance pattern more suggestive of halo stars than of stars in the solar neighborhood.

Why is this pertinent? Abundance ratio changes add a new level of complexity to age determinations because elemental mixture determines isochrone shape and stellar lifetimes, which directly affect age estimates. Furthermore, the two most important elements, Oxygen and Helium, don't have observable absorption features in the spectra of old populations, and so promise to be extremely difficult to calibrate. We already know that we cannot simply extrapolate the halo mix to higher metallicities because Na is enhanced in Es, but not in the halo. Furthermore, our Galactic bulge appears to have [O/Fe] nearly solar, again bucking the halo trend (Rich, this volume).

Even more pertinent, abundance ratio changes mean that a derived "mean age" will be a function of the elemental species used as the metal indicator. For example, using Lick/IDS data and Worthey models in Balmer-metal diagrams, ellipticals plotted versus Fe features look old and of nearly solar composition. However, plotted versus Mg_2, they look extremely metal-rich (maybe +0.75 to +1.0) and only 1-2 Gyr old. I am eager to see how the challenge of the abundance ratio dilemma will be met in years to come.

5. Balmer absorption is often filled by nebular emission

An observational difficulty with the current-best age indicators is that the stellar Balmer absorption can be partially or completely filled in by nebular emission.

6. populations that are composite in age and metallicity

Although it is now possible to invert a Balmer-metal diagram via models to obtain a "mean age" that doesn't mean it is a particularly good idea, since the "mean age" thus obtained is very strongly weighted toward any young populations that might be present. A 1-Gyr population has Balmer lines a factor of 5 stronger and a luminosity more than 6 times greater than a 15 Gyr population, so only a few percent by mass of a young population will dominate the age estimate.

Another aspect of composite populations that will cause trouble in the future is that the more sensitive metal lines probe the more metal-rich populations (since the lines are stronger for those populations) but Balmer lines and other age indicators sensitive to turnoff color will be more sensitive to the metal-poor subpopulations because they contain more warm stars. This is calibratable if one *assumes* a chemical enrichment history, but it looks like a difficult chore to derive it from the galaxy spectrum itself.

In summary, we find ourselves at the "mean age" stage, but beset by fairly serious problems, the most serious of which are instrinsic model unreliability and the whole new world of altered abundance ratios in Es. The next steps are (1) to explore age sensitivity in the spectrum to a greater extent, (2) to start a program of evolutionary models which take into account *many* different abundance mixes, (3) to begin to explore compositness, and (4) to utilize empirical feedback to improve overall model reliability.

References

Aaronson, M., Cohen, J. G., Mould, J. R., & Malkan, M. 1978, ApJ, 223, 824
Charlot, S., Worthey, G., & Bressan, A. 1996, ApJ, in press, Feb. 1
González, J. J. 1993, Ph.D. Thesis, Univ. of California, Santa Cruz
O'Connell, R. W. 1976, ApJ, 206, 370
Worthey, G. 1994, ApJS, 95, 107

DISCUSSION

O. Gerhard: Could you give me an indication of the spread of main sequence turnoff ages for 1 $M_\odot$ stars as a function of metallicity and elemental ratios?

G. Worthey: A 1 $M_\odot$ turnoff at [Fe/H] $\approx$ –1.0 is about 8 Gyr, at [Fe/H] $\approx$ 0 is about 12 Gyr. To a first approximation, if you change the elemental ratios the stellar lifetimes are approximately what you would get if you added all the metals up and forced a solar mix to equal that total. The details here are crying out for some good modeling. Above solar metallicity, helium abundance has a large impact on lifetime.

O. Gerhard: In some ellipticals, two mean metallicities have been assigned to their globular clusters, suggesting a merger history. If you took two stellar populations with these metallicities and related SF peaks what mean metallicity and age would you find?

G. Worthey: As a first cut, weight by V-band flux.

B. Dorman: Your plot [mean age versus mean abundance with galaxy index data from González 1993 inverted via Worthey models] puts M31 in a very strange location, with a similar age for M32, but much more metal-rich. My poster has a fit to M31 with age 10 Gyr and solar metallicity. The mid-UV flux strength (not dependent on any assumptions about the origin of the UV upturn) excludes a high metallicity except for very young ages.

G. Worthey: First, the age and metallicity zeropoints from the Balmer-metal diagrams are unknown, so from that perspective our conclusions about M31 are in accord. For large Es, about solar metallicity is what one derives from broad band colors assuming old ages. Second, the IUE data you use has a 10×20" aperture compared to ground based spectrographs which have something like a $1 - 2$" aperture. The González aperture is just large enough to get both the dynamical nucleus and the higher surface brightness interloper nucleus, so there is probably a bigger problem than just an aperture effect to remove before we are talking about the same target population. Third, your age-metallicity pair is along the 3/2 degeneracy slope from M31 in my figure. Fourth, I do not think that we can yet discount a youthful component or an underlying ordinary metal-poor-with-hot-horizontal branch component in (the *spiral* galaxy) M31 as significant contributors to the 1500-2500 Å flux. Fifth, based on abundances in our own Galactic bulge, one would naively expect to find nuclear populations in M31 which are greater than solar Z, if not [Fe/H].

B. Dorman: The galaxies with the strongest UVX appear to have a large age spread.

B. M. Poggianti: Studying Hδ strength in single stars and simple stellar populations using Kurucz's models we found that only warm stars have strong Balmer lines. For solar metallicities, only main sequence stars are placed in this "high Balmer region," but in metal poor populations, some horizontal branch stars populate that region. Balmer lines do *not* depend on metallicity for a single star, but in integrated spectra they *do* depend on it. Can you exclude that this effect affects your interpretation of the Hβ line for ages of galaxies and for radial gradients within galaxies?

G. Worthey: Yes, excluding bizarre composite population effects.

ELLIPTICAL GALAXIES:
THE AGE - METALLICITY DILEMMA

CESARE CHIOSI

Department of Astronomy
Vicolo dell'Osservatorio 5, 35122 Padova, Italy

Abstract. In this paper, we briefly report on the study by Bressan et al. (1995, BCT) who examine the properties of elliptical galaxies in the space of the parameters H_β, [MgFe], (1550-V), and velocity dispersion Σ, try to infer the age of these systems, and cast light on the age-metallicity dilemma.

1. Introduction

The merge of the recent work on population synthesis for the line strength indices H_β and [MgFe] by Worthey et al. (1994) and the spectro-photometric models for elliptical galaxies by Bressan et al. (1994, BCF) and Tantalo et al. (1995, TCBF) allows us to tackle the question whether or not these systems span a large range of ages. The indices H_β and [MgFe] are particularly suited to this purpose because H_β is a measure of the turn-off colour and luminosity, and age in turn, whereas [MgFe] is more sensitive to the RGB colour and hence metallicity. Gonzales (1993) analyzing the distribution of a sample of galaxies in the H_β - [MgFe] plane of single stellar populations with different metallicity and age concluded that a large spread in age seems to exist. BCT re-analyzed the same galaxies adding two more dimensions to the problem, i.e. the intensity of the UV excess as measured by the (1550–V) colour (Burstein et al. 1988) and the velocity dispersion Σ. No details are given here for the sake of brevity. Suffice it to recall that the source of the UV flux in the models is an admixture of old stars in various evolutionary stages, i.e. the classical P-AGB stars and the so-called hot HB and AGB-manqué stars of high metallicity, whose proportions and characteristics vary with the mean and maximum metallicity of the stellar mix (see BCF and TCBF). They also made use of the constraint imposed

R. Bender and R. L. Davies (eds.), New Light on Galaxy Evolution, 75–79.
© 1996 IAU. Printed in the Netherlands.

by the colour-magnitude relation (CMR) for elliptical galaxies of Bower et al. (1992) for the Virgo and Coma clusters. The small scatter of the CMR in the (U–V) colour (0.04 mag) is understood as the signature of a small scatter in the age of elliptical galaxies (13.5 ÷ 15 Gyr). Since Bower's et al. (1992) and Gonzales' (1993) samples refer to different groups of galaxies, one may wonder whether they can simultaneously be used. The two samples have five objects in common, namely NGC 4374, 4472, 4478, 4552, and 4697, whose positions in the H_β - [MgFe] plane is not particularly distinct from that of all remaining galaxies, perhaps suggesting that the analysis below will not depend on the particular CMR in usage.

2. Galaxies in the four dimensional space

The model galaxies of BCT in the H_β - [MgFe] plane are shown in Fig. 1 (left panel) together with the Re/8 data of Gonzales (1993). The solid lines are loci of constant mass (in units of $10^{12} M_\odot$) and nearly constant mean and maximum metallicities, the dotted lines are loci of constant age (in Gyr). The locus named CMR-strip is the expected trend for galaxies obeying the CMR of Bower et al. (1992) on the notion that this is a mass-metallicity sequence of old (13 ÷ 15 Gyr), nearly coeval objects. There are two important hints emerging from the H_β - [MgFe] data:

(1) In spite of their different luminosity and mass, most galaxies seem to possess nearly identical chemical structures.

(2) Galaxies do not distribute along the locus expected for coeval old objects following the mass-metallicity sequence of the CMR. In contrast, they seem to follow a sequence of about constant metallicity and varying age, which is the conclusion reached by Gonzales (1993).

In order to reconcile things, one may argue that the large spread along the H_β direction could be the result of a recent episode of star formation shifting the galaxies from their natural location (the CMR-strip) to the one we actually see. Deciphering the true age is not possible (see BCT for details). The major drawback with this idea is that a mechanism synchronizing the bursts in different galaxies is required. Using the simultaneous inspection of the data in the H_β - [MgFe] - (1550–V) - Σ space we get the following results: 1) There is a group of galaxies (NGC 4649 as a prototype) with ages confined in the range 13 ÷ 15 Gyr whose (1550 − V) colour is normal, i.e. fully compatible with the theoretical expectation for old objects containing a certain fraction of high metallicity stars. Differences from object to object can be accounted for by differences in mass and hence mean metallicity and perhaps in age by as much as 3 Gyr. 2) There is another group of galaxies (M32 and NGC 584 as prototypes) whose H_β is too blue for their [MgFe] and (1550 − V). While these latter parameters suggest an

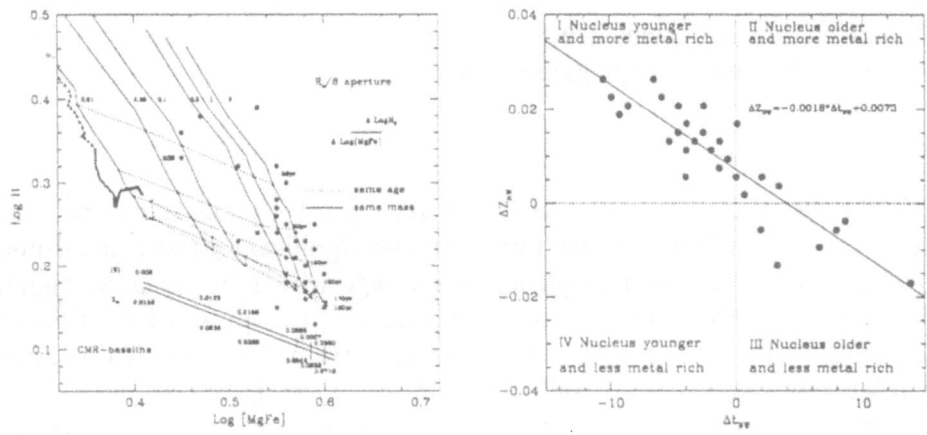

Figure 1. Left Panel: The H_β - [MgFe] plane of theoretical models and observational data. Right panel: the relation between Δt_{WN} and ΔZ_{WN}, i.e. the age and metallicity difference between the whole galaxy and its central regions

old age, H_β indicates a young age. It looks as if old objects suffered from a recent episode of of star formation that changed H_β leaving [MgFe] and $(1550 - V)$ unaltered.

Although the bursting hypothesis seems to be inevitable, before accepting this conclusion we examine the information provided by the gradients in H_β and [MgFe] observed across the galaxies under examination.

3. Gradients in age and metallicity

The quantities $\Delta H_{\beta NW} = \Delta \log H_{\beta N} - \Delta \log H_{\beta W}$ and $\Delta[\text{MgFe}]_{NW} = \Delta \log [\text{MgFe}]_N - \Delta \log [\text{MgFe}]_W$ where N and W stand for the Re/8- and R/2–data set of Gonzales (1993), i.e. for the central regions and the whole galaxy, measure the gradients within each individual galaxy. Translating $\Delta H_{\beta NW}$ and $\Delta[\text{MgFe}]_{NW}$ into Δt_{NW} and ΔZ_{NW}, i.e. in the age and metallicity difference between the nucleus and the whole galaxy, respectively, (see BCT for all details), one gets the correlation between Δt_{NW} and ΔZ_{NW} shown in the right panel of Fig. 1. Of the four quadrant's, the one characterized by $\Delta t_{NW} < 0$ and $\Delta Z_{NW} > 0$ is particularly rich of galaxies. In this region the nucleus is younger and more metal-rich than the external regions of the galaxy. This corresponds to a sort of out-inward process of

galaxy and star formation that in some cases continued for significantly
long periods of time. Whether the star formation process took place continuously or in a series of episodes we cannot say. BCT find the correlation
$\Delta t_{NW} = 3.327 \times \log \Sigma - 10.65$, where Δt_{NW} is in Gyr, Σ is in km sec[1], and
the correlation coefficient is 0.43. Δt_{NW} gets shorter at increasing velocity
dispersion Σ (and perhaps galactic mass).

4. Conclusions

(1) The distribution of galaxies in the H_β - [MgFe] plane is neither a sequence of sole metallicity with bluer galaxies significantly more metal-poor
than the red ones, nor a sequence of sole age with bluer galaxies significantly younger than the red ones. Furthermore, the observed distribution
seems to be in conflict with the expectation from the CMR if this latter
is interpreted as a mass-metallicity sequence of nearly coeval, old objects.
The hypothesis of a random burst of star formation in order to reconcile
the interpretation of the H_β, [MgFe], $(1550 - V)$ and Σ data and the CMR
requires a sort of *ad hoc* mechanism of synchronization. We rather favour
the idea that the overall duration of the star forming activity is somehow
related to Σ and perhaps galactic mass.
(2) The scenario emerging from the analysis of the gradients in H_β and
[MgFe] is consistent with current data on abundance ratios in elliptical
galaxies (Carollo et al. 1993, Carollo & Danziger 1994, Matteucci 1994)
and the expected trend for the ratio [Mg/Fe] inferred from the narrow band
indices (Faber et al. 1992, Davies et al. 1993). [Mg/Fe] seems to increase
with the galactic mass up to a value exceeding that of the most metal-rich
stars in the solar vicinity by $0.2 \div 0.3$ dex. According to Matteucci (1994)
this trend hints that the efficiency of star formation increases with the
galactic mass and that massive ellipticals should have formed the bulk of
their stars on shorter time scales than smaller ellipticals.

References

Bower, R. G., Lucey, J. R., & Ellis, R. S., 1992, MNRAS, 254, 601
Bressan, A., Chiosi, C., Fagotto, F., 1994, ApJS, 94, 63, BCF
Bressan, A., Chiosi, C., Tantalo, R., 1995, A&A, submitted, BCT
Burstein, D., Bertola, F., Buson, L. M., Faber, S. M., & Lauer, T.R., 1988, ApJ, 328 440
Carollo, C. M., & Danziger, I. J., 1994, MNRAS 270, 523
Carollo, C. M., Danziger, I. J., & Buson, L., 1993, MNRAS, 265, 553
Davies, R. L., Sadler, E. M., & Peletier, R. F., 1993, MNRAS, 262, 650
Faber, S. M., Worthey, G., Gonzales, J. J., 1992, in The Stellar Populations of Galaxies,
    B. Barbuy & Renzini eds., Kluwer Academic Publishers, Dordrecht, 225
Gonzales, J. J., 1993, Ph.D. Thesis, Univ. California, Santa Cruz
Matteucci, F., 1994, A&A, 288, 57
Tantalo, R., Chiosi, C., Bressan, A., & Fagotto, F., 1995, A&A, submitted, TCBF
Worthey, G, Faber, S. M., Gonzales, J. J., & Burstein, D., 1994, ApJS, 94, 687

Dickinson: We are now at the point where we can observe clusters of galaxies directly out to z=1 and beyond, select ellipticals morphologically from HST images, and study the evolution of the CMR. So far, these data show that both the slope and the very small scatter in the CMR are preserved out to z=1, i.e. to lookback times of $\simeq 10 \times H_{50}^{-1}$ Gyr. Can any of the models invoking wide age variations among elliptical galaxies account for these observations ?

Chiosi: By age of a galaxy we mean the age of the constituent stellar populations. In this context we tried to show that indeed the bulk stars are old objects (as indicated by their [MgFe] and (1550-V) data). Superposed to the old component, in some cases there might be a younger one (as suggested by the H_β and gradients in H_β and [MgFe]), which can mimic an age sequence. We do not expect the CMR to be much affected by the presence of this younger popualtion, considering the very short time scale on which a galaxy after a minor period of stellar activity recovers the previous situation. This can be inferred from comparing isochrones with the observational CMR (and blue luminosity to mass ratio).

Lee: If there is an age spread among elliptical galaxies, we do not need super metal-rich hot HB stars (see Lee & Park, this volume) to produce systematic variation of UV upturn. Then, why do we have to add completely ad-hoc assumptions required to produce hot super metal-rich HB stars in population models in addition to age spread ?

Chiosi: First excluding ongoing star formation as source of UV radiation, past the period of star formation the color (1550-V) gets very red (cf. BCF and TCB). Second, let us suppose that the the source of UV flux are the normal P-AGB and HB stars. In such a case, it would be difficult to match the observational dependence of the UV upturn with the galaxy mass, Mg2 index (metallicity ?) and luminosity (cf. BCF fo details). The hot HB and AGB-manqué channels are the natural consequence of the chemical evolution $(\Delta Y/\Delta Z)$ inside a galaxy. This is suggested by the detailed shape of the spectrum in the UV, the above relation, and the mean metallicities (cf. for instance the data in H_β - [MgFe] plane), only to mention a few. The kind of age spread we are talking about has little relationship with the UV properties.

Pepi Fabbiano and Guy Worthey

THE RATE OF SUPERNOVAE IN NORMAL GALAXIES

ENRICO CAPPELLARO
Osservatorio Astronomico di Padova
vicolo dell'Osservatorio 5, Padova, I-35122, Italy

1. Introduction

The rate of supernovae (SNe) is a key number linking stellar evolution with galaxy evolution models. Stellar evolution theories predict life times, fates and nucleosysntesis yields of individual stars which are used to predicted the galaxy chemical evolution once the star formation history in the galaxy is known. Constraints to the models are the present chemical content of galaxies but also the present observed SN rate (Arimoto & Yoshi, 1987; Ferrini & Poggianti, 1993; Matteucci, 1994; Renzini *et al.*, 1993; Bressan *et al.*, 1994; Elbaz *et al.*, 1995).

Over 90% of the SNe for which adequate observations are available can be assigned to one of the three basic SN types: Ia, Ib/c or II (e.g. Harkness & Wheeler, 1990). The progenitors of the different SN types belong to different stellar population: SN II and SN Ib/c result from young massive progenitors ($M_i > 8 - 10M_\odot$), hence their rate is expected to be directly related to the star formation rate (SFR). Instead SN Ia derive from low mass progenitors in close binary systems: their rate depends mainly on the parameter of the binary population, in particular the distribution of the mass ratios and of the separations of the two components. Indeed SN Ia are found also in ellipticals galaxies, where star formation ceased several billion years ago. Therefore, the rate of the different SN types can be used as a probe of the star formation history in the different type of galaxies.

2. SN rate estimates

Observational estimates of the SN rate, can be obtained through two different approaches called respectively *fiducial sample* and *control time* methods.

The fiducial sample method is based on the assumption that, during the last 30-40 years, all nearby, bright galaxies have been throughout searched

R. Bender and R. L. Davies (eds.), New Light on Galaxy Evolution, 81–84.
© *1996 IAU. Printed in the Netherlands.*

for SNe (Tammann *et al.*, 1994). It must be stressed that using the fiducial sample method only relative SN rate can be derived because the surveillance time is unknown. Moreover, no account is made for the fact that different SN type have different brightness and therefore different discovery probability.

These problems are eliminated by using the control time method, which is based on the detailed analysis of the log of individual SN searches. Unfortunately only for a handful of the many SN searches which have been carried out in the past decades the galaxy sample and control time are known. Among these are the visual search conducted by Evans (van den Bergh & McClure, 1994), the CCD search by the Berkeley group (Muller *et al.*, 1992), and the Asiago and Crimea photographic surveys (Cappellaro & Turatto, 1988; Tsvetkov, 1987). Because the SN sample is only made by the SNe discovered in a particular survey, in general the main caveat of this method is the small SN statistics which, however, can be improved by combining the data of different searches (Cappellaro et al. 1993a, 1993b, Turatto el al. 1994).

A comparison between recently published estimates of the SN rate is reported in Table 1. The table is divided in three sections for different galaxy morphological types. In the first column is the search identification, in the second is the number of SNe on which the SN rate estimates of columns 3 to 5 are based. Since early seventies it has been demonstrated that the rate of SNe is proportional to the (blue) galaxy luminosity (Tammann, 1974). This is why SN rates are expressed per unit of $10^{10}L_\odot$.

For the combined data of the Asiago+Crimea surveys, a special effort was made to estimate the errors. There are three types of uncertainties which affect the estimates: the event statistics, the errors in the input parameters and the discovery biases. The latter are particularly important since they cause a systematic under-estimate of the SN rate. Two biases have been found of particular relevance for photographic searches: the loss of SNe in the central region of distant galaxies (Shaw, 1979) and in inclined spirals. Whereas for the Asiago+Crimean surveys the first one is not of great importance (only a 10% correction need to be applied, Turatto *et al.*, 1994), the rate in late spirals must be multiplied by a factor three to account for the loss of SNe in inclined spirals. The uncertainties on this factor gives a major contribution to the error-bar on the SN rate in late spirals reported in Table 1.

Considering the relatively large errors, there is a fair general agreement between the different estimates reported in Table 1, especially uncertainties, but also a few significant disagreements. In particular the rate of SN Ia in early type galaxies based on the fiducial sample is a factor 4 higher than that derived using the control time method. Possibly, this is related to a failure in the assumption that all galaxies of the fiducial sample have been

TABLE 1. Comparison between different estimates of the SN Rates.

E-S0	No. SNe	SN rate [SNu][*]	
		Ia	II+Ib/c
Asiago+Crimea[a]	8	0.13 ± 0.06	< 0.06
Evans [b]	2	0.14 ± 0.10	
fiducial sample[c]	14	0.51	

S0a-Sb		Ia	Ib/c	II
Asiago+Crimea[a]	17	0.17 ± 0.07	0.13 ± 0.11	0.30 ± 0.19
Evans[b]	12	0.17 − 0.32	0.00 − 0.22	0.00 − 0.81
fiducial sample[c]	19	0.27	0.02 − 0.16	0.00 − 0.77

Sbc-Sd		Ia	Ib/c	II
Asiago+Crimea[a]	35	0.39 ± 0.19	0.27 ± 0.18	1.48 ± 0.65
Evans[b]	9	0.10	0.20	0.60
Berkeley[d]	10	0.21 ± 0.13	0.88 ± 0.28	0.64 ± 0.28
fiducial sample[c]	61	0.27	0.43	2.21

[*] $1SNu = 1SNe \times (10^{10}L_\odot)^{-1} \times (100yr)^{-1}$. Rates in SNu scales as $(H/75)^2$
a – Cappellaro et al. 1993a,b; b – van den Bergh et al. 1994; c – Tammann et al. 1992;
d – Muller et al. 1992

searched for SNe at an equal intensity level (Turatto et al., 1994). Instead the surprisingly high rate of SN Ic in late spirals found by Muller et al. (1992) may be due to an underestimate of the peculiarity of their CCD search (Cappellaro et al., 1993b). Finally, the low rate of SN II in late spirals found by van den Bergh & McClure (1994) is mostly related to the claim that the visual search by Evans is unaffected by the spiral inclination bias which, due to the poor statistics, need further confirmation.

3. Conclusions

The rate of Ib/c+II SNe is strongly dependent on galaxy type, in late spirals being at least 30 times higher than in ellipticals. That is, the rate of SNe with massive progenitors is high in galaxies with high SFR.

The rate of SN Ia in late spirals is three times higher than in ellipticals. This is not a firm conclusion because only found using the Asiago+Crimea data. On the other side the Evans search has a small statistics and the

fiducial sample estimates is biased by the a priori assumption that the rate
of SN Ia is constant going from early to late spirals. We should note that
the initial mass of the SN Ia progenitor is more likely in the range 2–6 $M_\odot$.
Therefore, even if the complex evolution of binary systems may delay the
explosion for time of the order of one Hubble time and account for the SN Ia
in ellipticals, a high rate is expected some 10^8 years after a star formation
burst which is consistent with the Asiago+Crimea observations.

An estimate of the SN rate in our Galaxy can be obtained assuming that
it is similar to that in external galaxies of the same morphological type.
Adopting $Sb \pm 0.5$ for the Galaxy morphological type, $L_B = 2.0 \pm 0.6 \times$
$10^{10} L_\odot$ for the Galaxy luminosity and including all sources of uncertainties,
we expect in a millenium 3 ± 2 SN Ia, 2 ± 2 SN Ib/c and 12 ± 8 SN II. These
numbers are consistent, within the large errors, with the rates derived from
historical SNe, from SN remnants and from estimated pulsar birth rates.

Acknowledgments
I wish to thank Massimo Turatto with whom I shared most of the work
described in this contribution.

References

Arimoto, N., & Yoshii, Y.,1987, A&A, 173, 23
Bressan, S., Chiosi, C., Fagotto, F., 1994, ApJS, 94, 63
Cappellaro, E., Turatto, M., 1988, A&A 190,10
Cappellaro, E., Turatto, M., Benetti, S., Tsvetkov, D.Yu., Bartunov, O.S., Makarova,
    I.N., 1993a, A&A, 268, 472
Cappellaro, E., Turatto, M., Benetti, S., Tsvetkov, D.Yu., Bartunov, O.S., Makarova,
    I.N., 1993b, A&A, 273, 383
Elbaz, D., Arnaud, M., Vangioni-Flam, E., 1995, A&A, in press
Ferrini, F., & Poggianti, B.M., 1993, ApJ 410, 44
Harkness, R.P., & Wheeler, J.C., 1990, in Supernovae, A.G. Petschek, New York:
    Springer, 1
Matteucci, F., 1994, A&A, 288, 57
Muller, R.A., Newberg, H.J.M., Pennypacker, C.R., Perlmutter, S., Sasseen, T.P., Smith,
    C., 1992, ApJ, 384, L9
Renzini, A., Ciotti, L., D'Ercole, A., Pellegrini, S., 1993, ApJ, 419, 52
Shaw, R.L., 1979, A&A, 76, 188
Tammann, G.A., 1974, Supernovae and Supernova Remnants, C.B. Cosmovici, Dor-
    drecth: Reidel, 95
Tammann, G.A., Löffler, W., Schröder, A., 1994, ApJS, 92, 487
Tsvetkov, D.Yu.,1987, AZh, 64, 79
Turatto, M., Cappellaro, E., Benetti, S., 1994, AJ, 108, 202
van den Bergh, S., McClure, R.D., 1994, ApJ, 425, 205

SMALL STELLAR SYSTEMS
AND GALAXY CORES

THE FORMATION OF GLOBULAR CLUSTER SYSTEMS: HOW, WHEN, AND WHERE?

WILLIAM E. HARRIS
Dept of Physics & Astronomy, McMaster University
Hamilton ON L8S 4M1 Canada

Globular clusters, as fossil remnants of the protogalactic era, provide unique traces of the earliest events of galaxy formation. However, new observations – especially from HST – are showing that massive, globular-like star clusters belong not only to the pregalactic era but can form right up to the present day under the right circumstances. Appropriate interpretation may now let us learn *simultaneously* about the process of cluster formation as well as the nature of the gaseous fragments from which the galaxies were assembled.

Let us first briefly visit some of the relevant observations that have been accumulated for globular cluster systems (GCSs) in a wide range of galaxies, and which seem to constrain the formation parameters the most strongly. These include the cluster metallicity distributions, their specific frequencies (relative numbers), and their distribution by mass.

Metallicity distributions are now available for GCSs in galaxies of all types (though most often in giant ellipticals, which hold by far the largest cluster populations). Most such data come from metallicity-sensitive broadband photometric indices ($C - T_1, B - R$, etc.) that have been calibrated spectroscopically. These results show unambiguously that globular clusters can form at very much the same mass range over more than two orders of magnitude in heavy-element enrichment, from [Fe/H] $\simeq -2.3$ up to above-solar metallicity. In dwarf ellipticals the clusters are almost uniformly metal-poor (Harris 1991; Durrell 1995). The Milky Way clusters, with their well known bimodal metallicity distribution corresponding to halo and bulge clusters (Zinn 1985; Armandroff 1989; Minniti 1995) cover most of this [Fe/H] range, and the distribution in M31 is basically quite similar (e.g. Reed et al. 1994). In the giant ellipticals, the mean abundance shifts to higher levels, with mean values typically $<$[Fe/H]$> \gtrsim -1$ even far out into their halos. The most extreme case discovered so far is in the giant

R. Bender and R. L. Davies (eds.), New Light on Galaxy Evolution, 87–95.
© *1996 IAU. Printed in the Netherlands.*

cD NGC 3311 (Secker et al. 1995), in which a large fraction of the clusters are probably above solar metallicity and the mean is $<[Fe/H]> \simeq -0.3$. The gE's can also have bimodal or multimodal [Fe/H] distributions that strongly suggest several distinct enrichment phases during formation (e.g. Ashman & Zepf 1992; Zepf et al. 1995a). But many differ widely among themselves, exhibiting a menagerie of unimodal, multimodal, narrow, or wide distributions, with or without radial metallicity gradients (see Ajhar et al. 1994). These large galaxies may be the end results of complex and different formation histories, and it seems that any successful theory for globular cluster formation must be virtually immune to the degree of chemical enrichment or the type of the host galaxy. Additional evidence for this conclusion comes from the observation that one of the young, massive star clusters in NGC 4038/39 has near-solar metallicity (Zepf et al. 1995b).

Specific frequency S_N is the number of globular clusters per unit galaxy luminosity (Harris 1991), and represents the *global cluster formation efficiency* over the lifetime of the galaxy. Figure 1 shows the measured S_N values for E galaxies of all sizes and locations. Clearly, E's and dE's display very much the same range of specific frequencies (except for the occasional cD galaxies with anomalously high S_N; see Harris et al. 1995). This information for the dE's is especially valuable because these small galaxies must have formed in quieter, more isolated environments than the giants. Yet they have been just as efficient, on average, at generating rather normal globular clusters. This suggests, perhaps, that collisions and mergers which build up large galaxies may *not* be especially favorable places for globular cluster formation.

The **Mass Distribution Function** (MDF) is turning out to be a major new clue to the process of globular cluster formation. The MDF is, surprisingly, the most nearly constant phenomenon in the observations (Harris 1991, 1995; Harris & Pudritz 1994 [HP94]). The GCSs in all galaxies have near-identical MDFs, with a number distribution by mass following a power-law form $dN/dM \sim M^{-\gamma}$ for $M \gtrsim 10^5 M_\odot$. In the giant ellipticals, we generally find $\gamma = 1.6 \pm 0.1$, while for spiral galaxies and dE's $\gamma \simeq 1.9$ (HP94; McLaughlin 1994; Durrell et al. 1995; Durrell 1995). An exciting new result which reinforces the idea that the MDF is mostly determined by formation, rather than later dynamical evolution, is the discovery by Whitmore & Schweizer (1995) that the hundreds of recently formed star clusters in the merging NGC 4038/39 galaxy pair follow a MDF of just the same shape ($\gamma = 1.78$), simply shifted to higher luminosity in accord with their much younger ages. In short, the shape of the MDF is *almost* independent of *almost* everything: (a) parent galaxy size and Hubble type, (b) parent galaxy location and environment, (c) metallicity, (d) total cluster population S_N, and (e) even the epoch of formation, if the new HST

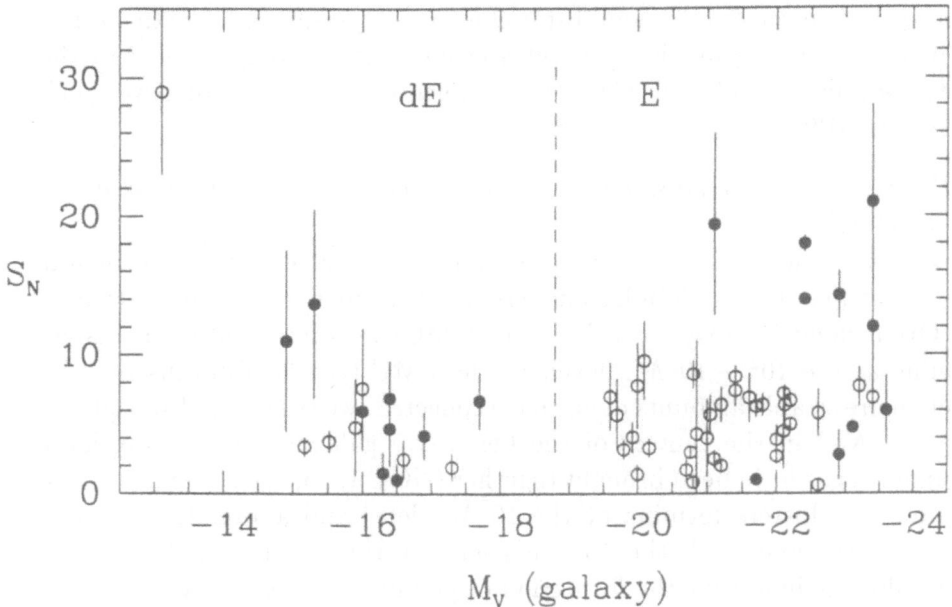

Figure 1. Specific frequencies for globular cluster systems in E and dE galaxies, where $S_N \equiv const \times N_{cl}/L_{V,gal}$. For the dE's, *filled symbols* denote nucleated dE's; while for the giant ellipticals, filled symbols denote central-giant cD galaxies. Most of the information from the dE's is from the new survey of Virgo dwarfs by Durrell (1995).

observations are to be believed.

Can all of this be put into a coherent framework? Our accumulated data add up to the view that globular cluster formation is an extremely robust process, and it seems time to put forward a new set of *observationally based* precepts which will allow us to construct a quantitative model. I suggest that we need the following three:

1. *Globular clusters in the early galaxies formed the same way we see star clusters form today:* that is, out of the rare, very dense gas cores that build up within giant molecular clouds (GMCs).

This view – that globular clusters do *not* come from some highly specialized formation mechanism but that they can form anywhere that gaseous raw material can collect into GMCs – is very close to the ideas that have frequently been expressed by Larson (e.g. 1990a,b), and is made explicit in HP94. As supporting evidence, we can note that virtually every part of the parameter space of cluster age, metallicity, and mass (τ, Z, M) is occupied by star clusters that can be found in *some* galaxy, and there is no compelling reason to distinguish any of these star clusters from one another.

Cluster formation within GMCs is a notably *inefficient process*: only a small fraction of the gas finds its way into the dense cores which eventually transform themselves into bound star clusters. Typically, in Galactic GMCs we observe that the median protocluster core mass is only $\sim 10^{-3}$ of the host GMC mass, and each GMC might form of order ~ 10 or fewer such cores (see HP94).

2. *The typical cluster mass goes up in direct proportion to the mass of the parent GMC.*

This statement implies that we need a *big* reservoir of gas to form star clusters as massive as globular clusters are. Consider, for example, that in the Orion cloud (a fairly typical, nearby GMC) we see a handful of clusters forming at the $10^2 - 10^3 M_\odot$ level. In the LMC (the 30 Doradus region), where more sizable amounts of gas have collected, we see a $\sim 10^4 M_\odot$ cluster forming. And in the centers of the big active galaxies where truly large amounts of gas have been brought together (NGC 1275, the Antennae, etc.), we now see clusters forming at the $10^5 M_\odot$ level and above. If we couple these observations with the 10^{-3} efficiency ratio mentioned above, then we would conclude that the formation of globular clusters of $10^5 - 10^6 M_\odot$ required host GMCs of $10^8 - 10^9 M_\odot$ – i.e., gas clouds that were far larger than any existing in the Milky Way today. HP94 define these as 'supergiant' molecular clouds (SGMCs).

Points (1) and (2) together suggest a rather straightforward answer to an old question: Why doesn't the Milky Way know how to make globular clusters any more? Our reply is that it *does* know how; but it simply does not have enough raw material left in the right form to build anything larger than the relatively tiny objects we call open clusters. The new HST observations demonstrate that any galaxy is capable of constructing massive clusters, as long as we can feed enough low-temperature gas into its potential well.

3. *The mass distribution function (MDF) of globular clusters is primarily a result of their formation process* and not dynamical evolution.

Numerous recent papers (see, e.g., Aguilar et al. 1988; Okazaki & Tosa 1995, for overviews) argue the importance of several dynamical effects on clusters (principally bulge and disk shocking, dynamical friction, and stellar evaporation). For low-mass clusters or ones very near the galactic nucleus, these destructive processes are unquestionably severe. But for most of the cluster mass range (i.e. $\gtrsim 10^5 M_\odot$) and for the majority of the halo ($R_{gc} \gtrsim$ 3 kpc), the MDFs described above are so similar in all galaxies that they seem unlikely to be the products of convergent evolution.

The first stages of a specific theory based on these three postulates are laid out in HP94 and in McLaughlin & Pudritz 1996 [MP96]. In this model, a given host GMC is assumed to contain a large supply of small cloudlets of mass m_0. These cloudlets collide and coalesce to build up larger clouds (and eventually protoclusters) over a range of masses. Once a protocluster gets above some critical point $m_\star$, it can begin star formation; soon after it does, it rapidly ejects its remaining gas by OB star winds and supernovae, thus partially repopulating the reservoir of cloudlets. A statistical equilibrium is thus set up, for which a rate equation can be written down and solved for the resulting distribution function $N(m)$. This agglomeration process is well known to yield a power-law MDF of the right approximate form (see, e.g., Field & Saslaw 1965; Kwan 1979, among others). MP96 generalize it by assuming that the *cloud lifetime against self-disruption* by star formation varies as $\tau_m \sim m^c$, for $m > m_\star$. Higher-mass clouds have lower mean densities (see HP94) and thus undergo relatively more disruptive star formation, so we expect $c < 0$; the actual model solutions (see below) give $c \simeq -0.6$. Now τ_m can be compared with the *collisional growth time*, which is simply $\tau_0 = m_0/\rho\sigma v$ for cloud velocities v, collision cross sections σ, and number densities ρ. The crucial ratio $\beta \equiv \tau_\star/\tau_0$ is what determines the steepness of the resulting mass distribution $N(m)$: the faster the growth time relative to the fiducial disruption time $\tau_\star$, the more clouds can build up to higher masses and the flatter the resulting exponent γ. At the very top end – near $5 \times 10^6 M_\odot$ if $m_\star \simeq 10^5 M_\odot$ – the buildup of still bigger protoclusters becomes so statistically improbable that the MDF finally cuts off quite abruptly.

This model has been used by MP96 to produce the first quantitative theory of the MDF for globular clusters. Model fits to the real galaxies with the best available data (M87, M31, and the Milky Way) are shown in Figure 2. The best-fit ratio β is noticeably higher for M87 than for the spiral galaxies, suggesting that the protocluster clouds built up much more rapidly in the giant proto-elliptical than in the proto-spiral halos.

A traditional problem with any cluster formation model has been to understand how the protoclusters can avoid rapid cooling through radiation from heavy elements or molecular hydrogen (e.g. Fall & Rees 1985; Murray & Lin 1992). That is, if the cloud is supported only by thermal pressure, then we expect $\tau_\star \sim \tau_0$ – essentially the free-fall timescale – and there would be no time for collisional growth to occur. However, GMCs and their embedded cores are strongly nonthermal, and appear to be supported primarily by turbulence and weak magnetic fields in the range of $10 - 100$ microgauss (e.g. Myers & Goodman 1988; Heiles et al. 1993). Thus the relevant quantity governing the cloud lifetime is the ambipolar diffusion time τ_{AD} for the magnetic field to leak out of the cloud, which is an order of magnitude larger than τ_{ff} (see HP94). This is the key factor which allows the clouds enough time to build up to high mass while still gaseous. For the Milky Way halo at $R_{gc} \simeq 8$ kpc, the expected growth time is only a few $\times 10^8$ yr (MP96), which sets a lower limit on the age dispersion among the globular clusters there.

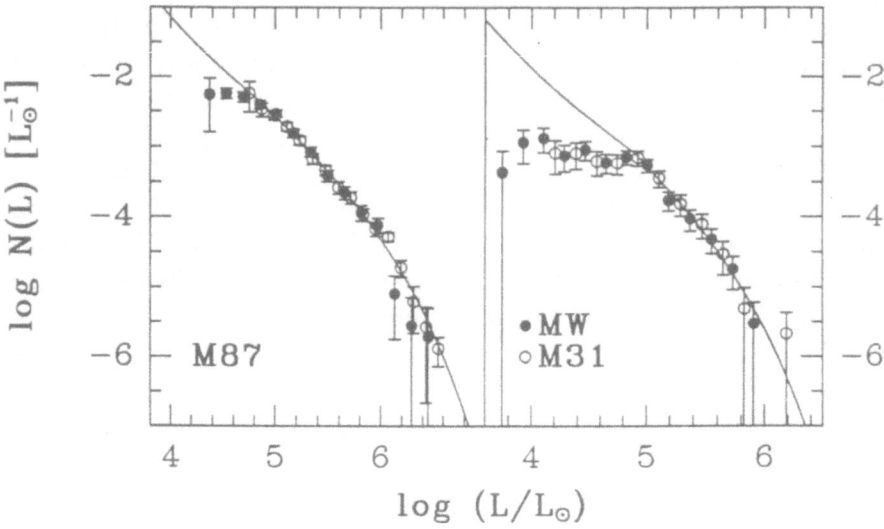

Figure 2. Observation vs. theory for the mass distribution functions of globular clusters in three galaxies, from McLaughlin & Pudritz 1996. Here $N(L)$ is the number of clusters per unit luminosity; the classic 'turnover' point (the maximum number of clusters per unit *magnitude*; see Harris 1991) is at $L_* \sim 10^5 L_\odot$. The solid line in each panel is the best-fit model for each set of data (see text): for M87, the deduced ratio β (the ratio of cloud lifetime to collisional growth time) is $\simeq 50$, while for M31 and the Milky Way, $\beta \simeq 10$ (a steeper MDF). For $L < L_*$ (the bottom 10% of the mass range), the collisional growth model predicts too many clusters, though these small objects may have been preferentially depleted by dynamical disruptive processes since their formation; see text.

For purposes of galaxy formation, the utility of this model is that we can use the known characteristics of globular clusters to gauge how large their parent clouds must have been, and thus to get a picture of the raw pieces that built galaxies. As is shown by HP94, the SGMC's need to be $10^8 - 10^9 M_\odot$ and up to ~ 1 kpc in size. It is obvious that the SGMCs can easily be identified with the protogalactic 'fragments' of Searle & Zinn (1978), which they postulated in order to explain the metallicity distribution of the Milky Way halo clusters. Although we invoked the SGMCs for a completely different reason (as host environments for generating the cluster MDF), they have exactly the desired properties for reproducing several other observations too. For example, the MDF will be independent of metallicity as long as magnetic field is the principal support mechanism. The long cloud lifetimes $\tau_{AD} \sim \tau_*$ also allow each cluster to be well mixed and thus chemically homogeneous by the time it forms stars. In addition, since *both* the cloud lifetime τ_* and growth time τ_0 scale with the ambient gas density as $\rho^{-1/2}$ (the free-fall time; see MP96 for derivation), the critical ratio β is independent of galactocentric distance R_{gc}. Thus the MDF should also be independent of position in the halo. In other words, in the outer parts of the protogalaxy everything simply happens more slowly: the clouds grow more gradually because of the lower density, but their self-disruption lifetimes are longer in the same proportion, so the resulting MDF is the same (as long as they are not interrupted by external events such as mergers or tidal breakup).

In summary – to return to the three questions posed in the title – if this picture for cluster formation is valid then we can say that massive (globular) clusters will form anytime and anywhere a galaxy can accumulate gas clouds the size of SGMCs (Searle-Zinn fragments): in protohalos *and* protobulges of large galaxies; in dwarf galaxies; in mergers; and by late infall and accretion. But as a bonus, we have caught a better glimpse of the original building blocks of the galaxies themselves, inside of which the globular clusters assembled.

References

Ajhar, E.A., Blakeslee, J.P., & Tonry, J.L. 1994, *AJ*, **108**, 2087
Armandroff, T.E. 1989, *AJ*, **97**, 375
Ashman, K.M., & Zepf, S.E. 1992, *ApJ*, **384**, 50
Durrell, P.R. 1995, in preparation
Durrell, P.R., McLaughlin, D.E., Harris, W.E., & Hanes, D.A. 1995, *ApJ*, submitted
Fall, S.M., & Rees, M.J. 1985, *ApJ*, **298**, 18
Field, G.B., & Saslaw, W.C. 1965, *ApJ*, **142**, 568
Harris, W.E. 1991, *ARAA*, **29**, 543
Harris, W.E. 1995, in *Stellar Populations*, IAU Symposium No.164, in press
Harris, W.E., Pritchet, C.J., & McClure, R.D. 1994, *ApJ*, **441**, 120
Harris, W.E., & Pudritz, R.E. 1994, *ApJ*, **429**, 177
Heiles, C., Goodman, A.A., McKee, C.F., & Zweibel, E.G. 1993, in *Protostars and Planets*, ed. E.H.Levy & J.I.Lunine (Tucson: University of Arizona Press), p.279
Kwan, J. 1979, *ApJ*, **229**, 567
Larson, R.B. 1990a, *PASP*, **102**, 709
Larson, R.B. 1990b, in *Physical Processes in Fragmentation and Star Formation*, ed. R.Capuzzo-Dolcetta, C.Chiosi, & A.DiFazio (Dordrecht: Kluwer), p.389
McLaughlin, D.E. 1994, *PASP*, **106**, 47
McLaughlin, D.E., & Pudritz, R.E. 1996, *ApJ*, in press
Minniti, D. 1995, *AJ*, **109**, 1663
Murray, S.D., & Lin, D.N.C. 1992, *ApJ*, **400**, 265
Myers, P., & Goodman, A.A. 1988, *ApJ*, **329**, 392
Reed, L.G., Harris, G.L.H., & Harris, W.E. 1994, *AJ*, **107**, 555
Searle, L., & Zinn, R. 1978, *ApJ*, **225**, 357
Secker, J., Geisler, D., McLaughlin, D.E., & Harris, W.E. 1995, *AJ*, **109**, 1019
Whitmore, B.C., & Schweizer, F. 1995, *AJ*, **109**, 960
Zepf, S.E., Ashman, K.M., & Geisler, D. 1995a, *ApJ*, **443**, 570
Zepf, S.E., Carter, D., Sharples, R.M. & Ashman, K.M. 1995b, *ApJ*, **445**, L19
Zinn, R. 1985, *ApJ*, **293**, 424

DISCUSSION

ZEPF: (i) A comment about relating the color distributions of globular clusters to yesterday's discussion of the stellar populations of elliptical galaxies: The broad and often multi-peaked color histograms indicate complex formation histories, in agreement with those suggested yesterday. With deeper images going past the turnover, or spectroscopy, it may be possible to read the formation history of the galaxy in its globular clusters. (ii) How much mass is there in the difference between the predicted and observed LF for the Milky Way and M31 GCS's?

HARRIS: If our simple LF formation model is correct and there were originally many more low-mass clusters, most of them have now been dissolved into the field stellar population. But they would only add up to $\sim 10\%$ of the total mass of the GCLF, and much less than 1% of the whole halo mass. In other words, you can rip up huge numbers of these small clusters without affecting the halo at large.

MEURER: You emphasized cluster formation in merging systems. We have done UV imaging of several starburst galaxies and find that all have UV-luminous clusters. The fact that we see them in mergers may just be a secondary effect resulting from the fact that starbursts occur in merging systems. In addition, although cluster formation may be inefficient relative to the gas content, on average they make up $\sim 20\%$ of the UV flux of starbursts, so they make a significant contribution to the total star formation.

HARRIS: At early stages in a starburst it may look like the clusters or clumps are dominating the light output, but that may be because they form sooner than the field stars. In fact, only 1% or so of the total gas mass may be involved in the cluster phase. Thus what we really need to know for these active regions is, *after* the starburst has finished and everything has died down, what fraction of the mass ends up in bound clusters?

GERHARD: If clusters form from dense cloud cores, their process may require a minimum metallicity to cool the gas down to sufficiently low temperature and high density, something like $Z \simeq 0.01$ or so. For comparison, what is the minimum metallicity you observe in globular clusters?

HARRIS: The lowest observed metallicity for any globular cluster is very near $Z \simeq 0.005 Z_\odot$. However, there are various papers in the literature which suggest that nonequilibrium formation of H_2 can cool the clouds rapidly even at low metallicity. Once the magnetic field support has gone, there doesn't seem to be any barrier to rapid cloud collapse.

BRINKS: You might be pleased to hear that 'SGMCs' are now seen in *non*-destructively interacting galaxies, such as the 'ocular' galaxy IC2163 which is tidally disturbed by NGC 2207. Apparently, the interaction boosts the velocity dispersion in the neutral gas and $10^8 - 10^9 M_\odot$ clumps of neutral gas are formed (Elmegreen et al. 1995, in press).

MINNITI: Does the observed luminosity distribution depend on metallicity?

HARRIS: To first order it doesn't, as far as the available data show. To second order, there may be differences in (say) the turnover luminosity at the quarter-magnitude level in the sense that higher-metallicity systems have slightly fainter turnovers (Ashman et al. 1995, in press).

FABER: Your picture invokes a turnover caused by some kind of destruction effects. Those are likely to vary greatly from galaxy to galaxy and within a galaxy. How does $m_\star$ manage to be so constant?

HARRIS: I completely agree that dynamical effects such as tidal shocking, dynamical friction, and disk shocking seem unlikely to govern the present-day level of $m_\star$; if they did, then the turnover luminosity should differ strongly from place to place among galaxies. So if $m_\star$ actually *is* a result of dynamical evolution rather than formation, then it's probably just caused by the slow process of evaporation of stars from the cluster within the overall tidal field of the galaxy. However, the more correct answer to your question is that we don't really have any good, quantitative theory for $m_\star$ at present, nor does anyone else. Our opinion is that formation is still the dominant role in determining it, but it's not clear yet just how.

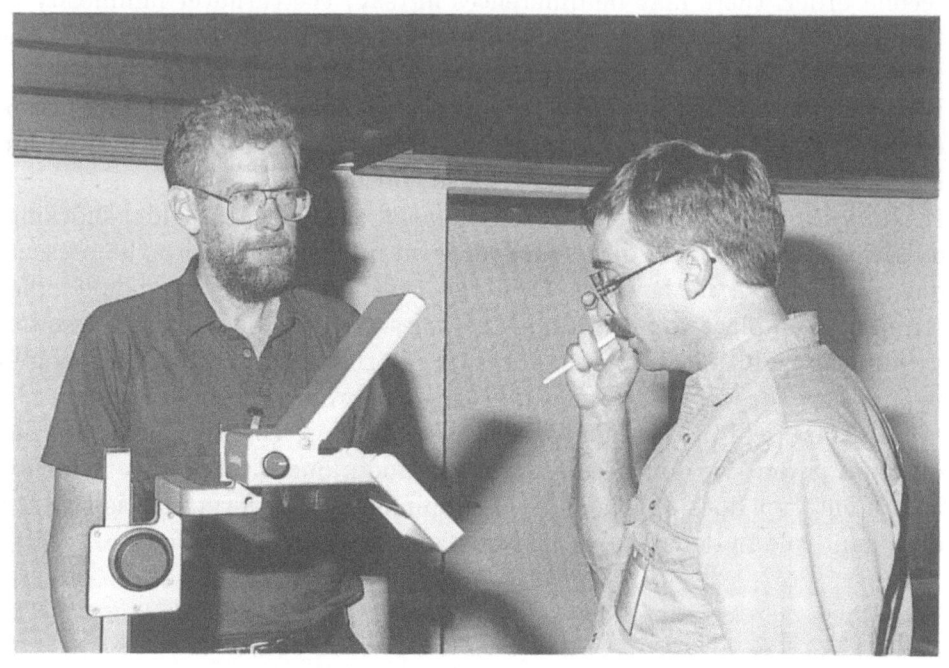

Bill Harris and Simon Lilly

DWARF AND LOW SURFACE BRIGHTNESS GALAXIES

Field Populations

STACY S. MCGAUGH
Institute of Astronomy
The Observatories
Cambridge, England

1. Introduction

The terms 'dwarf' and 'low surface brightness' are commonly used to mean a variety of different things, and are sometimes used interchangeably. It is thus necessary to explicitly quantify the definitions which are adopted. Here I shall limit the discussion to field galaxies which are [exponential] disks characterized by a scale size h and central surface brightness μ_0.

The word dwarf implies things which are very small in linear extent, so one definition might be $h < 1$ kpc. More commonly, dwarf is used to mean things which are intrinsically faint, $L < 0.01L^*$. By either criterion, such things are extremely rare in field samples (Fig. 1), simply because they are so faint that surveys are sensitive to them only over very limited volumes.

On the other hand, there exist many low surface brightness (LSB) disk galaxies which are not small, and do not satisfy either of the above definitions of 'dwarf'. Indeed, these exist right up to L^* (Fig. 1), and typically exhibit spiral structure (McGaugh et al. 1995a; de Blok et al. 1995a). So I use the term LSB to refer to disks with $\mu_0 > 22.5\ B$ mag arcsec^{-2} (about the brightnesses of the moonless sky) which are too large and luminous to be considered dwarfs. This population constitutes $\sim 1/2$ of all disks by number (McGaugh et al. 1995b; Sprayberry et al. 1995a).

2. Dwarf Galaxies

While a great deal is known about dwarfs in the local group (Hodge 1971; 1989) and in nearby groups and clusters (Ferguson & Binggeli 1994), little is known about true dwarfs in the field for the simple reason that their faintness makes them very rare in flux limited samples. I will therefore further limit the discussion to the population of dwarfs discovered by their

R. Bender and R. L. Davies (eds.), New Light on Galaxy Evolution, 97–104.

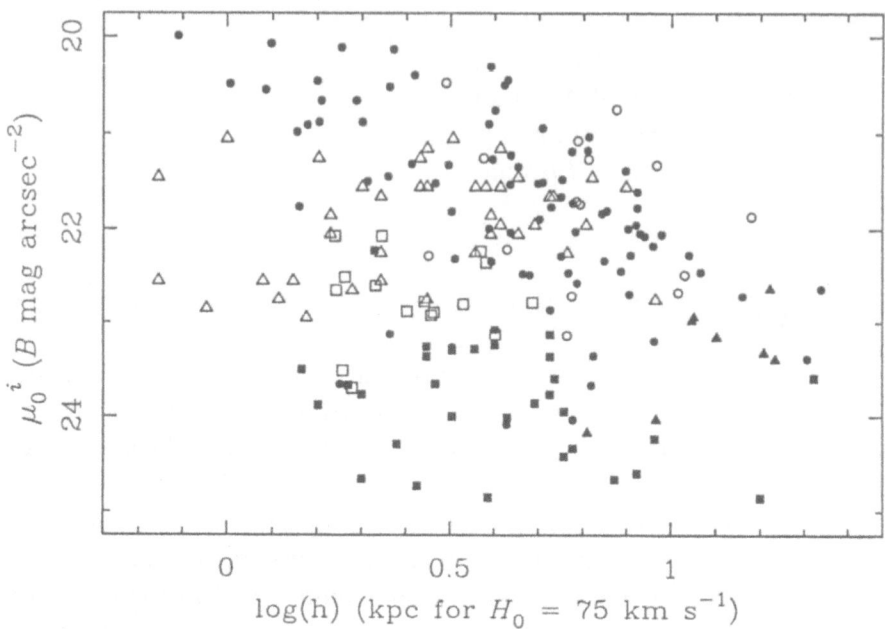

Figure 1. Central surface brightness vs. scale length for disk galaxies [data from Boroson (1981; open circles), Romanishin et al. (1983; open squares), van der Kruit (1987; open triangles), McGaugh & Bothun (1994; solid squares), de Blok et al. (1995a; solid squares), Sprayberry et al. (1995b; solid triangles), and de Jong (1995; solid circles)]. Galaxies fill this plane up to maxima in both luminosity and surface brightness.

strong emission lines. Variously known as HII galaxies, BCDs, & BCGs (see Salzer et al. 1989), these are intrinsically small galaxies (though of course with some distribution of sizes) undergoing a strong burst of star formation involving typically $\sim 10^4$ O stars.

These galaxies have been extensively studied both for their remarkable star formation activity, and because they dominate objective prism surveys (Salzer et al. 1989; Terlevich et al. 1991). They are typically low metallicity ($Z \approx 0.2Z_\odot$), gas rich objects, and as such are relatively unevolved. This led to the hope that some might be true protogalaxies undergoing their first episode of star formation, but with the famous exception of I Zw 18, so far all do have older underlying stellar populations (Salzer, private communication; Telles 1995).

Since this conference focuses on evolution, which primarily means the star formation history (Kennicutt 1995), I would like to review an important outstanding problem these objects pose: the progenitor problem (e.g., Tyson & Scalo 1988). This arises because the inferred star formation history is one of brief episodic bursts interspersing lengthy quiescent periods. Therefore, there must be a vast reservoir of progenitors for each individual H II galaxy currently undergoing a burst:

$$n_{prog} = \frac{\tau_{off}}{\tau_{on}} n_{burst}$$

where n_{burst} is the observed density of currently active galaxies, and τ represents the duty cycle for star formation, i.e., the period of time spent in bursting and quiescent phases. The quiescent phase is essentially a Hubble time less the bursts, which are generally inferred to be few and brief, $\tau_{on} \sim 10^7$ yr. Thus

$$n_{prog} \approx \frac{10^{10}}{10^7} n_{burst} \approx 1000 \; n_{burst}$$

which is an enormous problem since n_{burst} is observed to constitute $\sim 10\%$ of the total field galaxy population (Salzer 1989; see also Schade & Ferguson 1994). Thus, for any amount of fading after the burst, the inferred number of progenitors exceeds everything else we know about optically by a factor of ~ 100, but is undetected in 21 cm surveys (Weinberg et al. 1991).

One way to ease this problem is to increase τ_{on}, presumably with a concomitant decrease in burst strength consistent with recent estimates that the burst itself contributes $\lesssim 1$ mag. to the total luminosity (Salzer, private communication). This then leads to a qualitatively different picture for the star formation history with substantial peaks and troughs but not sharp δ-functions. However, one is limited in the degree to which the star formation rate can be smoothed out in this sense by the need to

1. not overproduce luminosity in long lived stars
2. not overproduce metallicity
3. not consume all the gas, and most crucially
4. provide enough ionizing photons to yield the observed Hα luminosities for long periods.

Item [1] could be avoided by truncating the IMF so that only high mass ($\gtrsim 10 \; M_\odot$) stars are formed in the burst. While this is appealing in some respects, tailor made IMFs can fit anything and there really is no evidence for variations in the IMF (McGaugh 1991). Metallicity [2] may be lost in preferentially enriched supernova driven winds, but note that there is no evidence that these galaxies 'explode' and lose all of their gas. Item [3] provides the ultimate constraint unless very substantial amounts of gas are subsequently accreted to replenish the supply. I think item [4] places the tightest constraints on the burst duration, but given the desperate lack of adequate model atmospheres for hot, low metallicity stars, it is conceivable that not quite so many O stars are required if low metallicity stars produce a lot more ionizing photons per unit mass than is usually assumed based on solar metallicity models.

Another, related puzzle is that the *underlying* stellar population is itself very blue (Telles 1995). For a star formation history consisting of a few intermittent bursts, the remnants of the preceding burst should have reddened substantially. This appears not to be the case, and the colors are so blue ($B - V \approx 0.4$) that I don't think that low metallicity can be the entire explanation. It is also hard to see how to address this by varying the IMF, since we are considering an underlying population which is presumably much older than the lifetimes of blue stars. Perhaps the mean age is implicated — either it has not been long since the previous burst, or the system as a whole formed late and is rather less than 10 Gyr old, or most likely some combination of these and metallicity effects.

A young mean age for the underlying population suggests a decrease in τ_{off}, but even if one takes $\tau_{on} \rightarrow$ a few $\times 10^8$ and $\tau_{off} \rightarrow$ a few $\times 10^9$, the entirety of the normal galaxy population fainter than $\lesssim \frac{1}{2}L^*$ is needed to serve as progenitors. A substantive progenitor population seems to be demanded by the intensity of the observed star formation, but ruled out by optical and 21 cm surveys. However, optical surveys are very insensitive to objects which are both small and low surface brightness, and these could be quite *numerous* without violating the 21 cm constraints on *mass* density.

3. Low Surface Brightness Galaxies

Now let us turn to the population of LSB galaxies which are comparable in size to the high surface brightness spirals which define the Hubble sequence. These galaxies are extremely blue ($B - V \approx 0.4$, $V - I \approx 0.7$), especially in the redder colors (McGaugh & Bothun 1994; Rönnback & Bergvall 1994; de Blok et al. 1995a). The colors of disks become generally bluer (with much scatter) as either size or surface brightness decrease. This suggests a connection between small, LSB galaxies and the underlying components of H II galaxies, but in this section I will discuss larger, Milky Way size objects which are unlikely to contribute to the progenitor population.

Understanding the very blue colors of LSB galaxies is challenging. In order to disentangle the effects of age and metallicity, it is useful to measure the latter. LSB galaxies are quite metal poor, with typical metallicities in the range $0.1 < Z < 0.3Z_\odot$ (McGaugh 1994; Rönnback & Bergvall 1995). Thus at least some of the blueness must be due to this. However, metallicity can not explain it entirely, as color and metallicity are not correlated (McGaugh & Bothun 1994). A rather low mean age is thus implicated, with a birth rate function weighted more heavily towards recent epochs than early ones. This is at once consistent and at odds with the trends along the

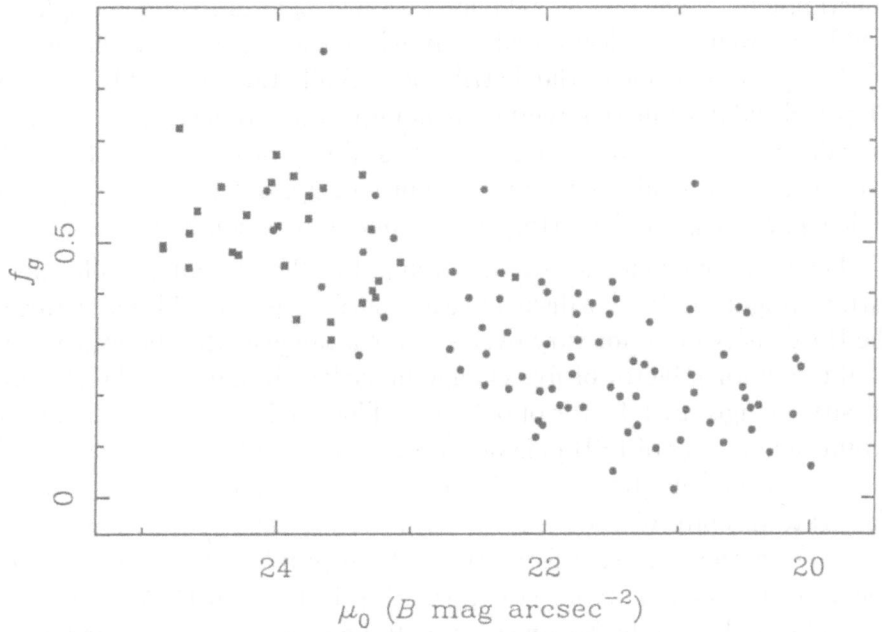

Figure 2. The correlation of gas mass fraction with surface brightness.

Hubble sequence (Kennicutt 1995): as morphological types typically later than Sc, one might expect LSB galaxies to have such birth rate functions. However, they also have low current star formation rates per unit area, as in very early type disks.

The inferred ages are typically a few Gyr less than those of high surface brightness disks, suggesting a late formation epoch and/or slow evolution. The latter is certainly indicated by the low metallicities, and also by the large gas mass fractions (Fig. 2). Given the observed ratio of 21 cm to optical flux (M_{HI}/L), the gas mass fraction $f_g = M_{gas}/(M_{gas} + M_*)$ can be calculated from

$$f_g = \left[1 + \frac{\Upsilon_*}{\eta}\frac{L}{M_{HI}}\right]^{-1}$$

with some reasonable assumption about the stellar mass to light ratio Υ_* and the fraction of gas in H I , η^{-1}. For simplicity, I take $\Upsilon_* = \eta$; though not exactly correct since LSB galaxies have little molecular gas (Schombert et al. 1990), this is not a bad approximation and the trend is clear in the raw data (de Blok et al. 1995b; Sprayberry et al. 1995c).

The correlation between f_g and μ_0 is remarkably strong, but there is no correlation between f_g and scale length. The lack of objects with high gas fractions at high surface brightnesses is certainly real and can not be a selection effect. The lack of galaxies with low f_g at low surface brightnesses could very well be a selection effect, in which case the correlation line would be the upper envelope of the distribution. While there certainly exist dwarf Elliptical galaxies in this regime, it is important to determine if these are causally connected populations. That is, do spirals evolve to lower f_g at fixed μ_0, perhaps also evolving in morphology, or along the sequence to higher μ_0? I suspect the latter, but of course some combination is possible.

The trend of global gas fraction seen in Fig. 2 is also mimiced locally: low surface brightness stellar disks have low surface density H I disks. However, the H I density does not vary over as large a range as that in optical surface brightness: for a factor of five change in surface brightness, the H I surface density changes by a factor of only ~ 2. This holds the key to the inhibited evolutionary rates of LSB galaxies: they exist close to the critical threshold for star formation (Kennicutt 1989), and as a result form stars at a very slow rate in spite of their enormous gas reservoirs. (Note that the usual assumption that the star formation rate is proportional to the gas mass, which leads to exponential star formation histories with $\dot{M} \propto e^{-t/\tau_s}$ can not hold in these galaxies unless $\tau_s \leq 0$, i.e., an *increasing* star formation rate or a constant one with a low age.)

Surface density is thus a critical parameter in governing a disk's evolution. So what determines the density? All lines of evidence, the low metallicities, blue colors, and large gas mass fractions, indicate slow evolution and relative youth. One expects a galaxy to form late if it arises from a low density peak in the initial field of fluctuations. That low initial density should lead directly to a low final density (Fig. 3), with the observed consequences. This simple picture, derived from the physical properties listed above (McGaugh 1992), makes a clear prediction about the spatial distribution of LSB galaxies: they should be less strongly clustered than higher surface brightness spirals. This prediction has been confirmed (Mo et al. 1994) with the additional observation that LSB galaxies are extremely isolated on small (< 2 Mpc) scales (see also Bothun et al. 1993). They have no bright companions, and have suffered no tidal perturbations which might clump their gas and induce star formation (presumably raising their surface brightnesses). They have endured no merging, being the poster children for galaxy formation by gradual collapse. They may, however, compose a population which, in hierarchical structure formation scenarios, is expected to fall into larger group and cluster structures at late times (Rakos & Schombert 1995).

The evolution of disks is governed by their characteristic density as well as total mass. The surface brightness of a disk is intimately related to its evolutionary rate and collapse epoch. The star formation history is relatively stable in large disks, but tends increasingly towards episodic

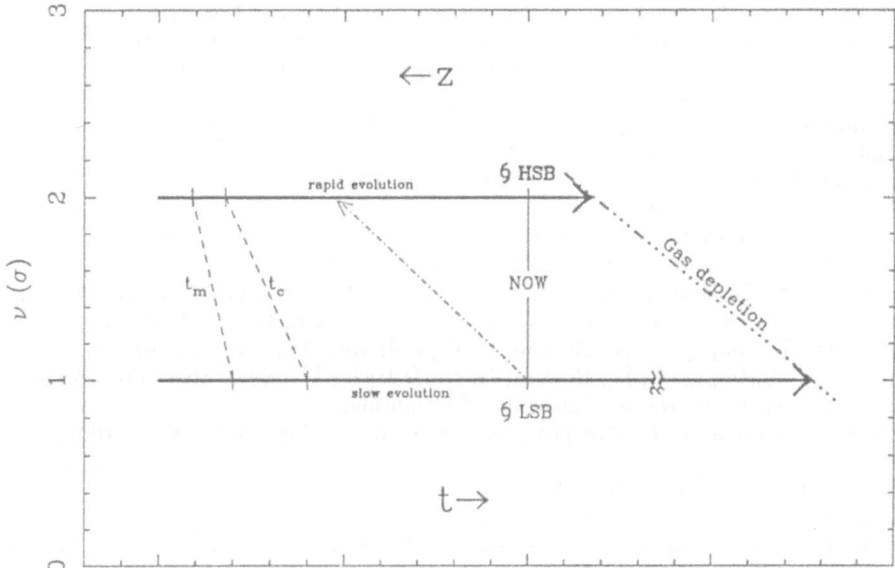

Figure 3. The surface brightness of a disk is related to the amplitude ν of the primordial density perturbation from which it arises. Low density perturbations collapse late (t_c), forming low density galaxies which evolve slowly. When observed at any given epoch, their gas content, metallicity, etc. will correspond to an earlier state of higher density galaxies (dash-dotted line).

bursts as size decreases. This may simply be a statement that star formation is inherently a local process, so that larger disks in effect average over larger numbers of discrete star forming events.

References

de Blok, W. J. G., van der Hulst, J. M., & Bothun, G. D. 1995a, MNRAS, 274, 235
de Blok, W. J. G., McGaugh, S. S., & van der Hulst, J. M. 1995b, these proceedings
Boroson, T. 1981, ApJS, 46, 177
Bothun, G. D., Schombert, J. M., Impey, C. D., Sprayberry, D., & McGaugh, S. S. 1993, AJ, 106, 530
Ferguson, H. C., & Binggeli, B. 1994, A&AR, 6, 67
Hodge, P. 1971, ARA&A, 9, 35
Hodge, P. 1989, ARA&A, 27, 139
de Jong, R. S. 1995, Ph.D. thesis, University of Groningen
Kennicutt, R. C. 1989, ApJ, 344, 685
Kennicutt, R. C. 1995, these proceedings
McGaugh, S. S. 1991, ApJ, 380, 140
McGaugh, S. S. 1992, Ph.D. thesis, University of Michigan
McGaugh, S. S. 1994, ApJ, 426, 135

McGaugh, S. S., & Bothun. G. D. 1994, AJ, 107, 530
McGaugh, S. S., Bothun, G. D., & Schombert, J. M. 1995b, AJ, in press
McGaugh, S. S., Schombert, J. M., & Bothun, G. D. 1995a, AJ, 109, 2019
Mo, H. J., McGaugh, S. S., & Bothun, G. D. 1994, MNRAS, 267, 129
Rakos, K. D., & Schombert, J. M., 1995, ApJ, 439, 47
Romanishin, W., Strom, K. M., & Strom, S. E. 1983, ApJS, 53, 105
Rönnback, J., & Bergvall, N. 1994, A&AS, 108, 193
Rönnback, J., & Bergvall, N. 1995, ESO preprint 1071
Salzer, J. J. 1989, ApJ, 347, 152
Salzer, J. J., MacAlpine, G. M., & Boroson, T. A. 1989, ApJS, 70, 479
Schade, D. J., & Ferguson, H. C. 1994, MNRAS, 267, 889
Schombert, J. M., Bothun, G. D., Impey, C. D., & Mundy, L. G. 1990, AJ, 100, 1523
Sprayberry, D., Bernstein, G. M., Impey, C. D., & Bothun, G. D. 1995c, ApJ, 438, 72
Sprayberry, D., Impey, C. D., Bothun, G. D., & Irwin, M. 1995b, AJ, 109, 558
Sprayberry, D., Impey, C. D., Bothun, G. D., & Irwin, M. 1995a, these proceedings
Telles, E. 1995, Ph.D. thesis, University of Cambridge
Terlevich, R., Melnick, J., Masegosa, J., Moles, M., & Copetti, M. V. F. 1991, A&AS, 91, 285
Tyson, N. D., & Scalo, J. M. 1988, ApJ, 329, 618
van der Kruit, P. C. 1987, A&A, 173, 59
Weinberg, D. H., Szomoru, A., Guhathakurta, P., & van Gorkom, J. H. 1991, ApJ, 372, L13
Zwaan, M. A., van der Hulst, J. M., de Blok, W. J. G., & McGaugh, S. S. 1995, MNRAS, 273, L35

DISCUSSION:

McCall: In your graphs of metallicity vs. color; were the colors integrated, or were they corrected for the star forming regions? I am concerned that the scatter in the colors might be in part due to the effect of the young component, which of course can vary substantially in luminosity fraction from galaxy to galaxy.

McGaugh: For H II galaxies a correction must very carefully be applied; it is not necessary for LSB galaxies.

Meurer: How do you know that the cycle time in BCDs is 10^7 years? From the size of the "starburst" region I would say this is more like a lower limit to the cycle time, although the clusters they contain should themselves be better analogs to true instantaneous bursts.

McGaugh: A burst duration of $\sim 10^7$ yr is the consensus number in the literature, though I have some sympathy for the case that it be longer.

Djorgovski: These objects obviously have fewer stars than high surface brightness galaxies, but for a given type (e.g., giant disks, true dwarfs, etc.), do they have fewer baryons, and do they have less dynamical mass?

McGaugh: Fewer stars per unit area, certainly. They do not have much lower dynamical masses; the mass to light ratio within the optical radius increases very systematically with decreasing surface brightness (Zwaan et al. 1995; de Blok et al. 1995b). Whether this is due to fewer baryons per unit mass or what is very hard to say, and poses a fundamental puzzle.

AN HST SURVEY OF CORES OF EARLY-TYPE GALAXIES [1]

JOHN KORMENDY [2] AND YONG-IK BYUN
Institute for Astronomy, University of Hawaii, 2680 Woodlawn Dr., Honolulu, HI 96822

E. A. AJHAR AND TOD R. LAUER
Kitt Peak National Observatory, National Optical Astronomy Observatories, P. O. Box 26732, Tucson, AZ 85726

ALAN DRESSLER
Carnegie Observatories, 813 Santa Barbara St., Pasadena, CA 91101

S. M. FABER AND CARL GRILLMAIR
UCO/Lick Observatory, University of California, Santa Cruz, CA 95064

KARL GEBHARDT AND DOUGLAS RICHSTONE
Dept. of Astronomy, Univ. of Michigan, Ann Arbor, MI 48109

AND

SCOTT TREMAINE
Canadian Institute for Theoretical Astrophysics, University of Toronto, 60 St. George St., Toronto M5S 1A7, Canada

Abstract. Photometry of the central parts of bulges and elliptical galaxies with the *Hubble Space Telescope* (HST) confirms and extends ground-based results. Most giant ellipticals have cuspy cores: at the "break radius" r_b (formerly the core radius r_c), the steep outer surface brightness profile turns down to a shallow inner power law $I(r) \propto r^{-\gamma}$, $0 \leq \gamma \lesssim 0.25$. The corresponding slope of the deprojected profile is derived; the flattest cores allow box orbits to survive. Cores continue to satisfy fundamental plane

[1] Based on observations with the NASA/ESA *Hubble Space Telescope*, obtained at the Space Telescope Science Institute, which is operated by AURA, Inc., under NASA contract NAS 5-26555.

[2] Visiting Astronomer at the Canada-France-Hawaii Telescope (CFHT), operated by the National Research Council of Canada, the Centre National de la Recherche Scientifique of France, and the University of Hawaii.

R. Bender and R. L. Davies (eds.), New Light on Galaxy Evolution, 105–116.
© *1996 IAU. Printed in the Netherlands.*

parameter correlations like those found from the ground. In particular, HST confirms that the luminosity sequence of elliptical galaxies (from cDs to M 32) is physically unrelated to spheroidal galaxies like Fornax. The latter are closely related to late-type dwarfs. Low-luminosity ellipticals do not show cores: $0.5 \lesssim \gamma \lesssim 1.3$. The most important new result is that global and core properties both show signs of a dichotomy between (i) low-luminosity ellipticals that rotate rapidly, that are nearly isotropic and oblate-spheroidal, that have disky-distorted isophotes, and that are *coreless* and (ii) giant ellipticals that are essentially nonrotating, anisotropic, and moderately triaxial, that are boxy-distorted, and that have *cuspy cores*.

Key words: Galaxies: Nuclei – Galaxies: Photometry – Galaxies: Structure

1. Introduction

The study of galaxy cores is a prime mission of HST. High-resolution photometry has now been published by a number of groups (Lauer *et al.* 1991; 1992a, b; 1993; 1995; Crane *et al.* 1993; Stiavelli *et al.* 1993; Kormendy *et al.* 1994; Grillmair *et al.* 1994; Forbes 1994; Forbes *et al.* 1994, 1995; Jaffe *at al.* 1994; van den Bosch *et al.* 1994; Ferrarese *et al.* 1994). This paper focuses on the work of our group; results from the other groups are similar. HST has enriched our understanding of galaxy cores; it has settled some outstanding issues, and it has provided a few surprises. But many results were already in place from ground-based photometry, and most of these have survived. We therefore begin with a brief review of ground-based work. We concentrate on one result that is particularly relevant at this meeting, i. e. the clear physical distinction between elliptical and spheroidal galaxies.

2. Ground-Based Results: Elliptical and Spheroidal Galaxies as Distinct Families of Stellar Systems

Ground-based work on galaxy cores is reviewed in Kormendy (1982, 1987a) and in Kormendy & Djorgovski (1989). The main results are:

1 – Cores: Most giant ellipticals have cores; i. e., central regions where the surface brightness profile $I(r)$ turns down from a steep $I \propto r^{-\beta}$ ($\beta \sim 2$) outer power law toward $I \simeq$ constant. The turndown is more gradual than in an isothermal sphere; this was demonstrated by the first CCD photometry (Young *et al.* 1978; Lauer 1985a, b) and is most convincingly seen in high-resolution photometry from the CFHT (Kormendy 1985a, 1987a) and NOT (Møller *et al.* 1995). The brightness profile is still rising where seeing becomes dominant, but ground-based photometry did not tell us the functional form of $I(r)$ at $r \ll r_c$. HST solves this problem (§ 3).

2 – <u>Fundamental plane (FP) correlations:</u> Lower-luminosity giant Es have smaller core radii r_c, higher central surface brightnesses I_0, and larger central velocity dispersions σ (Kormendy 1984; 1985b; 1987a,b; Lauer 1985a,b; see Kormendy & McClure 1993 and Kormendy & Bender 1994 for recent versions). Bulges of disk galaxies are consistent with these correlations; when we speak of "ellipticals" below, we include bulges.

3 – <u>Low-luminosity galaxies do not show resolved cores.</u> Limits on r_c are consistent with the FP relations, but there may be a dichotomy between coreless ellipticals with disky isophote distortions and boxy ellipticals with resolved cores (Nieto et al. 1991). This dichotomy is the subject of § 6.

4 – <u>Families of ellipsoidal stellar systems:</u> The FP correlations of elliptical galaxies are very different from those of spheroidal (Sph) galaxies. For example, spheroidal galaxies with lower luminosities L have lower core and effective surface brightnesses, while lower-luminosity ellipticals have higher surface brightnesses. The low-luminosity end of the E sequence is defined by M 32 and by similar $M_V \simeq -16$ ellipticals in Virgo, not by dwarf spheroidals like NGC 205, Fornax, or Draco. This was correctly postulated by Wirth & Gallagher (1984) from remarkably meager statistics and then demonstrated by Kormendy (1985b, 1987b) using CFHT photometry of galaxies with a wide range in luminosities. The difference between E and Sph galaxies is global, not just a core property (Ichikawa et al. 1986, 1988; Kormendy 1987b; Binggeli & Cameron 1991)[3]. Also, Wirth & Gallagher (1984) suggested and Sandage et al. (1985), Binggeli (1987), Binggeli et al. (1988), and Ferguson & Sandage (1991) showed that E and Sph galaxies have different luminosity functions. Ellipticals are bounded in luminosity. Objects like M 32 are rare; we are extremely fortunate to live so near a prototypical example. Spheroidals, on the other hand, begin to appear at $M_B \simeq -18$ and then have exponentially rising luminosity functions at faint magnitudes M_B.

Kormendy (1985b, 1987b) further showed that dwarf spheroidals are similar in global structure to dwarf spirals and irregulars. This almost certainly means that they are physically related. Binggeli (1994b) and Ferguson & Binggeli (1994) review the possibilities. The relationship is complex; more than one physical process is likely to be be important even at a single luminosity. However, it is worth noting that about half of the Galaxy's dSph companions have stellar subpopulations that are 3 – 7 Gy old (e. g., Da Costa 1992), so many dSph galaxies were Magellanic irregulars until relatively recently (Kormendy & Bender 1994).

---

[3] Caution: Binggeli and collaborators call the galaxies in the Sph family "dwarf ellipticals" or "dEs" even though they are not related to ellipticals; see Binggeli (1994a) and Kormendy & Bender (1994) for contrasting views on the terminology. We follow the Kormendy & Bender convention.

The "bottom line" is this: Sph galaxies are not ellipticals and probably formed differently from ellipticals. Compared to the difference between E and Sph galaxies, ellipticals are remarkably homogeneous in properties (Djorgovski & Santiago 1993; Bender et al. 1993, 1994; Saglia et al. 1993; Djorgovski, Pahre, & de Carvalho 1995), despite heterogeneous merger histories and even including the physical dichotomy discussed in § 6.

3. An HST Perspective on Galaxy Cores

HST work on galaxy cores began with the S0 galaxy NGC 7457 (Lauer et al. 1991) and with the ellipticals M 87 and M 32 (Lauer et al. 1992a,b). NGC 7457 and M 32 have coreless power-law profiles, although limits on r_c (0″.05 and 0″.11, respectively) are consistent with the FP correlations (Kormendy & McClure 1993). In contrast, the core of M 87 was already well resolved from the ground (Kormendy 1985a). With HST, it is so well resolved that the nature of the inner profile becomes clear: inside a break radius $r_b \simeq r_c$, the steep outer power law turns down to a shallow inner power law, $I \propto r^{-0.26}$. These two types of profiles – power laws and cuspy cores – characterize almost all ellipticals (references in § 1; Tremaine 1995).

Our own group obtained photometry for 45 galaxies in Cycles 1 and 2; Kormendy et al. (1994) present a preliminary report, and Lauer et al. (1995) publish the data in full. The images were Lucy – Richardson deconvolved; steep brightness profiles are accurate to ~ 0.1 mag arcsec^{-2}; core profiles are accurate to $\lesssim 0.05$ mag arcsec^{-2} (§ 4).

Almost all high-luminosity galaxies have resolved cuspy cores like that of M 87. At small r, the profiles are shallow power laws, $I(r) \propto r^{-\gamma}$, with $0 \leq \gamma \lesssim 0.25$. A convenient parametrization is

$$I(r) = I_b \, 2^{\frac{\beta-\gamma}{\alpha}} \left(\frac{r}{r_b}\right)^{-\gamma} \left[1 + \left(\frac{r}{r_b}\right)^{\alpha}\right]^{\frac{\gamma-\beta}{\alpha}}. \tag{1}$$

Here r_b and I_b replace the former parameters core radius r_c and central surface brightness I_0: r_b is the radius at which the steep outer $I \propto r^{-\beta}$ profile breaks into the shallow inner profile, and I_b is the surface brightness at r_b. The parameter α measures the sharpness of the break. Fits of Equation 1 to the profiles are calculated in Byun et al. (1995); resulting core parameters are discussed in Faber et al. (1995) and in § 5, below.

All of the low-luminosity galaxies except NGC 4486B are unresolved. Like M 32, they have power-law profiles that remain steep ($0.5 \lesssim \gamma \lesssim 1.3$) to radii $r < 0″.1$. The division between galaxies with and without resolved cores occurs at $M_V \simeq -21 \pm 0.5$ but is not completely sharp (§ 6).

A few galaxies show point sources added to core or power-law profiles. Some are active nuclei (NGC 6166: Filippenko, private communication). When we know or suspect that they are star clusters, we call them nuclei.

4. The Deprojected Brightness Profiles of Cuspy Cores

Many astrophysical questions about cores require us to know the slopes of the deprojected brightness profiles. At $r \ll 0.1r_b$ and $\gamma > 0$, this is $-(\gamma+1)$ for the fitting function in Equation 1, but at $0.1r_b \lesssim r < r_b$, it is considerably shallower than $-(\gamma + 1)$. Our observations do not reach $r \ll 0.1r_b$, so we cannot be sure that the slope is ever as steep as $-(\gamma + 1)$. In any case, as Merritt (private communication) has emphasized, small departures of the observed profiles from Equation 1 are greatly magnified in deprojection, so we can be misled if we merely deproject the fit of Eq. 1 to the data. More reliable is a nonparametric deprojection of the profiles. Therefore we ask: What is the relationship between the logarithmic slopes of the projected and deprojected profiles at the smallest radii we can reach, $r \simeq 0.1r_b$?

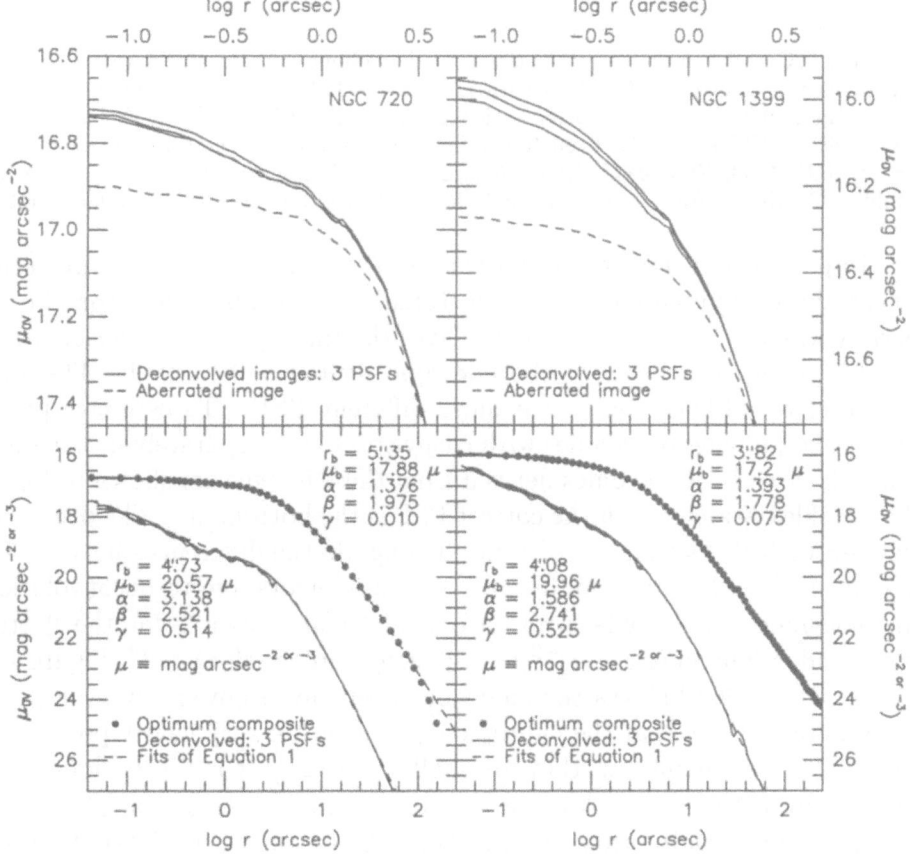

Figure 1. (*top*) Major-axis brightness profiles of NGC 720 and NGC 1399 before and after deconvolution with three PSFs. (*bottom*) The three deconvolved profiles from the top panels are shown before (*above*) and after (*below*) deprojection. Equation 1 has been fitted to the optimally deconvolved profiles (*dashed lines* and *tabulated major-axis parameters*).

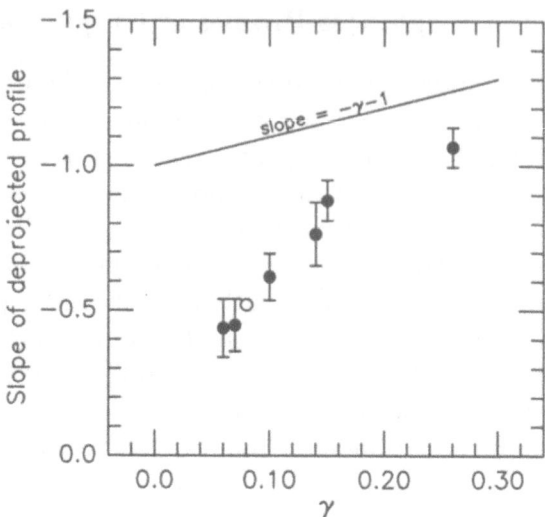

Figure 2.   Correlation between the slopes of the volume and surface brightness profiles for galaxies in Lauer *et al.* (1995) with $r_b = 2\rlap{.}''5 - 11\rlap{.}''1$, i. e., large enough so we can derive the profile slope at $r \simeq 0.1r_b$ with confidence. From left to right, the galaxies are NGC 4889, NGC 720, NGC 6166 (open circle: the active nucleus reduces our leverage on γ), NGC 1399, NGC 4874, NGC 4636, and M 87. Parameters are for the major-axis profiles. Merritt & Fridman (1995a) and Gebhardt *et al.* (1995) obtain similar results.

To proceed, we need to know the accuracy of the profiles. Deconvolution uncertainties dominate over photon statistics and calibration errors. So the easiest way to proceed is as follows. Over the time span of our observations (~ 1.5 years), the focus of the telescope drifted substantially. Therefore Lauer *et al.* (1995) used three quite different PSFs. To estimate profile errors here, images of galaxies with cuspy cores were separately deconvolved with all three PSFs. Profiles derived from these images are shown in Fig. 1. The profile obtained with the correct PSF is the bottom one. The others are derived with PSFs that are certainly wrong. So the differences between the profiles in Fig. 1 overestimate the systematic errors due to deconvolution. We conclude that profile errors are $\lesssim 0.02$ mag arcsec^{-2} for the flattest cores and $\lesssim 0.05$ mag arcsec^{-2} for all cuspy cores with $r_b \gtrsim 2''$. Figure 1 in Lauer *et al.* (1995) shows the analogous result for a power-law profile.

Figure 1 shows that the deprojected profiles are very nearly power laws at $r \simeq 0.1r_b$. As expected, they are shallower than $\rho \propto r^{-(\gamma+1)}$. Slopes were derived by fitting power laws or Eq. 1, allowing for errors of $\lesssim 0.05$ mag arcsec^{-2}. The results are in Fig. 2. For M 87, γ and r_b are large; then the slope after deprojection approaches $-(\gamma+1)$. But for flatter cores, the slopes depart more and more from $-(\gamma+1)$. The flattest profiles are quite shallow. One conclusion is that box orbits can survive: Merritt & Fridman (1995a, b) and de Zeeuw (1995) show that cuspy cores preclude box orbits unless the deprojected profile is sufficiently shallow (approximately $\rho \propto r^{-0.5}$).

5. Fundamental Plane Parameter Correlations

Figure 3 shows two projections of the FP correlations (Faber *et al.* 1987; Djorgovski, de Carvalho, & Han 1988). Resolved cores (*filled circles*) satisfy FP correlations like those seen from the ground. Lower-luminosity galaxies have smaller cores of higher surface brightness. Many unresolved galaxies are consistent with the extrapolation of these correlations, but we have only upper limits on any break radii, $r_b \lesssim 0\rlap{.}''1$. These objects may have *much* smaller cores or they may not have cores at all.

The faintest Virgo Es look much less compact than M 32. This is a resolution effect. Figure 3 shows that if M 32 were in the Virgo Cluster, the HST limits on its core parameters would be similar to those observed for the smallest ellipticals in the cluster. M 32 appears normal for its low L.

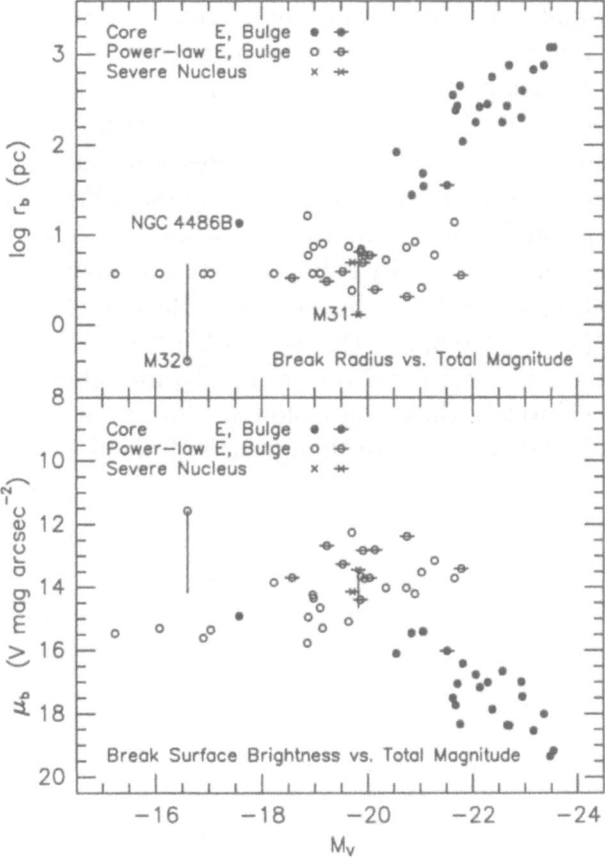

Figure 3.   Correlations of r_b and μ_b with absolute magnitude (from Faber *et al.* 1995). M 31 and M 32 are plotted twice; the symbols represent the galaxies as observed; the lines point to the parameters that we would observe if the galaxies were in the Virgo Cluster. Distances are based on a Hubble constant of $H_0 = 80$ km s^{-1} Mpc^{-1}.

6. A Dichotomy Between Two Kinds of Elliptical Galaxies?

At $M_V \simeq -21$ in Fig. 3, some galaxies have rather flat cores and others have power-law profiles. This illustrates the biggest surprise in our data. From ground-based FP correlations, we expected that marginally resolved cores would be well resolved with HST. At $M_V \simeq -21$, *this did not happen.* Most galaxies with marginal cores turned out to have power-law profiles.

As a result, the scatter in Fig. 3 is not random. Figure 4 shows the correlations of γ with M_V and r_b. There are signs of a dichotomy. Bright ellipticals have $0 \leq \gamma \lesssim 0.25$ (they are very well resolved); low-luminosity galaxies have $0.5 \lesssim \gamma \lesssim 1.3$ (they are very unresolved); between these, there is a gap (few galaxies are marginally resolved). The gap is especially clearcut for galaxies with large r_b (in arcsec) (*right panel*). When $\gamma \gtrsim 0.5$, these have power-law profiles. A fit of Equation 1 then seizes on any small curvature in the profile and spits out a value of r_b that has no physical meaning. Equation 1 is not well suited to deriving parameters for power-law profiles.

The possible dichotomy between two kinds of ellipticals was discovered by Nieto *et al.* (1991). In their sample, no elliptical with disky isophote distortions showed a core. All resolved cores were in boxy galaxies. We confirm and extend this conclusion. Symbol types in Fig. 4 encode isophote distortion (*left*) and the dynamical importance of rotation (*right*). In particular, filled squares identify galaxies that have boxy or neutral isophote distortions ($100 a_4/a \leq 0.4$; see Bender 1987; Bender *et al.* 1989) and that rotate slowly ($V/\sigma^* < 0.5$; see Davies *et al.* 1983). Almost all power-law galaxies are disky-distorted and rotate rapidly, and almost all cores are in boxy/neutral ellipticals that rotate slowly. Slow rotation implies velocity anisotropy and triaxial structure (Illingworth 1977; Binney 1976, 1978a, b).

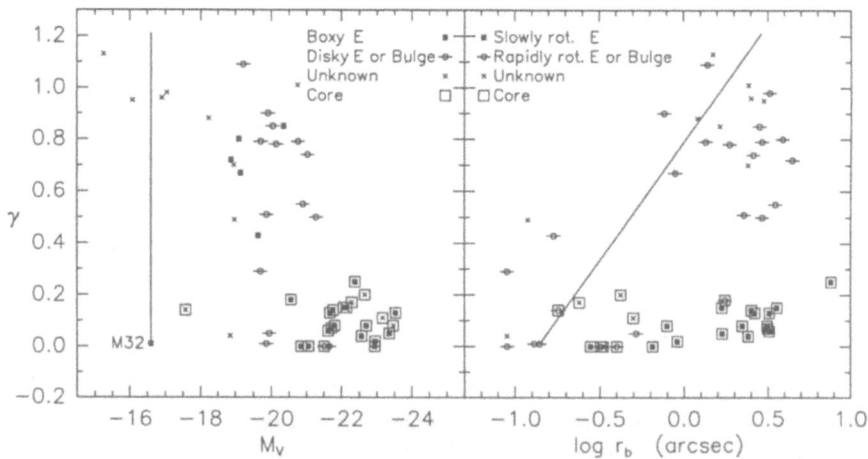

Figure 4. Inner profile slope *vs.* M_V (Kormendy *et al.* 1994) and r_b (Faber *et al.* 1995).

Isophote distortions also prove to be diagnostic of velocity anisotropy (Kormendy & Bender 1995). Figure 5 shows correlations of two kinematic diagnostics with $100a_4/a$. Here V/σ is the ratio of the maximum rotation velocity to the mean velocity dispersion near the center, and $(V/\sigma)^*$ is the ratio of V/σ to the value expected for isotropic oblate spheroids that are flattened by rotation. I. e., $(V/\sigma)^* \simeq 1$ implies a nearly isotropic velocity dispersion tensor, while $(V/\sigma)^* \lesssim 0.5$ implies substantial anisotropy. Similarly, minor-axis rotation implies triaxiality and hence anisotropy. So: Fig. 5 shows that disky-distorted galaxies are nearly isotropic, while essentially all of the anisotropic galaxies are boxy-distorted or neutral.

Given this result, global properties of ellipticals independently suggest the same dichotomy as do the core properties (Kormendy & Bender 1995; Kormendy & Djorgovski 1989). If we plot galaxy ellipticity *versus* $100a_4/a$,

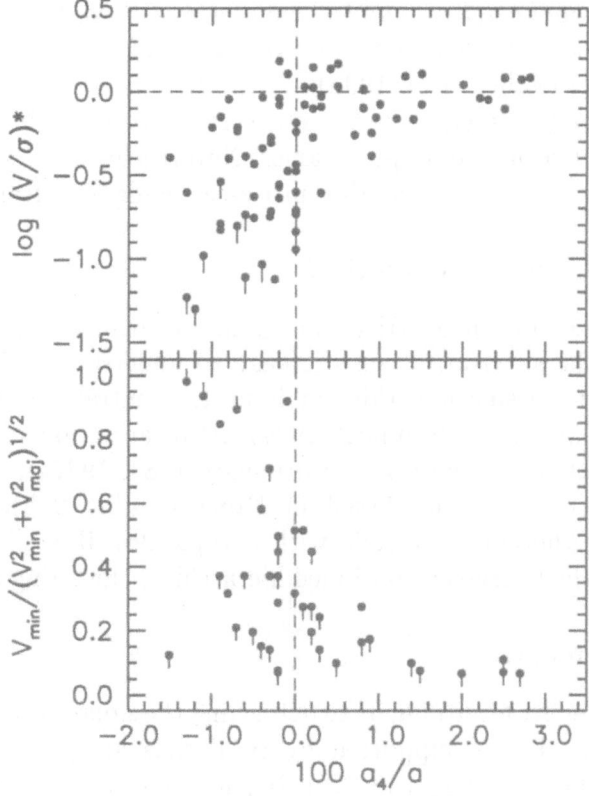

Figure 5.   Correlations with isophote shape of parameters that are diagnostic of velocity anisotropy (Kormendy & Bender 1995). Here $100\,a(4)/a$ is the percent inward or outward perturbation of isophote radii along the major axis; negative values indicate boxy isophotes; positive values indicate disky isophotes. The upper panel (first illustrated in Bender 1988) shows the rotation parameter $(V/\sigma)^*$. The lower panel shows minor-axis rotation velocity normalized by an indicative total rotation velocity.

we find a V-shaped distribution. E0 galaxies have $100a_4/a \simeq 0$. E4 galaxies have $100a_4/a \simeq -0.8$ or $\gtrsim +1$ but not ~ 0. If ellipticals are mostly oblate, then spherical galaxies are rare and the most common intrinsic shape is E4 (Sandage, Freeman, & Stokes 1970; Binney & de Vaucouleurs 1981). This suggests that the almost-round, almost-elliptical Es are almost face-on. Edge-on Es are either substantially disky or substantially boxy. Ellipticals divide themselves into boxy, anisotropic and disky, isotropic subgroups. This led Kormendy & Bender to propose that the Hubble sequence be revised as follows: boxy E – disky E – S0 – Sa – Sb – Sc galaxies. Then Es continue the sequence (right to left) of decreasing importance of rotation and increasing importance of random motions and velocity anisotropy.

We conclude: *Core and global properties both suggest that there are two different kinds of elliptical galaxies, (i) average- and low-luminosity Es that rotate rapidly and that are nearly isotropic, approximately oblate-spheroidal, disky-distorted, and* <u>*coreless*</u>*, and (ii) giant ellipticals that essentially do not rotate, that are anisotropic, moderately triaxial, and boxy-distorted, and that have* <u>*cuspy cores*</u> (Faber *et al.* 1995).

The dichotomy is suggestive but not certain. It is a subtler distinction than the one between E and Sph galaxies. Nevertheless, a dichotomy would suggest that two different formation processes made elliptical galaxies.

7. Why Do Cuspy Cores Exist?

As we explored core properties, we came to realize that neither the existence nor the survival of cores is easy to understand. Galaxy centers are vulnerable to dissipation; this builds up the central density and makes steep profiles (Mihos & Hernquist 1994). Also, accretion of small, dense Es tends to destroy the core FP (Kormendy 1984, 1987a). Possible ways to understand cores are discussed in Faber *et al.* (1995). None is yet convincing. We therefore conclude with two puzzles. How did cores form? And how have they survived continued hierarchical clustering and merging?

Acknowledgements

We thank D. Merritt for urging us to determine the slopes of the deprojected profiles. This work was supported by HST data analysis funds through grant GO–02600.01–87A and by NSERC. Kormendy's ground-based work was supported by NSF grant AST 9219221.

References

Bender, R.: 1987, *Mitt. Astr. Gesellschaft*, No. 70, 226
Bender, R.: 1988, *Astr. Ap.* **193**, L7
Bender, R., Burstein, D., & Faber, S. M.: 1993, *Ap. J.* **411**, 153

Bender, R., Burstein, D., & Faber, S. M.: 1994, in *Panchromatic View of Galaxies – Their Evolutionary Puzzle*, ed. G. Hensler *et al.*, Edit. Frontières: Gif-sur-Yvette, 99
Bender, R., *et al.*: 1989, *Astr. Ap.* **217**, 35
Binggeli, B.: 1987, in *Nearly Normal Galaxies: From the Planck Time to the Present*, ed. S. M. Faber, Springer-Verlag: New York, 195
Binggeli, B.: 1994a, in *ESO/OHP Workshop on Dwarf Galaxies*, ed. G. Meylan & P. Prugniel, ESO: Garching, 13
Binggeli, B.: 1994b, in *Panchromatic View of Galaxies – Their Evolutionary Puzzle*, ed. G. Hensler *et al.*, Editions Frontières: Gif-sur-Yvette, 173
Binggeli, B., & Cameron, L. M.: 1991, *Astr. Ap.* **252**, 27
Binggeli, B., Sandage, A., & Tammann, G. A.: 1988, *Ann. Rev. Astr. Ap.* **26**, 509
Binney, J.: 1976, *M. N. R. A. S.* **177**, 19
Binney, J.: 1978a, *M. N. R. A. S.* **183**, 501
Binney, J.: 1978b, *Comments Ap.* **8**, 27
Binney, J., & de Vaucouleurs, G.: 1981, *M. N. R. A. S.* **194**, 679
Byun, Y.-I., *et al.*: 1995, *A. J.*, in preparation
Crane, P., *et al.*: 1993, *A. J.* **106**, 1371
Da Costa, G. S.: 1992, in *IAU Symposium 149, The Stellar Populations of Galaxies*, ed. B. Barbuy & A. Renzini, Kluwer: Dordrecht, 191
Davies, R. L., *et al.*: 1983, *Ap. J.* **266**, 41
de Zeeuw, T.: 1995, in *IAU Symposium 171, New Light on Galaxy Evolution*, ed. R. L. Davies & R. Bender, Kluwer: Dordrecht, in press
Djorgovski, S., de Carvalho, R., & Han, M.-S. 1988: in *The Extragalactic Distance Scale*, ed. S. van den Bergh & C. J. Pritchet, ASP: San Francisco, 329
Djorgovski, S. G., Pahre, M. A., & de Carvalho, R. R.: 1995, in *Fresh Views of Elliptical Galaxies*, ed. A. Buzzoni *et al.*, ASP: San Francisco, in press
Djorgovski, S., & Santiago, B. X.: 1993, *ESO/EIPC Workshop: Structure, Dynamics, and Chemical Evolution of Early-Type Galaxies*, ed. J. Danziger *et al.*, ESO: Garching, 59
Faber, S. M., *et al.*: 1987, in *Nearly Normal Galaxies: From the Planck Time to the Present*, ed. S. M. Faber, Springer-Verlag: New York, 175
Faber, S. M., *et al.*: 1995, *A. J.*, in preparation
Ferguson, H. C., & Binggeli, B.: 1994, *Astr. Ap. Rev.* **6**, 67
Ferguson, H. C., & Sandage, A.: 1991, *A. J.* **101**, 765
Ferrarese, L., *et al.*: 1994, *A. J.* **108**, 1598
Forbes, D. A.: 1994, *A. J.* **107**, 2017
Forbes, D. A., Franx, M., & Illingworth, G. D.: 1994, *Ap. J.* **428**, L49
Forbes, D. A., Franx, M., & Illingworth, G. D.: 1995, *A. J.* **109**, 1988
Gebhardt, K., *et al.*: 1995, *A. J.*, in preparation
Grillmair, C. J., *et al.*: 1994, *A. J.* **108**, 102
Ichikawa, S.-I., *et al.*: 1988, *A. J.* **96**, 62
Ichikawa, S.-I., Wakamatsu, K.-I., & Okamura, S.: 1986, *Ap. J. Suppl.* **60**, 475
Illingworth, G.: 1977, *Ap. J.* **218**, L43
Jaffe, W., *et al.*: 1994, *A. J.* **108**, 1567
Kormendy, J.: 1982, in *Morphology and Dynamics of Galaxies, Twelfth Saas-Fee Course*, eds. L. Martinet & M. Mayor, Geneva Observatory: Sauverny, 113
Kormendy, J.: 1984, *Ap. J.* **287**, 577
Kormendy, J.: 1985a, *Ap. J.* **292**, L9
Kormendy, J.: 1985b, *Ap. J.* **295**, 73
Kormendy, J.: 1987a, in *IAU Symposium 127, Structure and Dynamics of Elliptical Galaxies*, ed. T. de Zeeuw, Reidel: Dordrecht, 17
Kormendy, J.: 1987b, in *Nearly Normal Galaxies: From the Planck Time to the Present*, ed. S. M. Faber, Springer-Verlag: New York, 163
Kormendy, J., & Bender, R.: 1994, in *ESO/OHP Workshop on Dwarf Galaxies*, ed. G. Meylan & P. Prugniel, ESO: Garching, 161
Kormendy, J., & Bender, R.: 1995. *A. J.*, submitted

Kormendy, J., & Djorgovski, S.: 1989, *Ann. Rev. Astr. Ap.* **27**, 235
Kormendy, J., *et al.*: 1994, in *ESO/OHP Workshop on Dwarf Galaxies*, ed. G. Meylan & P. Prugniel, ESO: Garching, 147
Kormendy, J., & McClure, R. D.: 1993, *A. J.* **105**, 1793
Lauer, T. R.: 1985a, *Ap. J. Suppl.* **57**, 473
Lauer, T. R.: 1985b, *Ap. J.* **292**, 104
Lauer, T. R., *et al.*: 1991, *Ap. J.* **369**, L41
Lauer, T. R., *et al.*: 1992a, *A. J.* **103**, 703
Lauer, T. R., *et al.*: 1992b, *A. J.* **104**, 552
Lauer, T. R., *et al.*: 1993, *A. J.* **106**, 1436
Lauer, T. R., *et al.*: 1995, *A. J.* in press
Merritt, D., & Fridman, T.: 1995a, in *Fresh Views of Elliptical Galaxies*, ed. A. Buzzoni *et al.*, ASP: San Francisco, in press
Merritt, D., & Fridman, T.: 1995b, *Ap. J*, submitted
Mihos, J. C., & Hernquist, L.: 1994, *Ap. J.* **437**, L47
Møller, P., Stiavelli, M., & Zeilinger, W. W.: 1995, *M. N. R. A. S*, in press
Nieto, J.-L., Bender, R., & Surma, P.: 1991, *Astr. Ap.* **244**, L37
Saglia, R. P., Bender, R., & Dressler, A.: 1993, *Astr. Ap.* **279**, 75
Sandage, A., Binggeli, B., & Tammann, G. A.: 1985, *A. J.* **90**, 1759
Sandage, A., Freeman, K. C., & Stokes, N. R.: 1970, *Ap. J.* **160**, 831
Schweizer, F.: 1980, *Ap. J.* **237**, 303
Schweizer, F.: 1981, *Ap. J.* **246**, 722
Stiavelli, M., Møller, P., & Zeilinger, W. W.: 1993, *Astr. Ap.* **277**, 421
Tremaine, S.: 1995, in *Some Unsolved Problems in Astrophysics*, in press
van den Bosch, F. C., *et al.*: 1994, *A. J.* **108**, 1579
Young, P. J., *et al.*: 1978, *Ap. J.* **221**, 721
Wirth, A., & Gallagher, J. S.: 1984, *Ap. J.* **282**, 85

Discussion

A. Renzini: As you pointed out, faint galaxies have such high phase space densities that they should easily survive if accreted by bright ellipticals. The fact that bright ellipticals don't have power-law nuclei seems to argue against recent merging, doesn't it?

J. Kormendy: Yes, that is precisely our point. There are rare exceptions, like NGC 1316, in which a steep density profile is seen and in which there is evidence for a recent merger (Schweizer 1980, 1981; Kormendy 1987a). But in general, mergers of present-day ellipticals tend to destroy the core FP relations *unless some process can heat the core*. One possibility is binary black holes, perhaps themselves a result of the merger (Faber *et al.* 1995).

W. Dehnen: A way to solve the problem of forming giant Es from low-luminosity Es may be to make the cuspy cores later by secular evolution. Do you see any signs of secular evolution (e. g., barlike distortions)?

J. Kormendy: The secular process for which we see evidence is accretion of gas-rich fragments. We see central dust disks, stellar disks (sometimes made of young stars), and in one case, both a dust disk and a stellar disk (Kormendy *et al.* 1994). Accretion tends to increase the central density and fill in any cores. The only galaxy in which we see an elongated center (actually an asymmetric one, as in M 31) is NGC 4486B.

RECENT PROGRESS IN THE SEARCH FOR
BLACK HOLES IN GALACTIC NUCLEI

ROELAND P. VAN DER MAREL[1]

Institute for Advanced Study

Olden Lane, Princeton, NJ 08540, USA

Abstract. Massive nuclear black holes (BHs) of $10^6 - 10^9 M_\odot$ are believed to be responsible for the the energy production in quasars and active galaxies, and are thought to be present in many quiescent galaxies as well. Dynamical evidence for this can be sought by studying the dynamics of gas and stars in galactic nuclei at high spatial resolution. This paper reviews the current evidence, with emphasis on some recent developments and ongoing projects. The evidence from water masers and gas kinematics in the active galaxies NGC 4258 and M87 is compelling. In quiescent galaxies only stellar kinematics are generally available. One well-studied case is M32. Stellar dynamical $f(E, L_z)$ models with a few million solar mass BH fit the ground-based kinematical data remarkably well. N-body simulations of an edge-on $f(E, L_z)$ model for M32 show that this model is stable. HST spectra should soon provide new and improved constraints on the presence of BHs in quiescent galaxies.

1. Active Galaxies: Gas Kinematics

Rapid gas motions have long been known to exist in the narrow and broad emission line regions of active galactic nuclei. To use these to constrain the gravitational potential one must be able to spatially resolve the gas morphology and determine whether the motions are gravitational (or due to inflow, outflow, turbulence, etc.). This has proved very difficult with ground-based optical techniques, but space based and radio observations have recently yielded important progress.

[1]Hubble Fellow.

R. Bender and R. L. Davies (eds.), New Light on Galaxy Evolution, 117–120.

In the 1970's M87 was the first galaxy for which stellar kinematical data hinted at the presence of a nuclear BH, but unambiguous interpretation remained difficult due to the unknown velocity dispersion anisotropy. The existence of rapid gas motions near the nucleus of M87 was pointed out by van der Marel (1994), who argued that these motions are gravitational. Ford et al. (1994) spatially resolved the gas morphology with HST, and showed the gas to be in a nuclear disk. Harms et al. (1994) measured gas rotation velocities in the disk of $\pm 500 \, \mathrm{km \, s^{-1}}$ at $R = \pm 0.3''$, implying a nuclear dark mass of $\sim 2.4 \times 10^9 M_\odot$, probably in a BH.

Even more impressive are the VLBA radio observations of water masers in the nucleus of the active galaxy NGC 4258 (Miyoshi et al. 1995). The individual maser sources reside in a nearly edge-on torus with inner and outer radii of 4 and 8 milliarcsec. The rotation curve of the sources is Keplerian to 1% accuracy and implies a mass within the torus of $3.6 \times 10^7 M_\odot$, most likely in a BH. The high spatial resolution and small deviations from Keplerian rotation rule out most alternatives (Maoz 1995).

NGC 4258 might well remain a unique case for several years to come: only a handful of (active) galaxies has nuclear water masers, and a torus (or disk) will only maser towards the observer if seen nearly edge on. The chances of finding more evidence for BHs from the kinematics of nuclear ionized gas disks appear more promising. Several groups are now taking HST data to this extent. One interesting galaxy is the E4 radio LINER NGC 7052 (van den Bosch & van der Marel 1995). HST images show a beautiful nuclear disk of dust and ionized gas (van der Marel, van den Bosch & de Zeeuw 1996, in preparation). Ground based spectra in 0.6'' seeing show rapid gas rotation ($250 \, \mathrm{km \, s^{-1}}$ at $R = 1.5''$) and an increasing line width towards the nucleus (nuclear FWHM $550 \, \mathrm{km \, s^{-1}}$). The latter can be due to seeing broadening of Keplerian rotation around a $5 \times 10^8 M_\odot$ BH, or to non-gravitational motions in an unresolved nuclear component. In Cycle 5 we will use HST to measure the gas rotation velocities at $R = 0.1''$ and $0.2''$, to obtain improved constraints on the mass of a possible BH.

2. Quiescent Galaxies: Stellar Kinematics

In quiescent galaxies only stellar kinematics are generally available. Unambiguous interpretation is complicated because the orbital structure is unknown, and difficult to derive from the data because of line-of-sight projection. Nonetheless, tentative evidence for nuclear BHs has been derived from ground-based data of a handful of nearby galaxies, most noticeably M32, M31, NGC 3115, NGC 3377, NGC 4594, and our own Galaxy (see Kormendy & Richstone 1995 for an extensive review). Here I focus on M32.

M32 has a steep central rotation velocity gradient, a central peak in the velocity dispersion, and a central surface brightness cusp, all hinting at the presence of a nuclear BH. van der Marel et al. (1994a) obtained kinematical data along five different slit position angles in $\sim 0.8''$ FWHM seeing to determine line-of-sight kinematics and velocity profile shapes. A flattened stellar dynamical model with $f = f(E, L_z)$ and a $\sim 1.8 \times 10^6$ BH (or other compact dark mass) provides a remarkably good fit to these data (van der Marel et al. 1994b; Qian et al. 1995; Dehnen 1995). Kormendy & Bender (1995, in preparation) recently obtained a major axis spectrum with the CFHT in $0.5''$ FWHM seeing, which appears to indicate a somewhat larger $(\sim 3 \times 10^6)$ BH mass. HST spectra of M32 will soon be obtained by van der Marel, Rix, de Zeeuw & White. To rule out models without a BH unambiguously requires the construction of flattened three-integral models. This has not been done yet, but several groups are working on this.

Kuijken & Dubinski (1994) found that flattened $f(E, L_z)$ models with much mean streaming can be unstable to the formation of a bar in their inner regions (see also Dehnen, this volume). To test whether $f(E, L_z)$ models for M32 are stable, we constructed an N-body realization of the edge-on model with a BH presented by Qian et al. (1995). The time-evolution of the system was studied with the 'self-consistent field' (SCF) N-body code of Hernquist & Ostriker (1992). Figure 1 summarizes the main results for a simulation with 2.5×10^4 equal mass particles (van der Marel et al. 1995). The axial ratios of the mass distribution, its radial profile, and the shape of the velocity ellipsoid are all constant with time. This indicates that the model is globally stable. Models for M32 that are not edge-on are intrinsically flatter. We are currently testing the stability of models with lower inclinations.

Simulations with $N = 10^{4-5}$ adequately represent the dynamics over the radial range that contains $\gtrsim 99\%$ of the mass, and suffice to study global stability. More particles are required to study the secular evolution close to the BH, and to study in detail the influence of a nuclear BH on the onset of a bar instability. We are currently running simulations with $N \gtrsim 10^6$ on a parallel computer, using the SCF code implementation of Hernquist et al. (1995) and Sigurdsson, Hernquist & Quinlan (1995). This allows a detailed study of the effects of the presence of a BH on the nuclear structure of a realistic stellar system such as M32.

Support for this work was provided by NASA through a Hubble Fellowship, #HF-1065.01-94A, awarded by the Space Telescope Science Institute which is operated by AURA, Inc., for NASA under contract NAS5-26555.

References

Dehnen W., 1995, MNRAS, 274, 919
Dubinski J., Carlberg R.G., 1991, ApJ, 378, 496

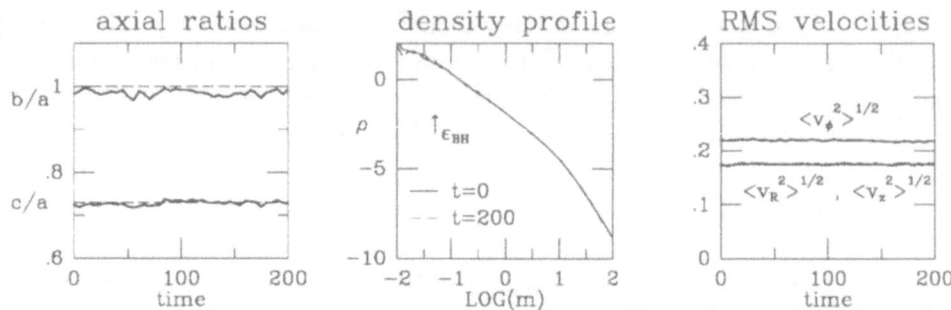

Figure 1. Time evolution of a self-consistent edge-on $f(E, L_z)$ model for M32 with a (softened) $1.8 \times 10^6 M_\odot$ BH, obtained from an N-body simulation with $N = 2.5 \times 10^4$. The left panel shows the 'mean' axial ratios b/a and c/a of the mass distribution as function of time, calculated as described in Dubinski & Carlberg (1991). Dashed lines indicate the analytical values $b/a = 1$ and $c/a = 0.73$. The middle panel shows the density profile $\rho(m)$ at the initial and final time, where $(m/a)^2 \equiv (x/a)^2 + (y/b)^2 + (z/c)^2$. The curves overlay each other. The arrow indicates the BH softening length $\epsilon_{BH} = 5 \times 10^{-2}$. The right panel shows the RMS velocities averaged over all particles, as function of time. The units of time, length and velocity used in the figures are: 1.1×10^5 yr; 30 pc $= 8.7''$; and 246 km s^{-1}. The period of a circular orbit in the equatorial plane is $T = 0.7$ at $m = \epsilon_{BH}$ and $T = 15$ at $m = 1$. The model is globally stable, and shows no sign of a bar instability. The RMS velocities remain constant with time, so that there is probably no secular instability that introduces a dependence of the DF on a third integral. The fractional energy conservation during the simulation was 5×10^{-4}. We are currently running simulations with larger N and smaller ϵ_{BH} to verify the accuracy of these results.

Ford H.C. et al., 1994, ApJ, 435, L27

Harms R.J. et al., 1994, ApJ, 435, L35

Hernquist L., Ostriker J.P., 1992, ApJ, 386, 375

Hernquist L., Sigurdsson S., Bryan G.L., 1995, ApJ, 446, 717

Kormendy J., Richstone D.O., 1995, ARA&A, in press

Kuijken K., Dubinski J., 1995, MNRAS, 269, 13

Maoz E., 1995, ApJ, 447, L91

Miyoshi et al, 1995, Nature, 373, 127

Qian E., de Zeeuw P.T., van der Marel R.P., Hunter C., 1995, MNRAS, 274, 602

Sigurdsson S., Hernquist L., Quinlan G.D., 1995, ApJ, 446, 75

van den Bosch F., van der Marel R.P., 1995, MNRAS, 274, 884

van der Marel R.P., 1994, MNRAS, 270, 271

van der Marel R.P., Rix. H–W., Carter D., Franx M., White S.D.M., de Zeeuw P.T., 1994a, MNRAS, 268, 521

van der Marel R.P., Evans N.W., Rix H–W., White S.D.M., de Zeeuw P.T., 1994b, MNRAS, 271, 99

van der Marel R.P., Sigurdsson S., Hernquist L., Quinlan G., 1995, ApJ, in preparation

Khachikian: Many (active) galaxies have a double nucleus. Does this speak for or against the BH hypothesis ?

van der Marel: This is not clear until we have a better understanding of what these double nuclei are. For the galaxy M31, which is a prime example, one can construct a viable model that has a BH and also explains the double nucleus (Tremaine S., 1995, AJ, 110, 628).

THE INTERSTELLAR MEDIUM

THE X-RAY EMISSION OF E AND S0 GALAXIES

G. FABBIANO

Harvard-Smithsonian CfA
60 Garden St., Cambridge MA 02138, USA

1. Introduction: Facts and Myths

It is well known by now that E and S0 galaxies are associated with hot, X-ray emitting gaseous halos, subject to central cooling flows, but otherwise homogeneous, which can be used to trace the galaxy potential and thus measure the total gravitational mass. Recent ASCA observations suggest that these halos have a surprisingly low metal abundance.

Although these 'facts' have entered the astronomical consciousness and do (or may) represent reality in some cases, if taken literally they may be more akin to myth. Although some galaxies are indeed associated with centrally cooling extensive X-ray halos, this is by no means true for all E and S0s. Although the X-ray data can be used to measure the binding mass of the galaxies, and in some cases the resulting M/L are indeed large, these measurements cannot be applied blindly to all galaxies. Although ASCA spectra are returning the unmistakeable signature of thermally emitting plasma, how well do we understand these data and the physical state of the hot ISM?

In this paper, I attempt to review the observations critically and give my assessment on our present state of knowledge of the X-ray properties of E and S0 galaxies and on what we have learned about the structure and possibly evolution of these galaxies by using these data.

There is significant overlap between this paper and a talk I gave at the meeting on early-type galaxies in Mexico earlier this year (Fabbiano 1995).

2. X-ray Bright and X-ray Faint E and S0's

Not all E and S0 galaxies retain their hot ISM. Some galaxies (e.g. NGC 4406 and NGC 4472 in Virgo; Forman et al 1979; Trinchieri, Fabbiano and Canizares 1986; Fabbiano, Kim and Trinchieri 1992) are associated with

R. Bender and R. L. Davies (eds.), New Light on Galaxy Evolution, 123–130.

extensive halos clearly displaced from the stellar component, and in some cases extending to very large radii (e.g. NGC 4636; Trinchieri et al 1994). Other galaxies (Fabbiano, Gioia and Trinchieri 1989; Fabbiano, Kim and Trinchieri 1994) in similar environments, display much fainter X-ray emission, typically coextensive with the stellar body, and generally consistent with the emission expected from a population of X-ray sources similar to that of M31 (Fabbiano, Trinchieri and VanSpeybroeck 1987; Trinchieri and Fabbiano 1991).

The X-ray spectra of the X-ray bright galaxies show clearly the presence of line emission (e.g. Awaki et al 1994) thus confirming the presence of an optically thin hot ISM. The temperature of the emitting gas also appears to decrease at smaller radii (e.g. Trinchieri et al 1994; Kim and Fabbiano 1995), as it would be expected in the presence of cooling flows. X-ray faint galaxies have significantly different spectral distributions from those of X-ray bright galaxies, suggesting the presence of multicomponent emission, with little – if any – hot ISM present (Fabbiano, Kim and Trinchieri 1994; Pellegrini and Fabbiano 1994).

The extent of the variation in the ability of retaining a hot gaseous halo by galaxies of a given optical luminosity is shown clearly when we plot the *Einstein* E and S0 sample in the $L_X - L_B$ plane. This diagram (Figure 1) shows an impressive amount of scatter. For a given L_B, one can find values of L_X differing by two orders of magnitude. The lower L_X galaxies have X-ray emission in the range expected from a population of bulge-type stellar X-ray sources.

3. Correlation Studies

We (Eskridge, Kim and Fabbiano 1995 a, b, c) explored the question of the important parameters affecting the presence of a hot ISM, and of the effect of the hot ISM on galaxy properties by studying the relations between *stellar* ($L_{12\mu m}, L_B, L_X$), *ISM* ($L_{HI}, L_{100\mu m}, L_X$), *nuclear* ($L_{6cm}$), and *structural* [potential shape: $a/b, a_4$, Bender at al 1989; potential depth: σ_v, Mg_2, Faber 1973] parameters. We used censored analysis methods, including regression analysis and partial-rank tests.

We find that morphology is important, in the sense that S0s are systematically X-ray fainter than E galaxies, for a given optical luminosity. Although in this case the increased fraction of rotational energy in the former may play a role, it appears more likely that this effect is related to the shape of the potential (see also Pellegrini 1994; Ciotti and Pellegrini 1995).

There are stronger and more direct indications that the galaxy potential is a major player. The *depth* of the potential is a key factor in promoting the retention of a hot ISM. There are strong correlations between L_X (and

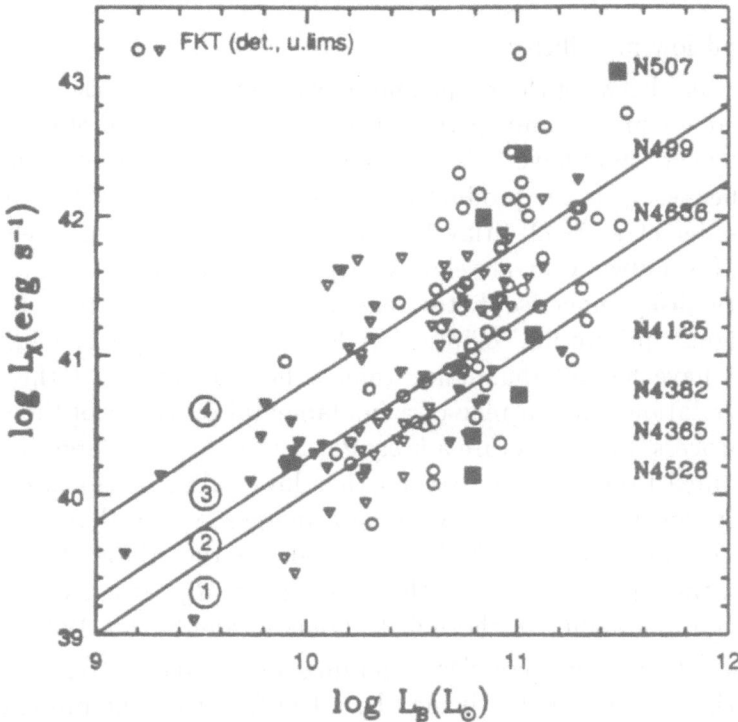

Figure 1. The distribution of the *Einstein* sample of E and S0 galaxies in the $L_X - L_B$ plane. The locus of Spirals and *bulge* dominated spirals overlaps regions 1 and 2 in this diagram.

the L_X/L_B ratio) and both σ_v and Mg_2, suggesting that the deeper the potential well, the better the hot ISM retention, and the better the retention of stellar ejecta from the early formation phase. Also, a certain depth appears to be critical for both retaining the hot ISM and fuelling an active nucleus: there are 'threshold' correlations between σ_v and both L_X/L_B and L_{6cm}.

The *shape* of the potential is also a key factor. Rounder (and boxier) galaxies have larger L_X, L_X/L_B, and L_{6cm}, with the strongest correlation being that between L_X/L_B and a/b. Also rounder galaxies tend to have boxier isophotes. These results persist when E and S0s are analyzed separately.

The potential shape and the depth parameters appear to be related, but not in a simple way. For instance, at a given value of a_4 the spread in σ_v appears directly related to the spread in Mg_2 such that the higher σ_v objects also tend to have higher Mg_2 (the $Mg_2 - \sigma_v$ relations). For a given σ_v in the range where substantial X-ray halos are retained, boxy galaxies tend to have *lower* Mg_2 values (and thus lower stellar abundances) than disky galaxies. However, the galaxies with the lowest σ_v tend to have disky isophotes and low metallicity.

The coupling between the shape and depth parameters is clearly an important input for understanding the formation and evolution of early-type systems. A galaxy may have a deep potential either because of initial conditions, or because it is the product of mergers of a number of less massive objects (Bender et al 1992; Stiavelli 1992). In the case of relatively high σ_v galaxies, for a given central velocity dispersion (depth of the potential), galaxies with disky inner isophotes tend to be more metal rich. If disky isophotes signal 'primordial' ellipticals, then a disky elliptical of a given mass would have higher abundance than a boxy elliptical, if the Mass-Abundance relationship is a relic of a fundamental contraint of the galaxy formation process. In this picture a boxy galaxy would be a merger product that was formed from lower mass (and thus lower abundance) precursors. Alternatively, disky isophotes may signal a merger system that has either formed a secondary nuclear disk (Hernquist and Barnes 1991), or captured a self-gravitating spiral disk. If so, then the higher Mg_2 at a given σ_v may be due to high abundance in those disks (Bender and Surma 1992).

The effect of merging is not clear. Merging was related to boxy isophotes and it was thus suggested that it may be a key factor for the presence of a hot ISM (Bender et al 1992). However, observations of galaxies believed to be remnants of recent mergers and found to be X-ray faint suggests that very recent merging may instead remove the hot ISM (e.g., NGC 4365, Fabbiano, Kim and Trinchieri 1994; NGC 3610 and NGC 4125, Fabbiano and Schweizer 1995). Perhaps the age of the merging event is a factor.

The link between L_X and potential depth is confirmed by the comparisons with the κ parameters of Bender, Burstein and Faber (1992), suggesting that L_X is more strongly correlated with inner mass to light ratio than any other luminosity indicators. This suggests that galaxies with central excesses of dark matter also have more massive extended dark matter halos, providing a mechanism for retaining larger amounts of hot ISM. We also find that galaxies with higher inner M/L for their mass also tend to be more metal rich. This may be due to the enhanced ability of such galaxies to mantain ongoing or episodic central star formation for extended periods, or to retain the enriched ejecta of early epochs of star formation, or both. This trend becomes stronger when tested for constant a_4, suggesting that this effect is not connected to the presence of metal-rich nuclear disks.

4. Measuring the Binding Mass

If E and S0 galaxies have hot gaseous halos in hydrostatic equilibrium in the galaxy potential, the X-ray data can be used to measure the mass of the galaxies (e.g. Fabricant and Gorenstein 1983). The use of the *Einstein*

X-ray observation for measuring the gravitational mass of E and S0 galaxies (Forman, Jones and Tucker 1985) has been controversial, mostly because of the quality of the available data (Trinchieri, Fabbiano and Canizares 1986; see review in Fabbiano 1989).

Where are we ten years later? It is clear that any attempt at applying the standard hydrostatic equilibrium approach to observations of X-ray faint E and S0 is meaningless and will result in erroneous results, because the X-ray emission of these galaxies does not appear to be dominated by a hot gaseous halo (see above).

X-ray bright E and S0s retain hot gaseous halos, which indeed dominate the X-ray emission in the ROSAT range. With the ROSAT PSPC one can measure the radial temperature profile of these halos with reasonable accuracy (e.g. Trinchieri et al 1994; Trinchieri et al 1995, in preparation), a key measurement for contraining the binding mass (see Fabbiano 1989). These results point to M/L ~20-50 and thus to a considerable amount of dark matter in these galaxies. One must however remember that these mesurements are still dependent on assumptions on the spectral properties of the hot ISM, which in some cases are not unique (e.g. in NGC 4636, see Trinchieri et al 1994). One also assumes hydrostatic equilibrium.

A different approach was followed by Buote and Canizares (1994), who compared X-ray and optical isophotes of the flattened elliptical galaxy NGC 720, using ROSAT data. They found that the flattening of the X-ray contours at large radii can only be explained if the galaxy contains an extended halo of dark matter.

Central group galaxies tend to have huge hot gaseous halos extending hundreds of kiloparsecs and M/L ~100 or more (e.g David et al 1994; Kim and Fabbiano 1995). These huge halos contain a sizeable fraction of the total mass of the system. Interestingly, these results may conflict with the ratio of baryonic to total mass expected in a $\Omega = 1$ universe (see also Briel, Henry and Boehringer 1992). However, even here a note of caution is necessary, since clumpiness of the X-ray halo which may be undetected with the present instruments may lower the estimate of the baryonic mass. The ROSAT PSPC image of NGC 507 suggests indeed the presence of such clumps (Kim and Fabbiano 1995).

5. Hot Halos and Active Nuclei

A strong connection was found between the presence of a hot ISM and that of radio sources in E and S0 galaxies. In a statistical sense, there are clear strong correlations between L_X, L_X/L_B and both total and nuclear radio power. These suggest that the hot ISM, accreting to the nuclei through cooling flows may be the fuel of the active nucleus. Moreover, the thermal pressure of this medium in X-ray bright galaxies is enough to represent an effective mechanism for confining and/or disrupting radio jets, and confining radio lobes in the less powerful radio galaxies (Fabbiano, Gioia and Trinchieri 1989).

We now begin to have direct evidence of the interaction between hot ISM and nuclear radio sources. The galaxy NGC 1399 is associated with a double lobed radio source, which is entirely contained in an extensive X-ray halo undergoing a cooling flow (Killeen and Bicknell 1988). We have obtained a high resolution ROSAT HRI image of the central regions of this galaxy (Kim et al 1995). This image shows a remarkable inhomogeneity of the ISM in a variety of scales, including an arcmin scale spiral-arm like feature; an E-W flattened central region with a total extent of about 1' and cooler than the surrounding regions; a few unresolved clumps; and finally a remarkable tunnel-like structure enveloping the northern radio lobe, with a possible suggestion of a similar structure in the south as well (see figure in Fabbiano 1995). The thermal pressure we derive from the tunnel's walls is in excess of that in the radio lobes. Thermal confinement is therefore the likely mechanism in this galaxy. Interactions between radio jets and hot ISM are also suggested in NGC 1316 (Kim, Fabbiano and Mackie in preparation).

These X-ray images suggest complexity in the hot ISM of early-type galaxies, and give us a taste of what is to come when AXAF will be returning high spectral and spatial resolution data on these galaxies.

6. Different Phases of the ISM

The relations between different phases of the ISM of early-type galaxies (hot – X-ray; warm – optical emission line; cold – FIR, HI), have been the subject of a number of papers (e.g. Trinchieri and diSerego Alighieri 1991; Bregman, Hogg and Roberts 1992; Eskridge, Fabbiano and Kim 1995a). These works suggest, but do not prove, a possible connection between the presence of cooling flows and Hα emission. They also show that the presence of a hot ISM and of cold HI gas or dust (FIR) are largely uncorrelated, suggesting either different origins (in the case of the HI), or an adverse effect of the hot X-ray emitting environment (in the case of dust).

Detailed comparisons of X-ray images of galaxies and images at other wavelengths are now being pursued and are beginning to reveal a rich multiphase ISM. In NGC 1316, dust lanes and hot ISM coexist in the core, as well as Hα emitting gas (Kim et al, in preparation).

7. The Abundance Question

The analysis of the ASCA CCD observations of galaxies provide far greater resolution than available with proportional counter data, leading to the detection of line emission. Using these data, significant subsolar metal abundance have been reported in the hot ISM of E and S0 galaxies (Awaki et al 1994; Matsushita et al 1994). These reports are in definite contrast with the expectations from stellar measurements and theory, which would predict metallicities of at least solar in the ISM (see N. Arimoto's talk, this volume).

A discussion of the possible pitfalls of the present X-ray spectral analysis, that may lead to spurious results as far as the abundances are concerned, can be found in Fabbiano (1995). Here I will not repeat those argument, but just state that this question is generating a lively amount of debate in the X-ray community. The low-abundance problem is not limited to elliptical galaxies. Similar puzzling results have been reported from observations of stars, for which abundances are well known (e.g. AR Lac, observed with ASCA, White et al 1994; and other stars, observed with both ROSAT and ASCA, J. Bregman 1995, private communication). Also the ASCA observation of the Cygnus Loop has been reported to suggest subsolar abundances (Miyata et al 1994). Unless we want to accept that metals are depleted in the X-ray emitting regions in these systems (interestingly, the opposite -if anything- may happen in the Sun; Meyer 1991, Phillips et al 1994), we have to conclude that there is something wrong with either our choice of models or our fitting procedures.

I think that before subscribing to the low-abundance interpretation of the spectral results on elliptical galaxies, we need to understand better what is going on with other less controversial astrophysical plasmas.

This work was partially supported by NASA contract NAS8-39073 (AXAF Science Center), and by NASA grant NAGW2681 (LTSA).

References

Awaki, H. et al 1994, P. A. S. J., 46, L65.
Bender, R., Surma, P., Doebereiner, S., Moellenhoft, C., and Madejsky, R. 1989, Astr. Ap., 217, 35.
Bender, R. et al 1992, in "Structure, Dynamics, and Chemical Evolution of Elliptical Galaxies", I. J. Danziger, W. W. Zeilinger, and K. Kjar, eds., ESO Proc. No. 45, p. 3.
Bender, R. and Surma, P. 1992, Astr. Ap., 258, 250.
Bender, R., Burstein, D., and Faber, S. M. 1992, Ap. J., 399, 462.
Briel, U. G., Henry, J. P., and Boehringer, H. 1992, Astr. Ap., 259, L31.
Buote, D., and Canizares, C. R. 1994, Ap. J., 427, 86.
Ciotti, L. and Pellegrini, S. 1995, M. N. R. A. S., submitted.
David, L. P., Jones, C., Forman, W., and Daines, S. 1994, Ap. J., 428, 544.

Eskridge, P. B., Fabbiano, G., and Kim, D.-W. 1995a, Ap. J. Suppl., 97, 141.
Eskridge, P. B., Fabbiano, G., and Kim, D.-W. 1995b, Ap. J., 442, 523.
Eskridge, P. B., Fabbiano, G., and Kim, D.-W. 1995c, Ap. J., in press.
Fabbiano, G. 1989, Ann. Rev. Ast. Ap., 27, 87.
Fabbiano, G. 1995, in "Fresh Views on Elliptical Galaxies", A. Buzzoni, ed., Pub. of A. Soc. P., in press.
Fabbiano, G., Kim, D.-W., and Trinchieri, G. 1992, Ap. J. Suppl., 80, 531.
Fabbiano, G., Gioia, I. M., and Trinchieri, G. 1989, Ap. J., 347, 127.
Fabbiano, G., Kim, D.-W., and Trinchieri, G. 1994, Ap. J., 429, 94.
Fabbiano, G., and Schweizer, F. 1995, Ap. J., in press.
Fabbiano, G., Trinchieri, G., and VanSpeybroeck, L. 1987, Ap. J., 316, 127.
Faber, S. M. 1973, Ap. J., 179, 731.
Fabricant, D., and Gorenstein, P. 1983, Ap. J., 267, 535.
Forman, Jones, C. and Tucker, W., 1985, Ap. J., 293, 102.
Forman, W., Schwarz, J., Jones, C., Liller, W., and Fabian, A. C. 1979, Ap. J., 234, L27.
Hernquist, L., and Barnes, J. E. 1991, Nature, 354, 210.
Killeen, N.E.D. and Bicknell, G.V. 1988, Ap. J., 325, 169.
Kim, D.-W., and Fabbiano, G. 1995, Ap. J., 441, 182.
Kim, D.-W., Fabbiano, G., Mackie, G., and Norman, C. 1995, Ap. J., submitted.
Matsushita, K. et al 1994, Ap. J., 436, L41.
Meyer, J. 1991, Adv. Space Res., 11, 269.
Miyata, E., Tsunemi, H., Pisarski, R., and Kissel, S. E. 1994, P. A. S. J., 46, L101.
Pellegrini, S. 1994, Astr. Ap., 292, 395.
Pellegrini, S. and Fabbiano, G. 1994, Ap. J., 429, 105.
Phillips, K. J. H., Pike, C. D., Lang, J., Watanabe, T., and Zarro, D. M. 1994, Proc. of Kofu Symp., NRO Report No. 360, p. 301.
Stiavelli, M. 1992 in "Structure, Dynamics, and Chemical Evolution of Elliptical Galaxies", I. J. Danziger, W. W. Zeilinger, and K. Kjar, eds., ESO Proc. No. 45, p. 303.
Trinchieri, G., and di Serego Alighieri, S. 1991, A. J., 101, 1647.
Trinchieri, G., and Fabbiano, G. 1991, Ap. J., 382, 82.
Trinchieri, G., Fabbiano, G., and Canizares, C. R. 1986, Ap. J., 310, 637.
Trinchieri, G, Kim, D.-W., Fabbiano, G., and Canizares, C. R., 1994, Ap. J., 428, 555.
White, N. E. et al 1994, P. A. S. J., 46, L97.

Inge Thiering: Concerning the metallicity problem you reviewed, I would like to comment that for our ROSAT PSPC observations of the groups around the first ranking galaxies NGC 4104, NGC 6269, and NGC 6329 we obtain spectral fits with $\chi^2/\nu \sim 1$ for two different models of the hot gas: one is assuming a two-component gas of variable temperature and $z/z_o \sim 1$, and the other assumes a roughly isothermal gas whose metallicity decreases from the inside to the ouside from $z/z_o \sim 1$ to $z/z_o \sim 0.1$–0.3.

Answer: Thanks for your comment. I agree that with the PSPC one cannot find a unique model (see also Trinchieri et al 1994; Fabbiano, Kim and Trinchieri 1994).

Duncan Forbes: Do you see a morphological coincidence between soft X-rays (10^6K) and Hα emission (10^4K)?

Answer: In general, Hα is found in the cores of galaxies, where cooling may occur. However, I am not aware of detailed one to one spatial comparisons for a large enough sample of galaxies.

HOT GAS FLOWS IN ELLIPTICAL GALAXIES

ALVIO RENZINI
European Southern Observatory
Garching bei München, Germany

1. Introduction

Stars in elliptical galaxies lose mass at an overall present rate $\dot{M}_* \simeq 1.5 \times 10^{-11} L_B \quad M_\odot yr^{-1}$ (e.g., Faber & Gallagher 1976; Renzini & Buzzoni 1986). When allowing for the predicted increase back with cosmological time it turns out that over one Hubble time the stellar population of an elliptical galaxy has cumulatively lost 20-50% of its initial mass, the precise value depending on the IMF. This review focuses on two simple questions: what happens to the gas being lost by the stars? Where is it ultimately disposed?

Ellipticals are dynamically *hot* stellar systems, hence the cool stellar winds emanating from individual red giants are quickly heated to nearly the same dynamical *temperature* of the stars, as such winds collide with each other of with the ambient ISM. This implies $T \simeq m_p \sigma_*^2/k \simeq 10^7$K ($\sigma_*$ being the stellar velocity dispersion), and the gas will shine in X rays.

Fig. 1 shows the X-ray luminosity of the elliptical galaxies vs their optical luminosity. Data refer to the objects observed with the *Einstein* satellite that have been recently reanalysed in a homogeneous way by Fabbiano, Kim, & Trinchieri (1992). Galaxies at any given optical luminosity exhibit a very wide range of X-ray luminosities, and Fig.2 demonstrates that this is not a result of distance errors (i.e. this is a distance-independent diagram). So, galaxies with the same luminosity and/or the same σ_* may differ by up to a factor ~ 100 in their X-ray luminosity.

The lines in Fig. 1 show the predicted X-ray luminosity for the two classical stady state solutions for the gas flow. In a steady state supersonic wind (Mathews & Baker 1971) very little hot gas accumulates, and L_X reduces to $\sim L_{discr}$, the contribution of discrete sources. In a steady state "cooling flow" (e.g. Sarazin & White 1987) a dense ISM is in place, and $L_X \simeq L_{in}$. Clearly, neither of the two steady state solutions accounts for the

131

R. Bender and R. L. Davies (eds.), New Light on Galaxy Evolution, 131–138.

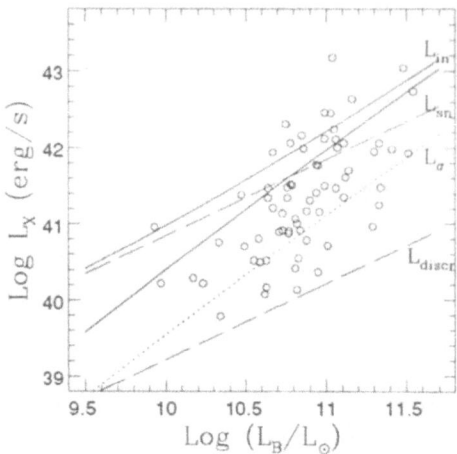

Figure 1.   The X-ray luminosity of elliptical galaxies observed with the Einstein Observatory vs their optical luminosity (From Fabbiano *et al.* 1992). L_{discr}, L_σ and L_{sn} are the expected contributions from discrete sources, from the thermalization of stellar winds, and from SN explosions (with $\vartheta_{SN} = 1$), respectively. The expected X-ray luminosity of galaxies in the inflow regime (L_{in}) is also shown, for $\vartheta_{SN} = 0$ and 1 (adapted from Ciotti *et al.* 1991).

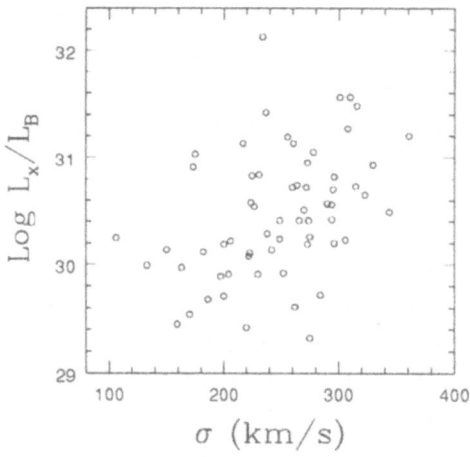

Figure 2.   The X-ray to optical luminosity ratio for the elliptical galaxies in Fig. 1 as a function of the central velocity dispersion. Luminosity units are as in Fig. 1.

observed X-ray properties of the galaxies. Some galaxies are faint enough in X-rays for their gas flow to be in a supersonic wind regime, and indeed their X-ray spectrum requires very little hot ISM contribution in excess of the emission from low mass X-ray binaries and other stellar sources (Pellegrini & Fabbiano 1994). Seemingly, some galaxies are bright enough in X-rays for being in a cooling flow regime. However, most galaxies exhibit X-ray luminosities that are intermediate between these two extremes.

By postulating a distibuted sink of gas from the hot ISM (e.g., Sarazin & Ashe 1989; Bertin & Toniazzo 1995), the X-ray luminosity of cooling flow models could be somewhat reduced, as only part of the flow would reach the center. Yet not enough to account for the galaxies for which $L_X < L_{sn} + L_\sigma$ (where L_{sn} and L_σ are the rates of energy deposition from supernova explosions and from the thermalization of stellar winds, respectively). Thus, steady state models of any kind can hardly account for the observed distribution of X-ray luminosities. There are also good theoretical reasons for dismissing steady state solutions. Indeed, both the rate of stellar mass return and the supernova (SN) heating are expected to exhibit a strong secular evolution, and therefore *evolutionary* – as opposed to time-independent – solutions are to be sought.

Life may get rather complicated when steady state solution are abandoned. Besides a time-dependent rate of stellar mass return a cornucopia of other time-dependent ingredients are likely to play a role, along with the very many parameters needed to describe them (see Renzini 1993 for a review). These include the evolution of the SN heating, the possible occurrence of star formation, the role of the intracluster medium (ram)pressure, recuTrent local thermal instabilities (Kritsuk 1992), and the interactions of the gas flow with a massive black hole (BH) that may exist at the center of galaxies. Here I concentrate on the role of two such ingredients, SN heating and BH-gas flow interactions.

2. What role for Type Ia Supernovae?

Evolutionary models of gas flows that include dark matter and SN heating meet several of the observed X-ray properties of ellipticals. There are basically two families of models, depending on the assumed past evolution of the Type Ia SN heating, that can be conveniently parameterized assuming a power-law evolution with time (Rate $\propto t^{-s}$). Also the evolution of the rate of stellar mass return is well represented by a power-law, with $\dot{M}_\star \propto\sim t^{-1.3}$ for a Salpeter IMF. Hence, for $s\gtrsim 1.3$ the specific heating of the stellar ejecta secularly *decreases* with time (the SN heating was more efficient in the past). Galaxy flows correspondingly evolve from early supersonic winds, to subsonic outflow, to an inflow regime when SN heating finally falls short of compensating for the radiative losses, and a central cooling catastrophe takes place (Ciotti *et al.* 1991, hereafter WOI models). Coversely, for $s\lesssim 1.3$ the specific SN heating *increases* with time, and flow models evolve in the reverse sequence of flow regimes, from early inflows, to subsonic outflows, to late supersonic winds (Loewenstein & Mathews 1987; David, Forman, & Jones 1991, hereafter IOW models).

The X-ray luminosity is very low during wind (W) phases, as the hot gas rapidly leaves the galaxy and does not accumulate. During inflow (I) phases galaxies are very bright in X-rays, as the hot gas remains confined to the galaxy and cools down. During outflow (O) phases L_X is intermediate between the two previous extremes, increasing with time in the WOI models as gas accumulates, decreasing with time in the IOW models. Thus, the large dispersion in the L_X/L_B ratio can be easily understood in the frame of either the WOI or the IOW models, and follows from relatively small differences in e.g. the depth of the potential well causing relatively large differences in the epoch of the O-I or I-O transition. In the frame of WOI models, in Fig. 1 galaxies evolve upward, from the lower dashed line when they are in the W phase, to the top line when they finally reach the I phase. IOW models instead drip from the top line, evolve downward, and finally land on the bottom line. In both cases, the observed X-ray luminosities indicate that most of present-day ellipticals are in the O phase.

How to distinguish between these two opposite options? The best, most direct way would be to look at high redshift ellipticals, with $z \gtrsim 1$, hence $t \lesssim 5$ Gyr. In the case of IOW models high-z E galaxies would be brighter in X-rays compared to low-z galaxies, while the reverse applies to WOI models. Unfortunately this is currently unfeasible; e.g. ROSAT can hardly see individual galaxies beyond the distance of the Coma cluster.

The present and past rate of SNIa's in ellipticals being the crucial ingredients, these can be constrained by the observed elemental abundances in their ISM and in the intracluster medium (ICM; Ciotti et al. 1991; Renzini et al. 1993). Within galaxies, one expects the iron abundance in the hot ISM to be $X_{Fe}^{ISM} \simeq < X_{Fe}^* > + 5\vartheta_{SN}$, thus reflecting the average stellar abundance ($< X_{Fe}^* >$) further enriched by Type Ia SNs, ϑ_{SN} being their present rate in units of 2.2×10^{-13} SNs $yr^{-1} L_\odot^{-1}$. The most extreme estimates of the Type Ia SN rate per unit light in ellipticals indicate $0.25 \lesssim \vartheta_{SN} \lesssim 1$ (Cappellaro, these proceedings), hence one expects $X_{Fe}^{ISM} \gtrsim 2 X_{Fe}^\odot$ (Ciotti et al. 1991). From the iron-L complex at ~ 1 keV very low values are instead inferred from ASCA data, $X_{Fe-L}^{ISM} \lesssim 0.3 X_{Fe}^\odot$ (Loewenstein et al. 1994, Matsushita, these proceedings). Formally, the ASCA results requires a vanishing SNIa activity in present day ellipticals. This *Iron-L Discrepancy* is extensively discussed by Arimoto (these proceedings).

The total amount of iron in clusters of galaxies (ICM plus galaxies) represent a record of overall past SN activity, hence it sets a constraint on the past evolution of the rate. The total iron mass to cluster optical luminosity ratio appears to be constant among clusters, irrespective of their luminosity, with $M_{Fe}/L_B \simeq 0.03 \pm 0.01$ $(M_\odot/L_\odot)$. To make all this iron requires either a much stronger SNIa activity in the past $(s \gtrsim 1.3)$, or a much stronger Type II SN activity at early times than expected from a Salpeter IMF (Ciotti et al. 1991; Arnaud et al. 1992; Renzini et al. 1993;

Elbaz, Arnaud, & Vangioni-Flam 1995). The elemental abundances in the ICM offer an attractive way to distinguish between these two alternatives (Renzini *et al.* 1993), hence providing indirect constraints on the history of gas flows in ellipticals. Again, ASCA observations seem to require very little SNIa contribution to the ICM (Mushotzky *et al.* 1995), which in turn would favor IOW over WOI models. If confirmed, this result would reduce SNIa's to a very marginal role in the chemical evolution of galaxies at a galaxy cluster scale, thus rising the puzzling question as to how the solar elemental proportions have been established in our own Galaxy, while non-solar proportions may prevail at galaxy cluster scales.

3. Gas Flows and Central Black Holes

If SNIa's are unimportant iron producers, they automatically become unimportant also as a heating source for galaxy gas flows (Ciotti *et al.* 1991). An alternative source of energy needs therefore to operate in order to power the outflows prevailing in present-day ellipticals. Nuclear activity seems to offer a very attractive option. If material being lost by stars is to flow to the center of a galaxy, then it can provide fuel to a central black hole (BH) that may be lurking there (e.g. Gisler 1976). The idea has been recently explored and expanded by Ciotti & Ostriker (1995), and a brief summary of some of their main results is given here.

During inflow phases $\sim 1 - 10 M_\odot yr^{-1}$ sink towards the center of the galaxy, and either IOW or WOI models simply postulate that this mass silently disappear. However, if instead this mass flow is accreted by a central BH the luminosity of the so powered AGN is expected to be in the range:

$$L_X^{AGN} \simeq \eta \dot{M}_{SINK} c^2 \simeq 10^{45} - 10^{47} \quad erg \, s^{-1}$$

for $0.01 < \eta < 0.1$, i.e., for 1 to 10% mass-energy conversion efficiency, with most of this luminosity being emitted in the hard X-ray range.

Moreover, during inflow phases a rather massive ISM is in place, and Ciotti & Ostriker estimate an optical depth for the Compton scattering of hard X-ray photons $\tau^{ISM} \simeq 10^{-3}$. Therefore, the rate of energy deposition in the hot ISM is of the order of $\sim \tau^{ISM} L_X^{AGN} \simeq 10^{42} - 10^{44}$ erg s^{-1}, much larger than the radiative losses of the hot ISM (i.e., of the X-ray luminosity of the galaxy before the activation of the AGN, see Fig. 1). A strong feedback is then extected on the gas flow itself, and in the Ciotti & Ostriker models the AGN energy deposition in the flow reverts the inflow back to a supersonic wind, thus discontinuing the fueling of the central BH.

Correspondingly, the evolution of the gas flow proceeds through a series of outflow phases separated by brief AGN phases during which the ISM of the galaxy is expelled. After each expulsion the AGN dies out of starvation, and stellar mass loss starts replenishing the ISM again until the next central cooling catastrophe takes place. The duration of the outflow phases depends on several model parameters, and can be as long as several Gyr provided SN heating delays the central cooling catastrophe. Also the duration of the AGN phase is highly model dependent, typically several 10^6 yr, leading to AGN phases duty cycle $\sim 10^{-3}$. These 1D models also predict high flickering activity during the active phases, as the BH accretion rate fluctuates and/or a transient cocoon of high optical depth ($\tau \simeq 1$) builds up in the immediate vicinities of the BH. The X-ray luminosity evolution of the galaxies (taking out the AGN phases) is confined between the upper and the lower lines in Fig. 1, and therefore this model quite naturally accounts for the observed large dispersion of the L_X/L_B ratios. After an AGN phase galaxies climb up slowly (in ~ 1 Gyr or more) starting from $L_X \simeq L_{discr}$, until the transition to the inflow takes place at $L_X \simeq L_{in}$ and the AGN is activated, then precipitously falling back to $L_X \simeq L_{discr}$ before starting a new cycle.

An attractive feature of this model is that it accounts for the fate of the cooling gas during the inflow phases, an aspect that remains obscure in other models. Just $\lesssim 1\%$ of the stellar ejecta eventually terminates inside the central BH, thus releasing enough energy to eject the other $\sim 99\%$ and enriching in metals the ICM. We may also wander if this same process could also work at the much larger scale of galaxy cluster "cooling flows", where a similar embarrassment exists as to where the cooling gas is eventually disposed. In other words, may central BHs in cD galaxies affect cluster scale "cooling flows"? Clearly a much brighter and hotter QSO would be required to compensate for radiative losses in cluster flows ($L_X \simeq 10^{45}$ erg s^{-1}), and temporarily revert inflows to outflows near the cluster center.

Central BH-gas flow interactions may also take place at a much smaller scale. At the deep bottom of the potential well of ellipticals stellar, hence gas density and radiative losses can be very high, and *mini-inflows* inside ~ 200 pc are common while the rest of the galaxy is in a wind or outflow regime (Ciotti *et al.* 1991). If central massive BHs are ubiquitous among ellipticals, mini-inflows with typical mass deposition rates $\sim 10^{-3} - 10^{-2} M_{\odot}$ yr^{-1} may feed a low-level, sporadic AGN activity, that may account for the observed $L_{RADIO} - L_X$ correlation (see Fabbiano, these proceedings), and producing occasional flares at the center of elliptical galaxies. A $\sim 10^6 L_{\odot}$ ultraviolet flare that might have been generated in this way has been recentlty detected with HST at the very center of the elliptical galaxy NGC 4552 (Renzini *et al.* 1995), one of the X-ray brightest galaxies in Fig. 1. Future HST observations – especially in the ultraviolet – should tell us how common these flare events are, and whether their occurrence correlates with other optical and X-ray properties of the host galaxies.

4. Conclusions

The study of gas flows in ellipticals is a fairly interesting field of research, as such flows connect to the central, active regions of galaxies to the one end, and to the intracluster medium at the other end, thus involving a rich variety of astrophysical phenomena. In summary, here stands the issue:

• Gas flows in ellipticals are not in a steady state.

• Outflows and winds appear to prevail over inflows in most ellipticals.

• The evolution of the flows over cosmological times is controlled by the evolution of the rate of Type Ia SNs, that is not yet constrained by direct observations.

• The iron and other element abundances in the ISM of ellipticals and in the ICM provide an indirect measure of the present and past SN rate, hence providing constraints for the evolution of the gas flows.

• Central BHs may play a main role in the evolution of the flows, with its interaction with the hot ISM leading potentially to a great variety of observable phenomena.

I am very grateful to Silvia Pellegrini for her preparation of the two figures shown in this paper and to her, Luca Ciotti and Annibale D'Ercole for stimulating discussions.

References

Arnaud, M., Rothenflug, R., Boulade, O., Vigroux, L., & Vangioni-Flam, E. 1992, A&A, 254, 49

Bertin, G., & Toniazzo, T. 1995, ApJ, 451, 111

Ciotti, L., & Ostriker, J.P. 1995, in preparation

Ciotti, L., D'Ercole, A., Pellegrini, S., & Renzini, A. 1991, 376, 380

David, L.P., Forman, W., & Jones, C. 1991, ApJ, 369, 191

Elbaz, D., Arnaud, M., & Vangioni-Flam, E. 1995, A&A, in press

Fabbiano, G., Kim, D.-W., & Trinchieri, G. 1992, ApJS, 80, 531

Kritsuk, A.G. 1992, A&A, 261, 78

Loewenstein, M., et al. 1994, ApJ, 436, L75

Loewenstein, M., and Mathews, W.G. 1987, 319, 614

Mathews, W.G., & Backer, J. 1971, ApJ, 170, 241

Mushotsky, R., et al. 1995, ApJ, in press

Pellegrini, S., & Fabbiano, G. 1994, ApJ, 429, 105

Renzini, A. 1993, in Panchromatic View of Galaxies, ed. G. Hensler, Ch. Theis, & J.S. Gallagher (Gif-sur-Yvette: Editions Frontières), p. 155

Renzini, A., Ciotti, L., D'Ercole, A.. & Pellegrini, S. 1993, ApJ, 419, 52

Renzini, A., Greggio, L., di Serego Alighieri, S., Cappellari, M., Burstein, D., & Bertola, F. 1995, Nature, 378, 39

Sarazin, C.L., & White, R.E.III 1987, ApJ, 320, 32

Sarazin, C.L., & Ashe, G.A. 1989, ApJ, 345, 22

Discussion

Forbes: Does mass drop out of the cooling flow at large radii in your model?

Renzini: In the Ciotti *et al.* (1991) models no distributed mass sink was included. Such an inclusion is mathematically trivial, yet physically *ad hoc* at this stage. In any event, even models with distributed mass sink cannot explain the distribution of L_X vs L_B that I have shown.

Elbaz: 1) Concerning the iron enrichment in the intracluster medium and the discrimination between a SNIa and a SNII origin, you did not mention the observed Si/Fe ratio by Mushowski with ASCA. 2) Since about half of the Fe mass in clusters is in the ICM and the other half in the galaxies themsemves, one may invoke a common origin and link it to the observed overabundance of Mg over Fe in E's (see poster from Barbuy *et al.*).

Renzini: 1) Yes, I had a transparency on that but flipped over for lack of time. I will mention the ASCA result in the paper for the proceedings. 2) I don't think that the iron share between galaxies and the ICM can discriminate between a SNIa or a SNII origin. The crucial test is instead in the elemental ratios.

Fabbiano: 1) You said that the Ciotti & Ostriker model can explain the scatter in L_X for a given L_B. However, the [average $L_X - L_B$] correlation points to the potential also, so the C&O effect cannot be all. 2) Is the C&O QSO duty cycle consistent with the fraction of QSOs vs galaxies?

Renzini: 1) I think the average trend of L_X with L_B is the easiest to reproduce. As I said, C&O models jump up and down between L_{in} and L_{discr}, so naturally account also for the observed slope. 2) It seems to me that a duty cycle $\sim 10^{-3}$ is broadly consistent with the QSO/AGN frequency at low z. doesn't it?

Ciotti: We expect the duty cycle to depend a lot on the galaxy parameters (especially luminosity), but so far we have not explored the whole parameter space.

Djorgovski: If the mechanism you describe is important for the fueling of quasars, how can you explain the very rapid evolution of the number density of quasars?

Renzini: Probably fueling central BHs from a hot ISM is not the most common way to activate QSOs. The observed peak of the QSO number density at $z = 2$ and beyond must have a very different origin, probably intimately connected to the very formation of galactic spheroids. Central cooling catastrophes in ellipticals would just relate to the tail of the QSO density - redshift distribution. Moreover, depending on how much SNIa heating is provided by nature, only some among the most massive ellipticals might have already experienced a central cooling catastrophe at late epochs.

IRON ABUNDANCE DISCREPANCY IN ELLIPTICALS AFTER ASCA

N. ARIMOTO
Institute of Astronomy, University of Tokyo
2-21-1 Osawa, Mitaka, Tokyo 181, Japan

1. Introduction

The chemical composition of the ISM is expected to reflect the average composition of the stellar component as established at the time of its formation, then further enriched by SN's of Type Ia that are known to currently explode in elliptical galaxies. Renzini et al. (1993) estimate the expected iron abundance in the hot ISM of elliptical galaxies to range from a minimum of ~ 2 times solar, to perhaps as much as ~ 5 times solar or more, depending on the adopted SNIa's rate.

2. The Iron Abundance in the Hot ISM

We have analysed the *ASCA* data for nine elliptical galaxies NGC 499, NGC 507, NGC 720, NGC 1399, NGC 1404, NGC 4374, NGC 4406, NGC 4472, and NGC 4636. We have integrated the SIS spectral data within a radius of 5' centered on each galaxy and have fitted the spectra with χ^2 method with the XSPEC spectral fitting package. Thin thermal plasma model (Raymond & Smith 1977) modified by interstellar absorption is fitted to each galaxy. The iron abundances are determined from the iron-L blends around 1keV.

Figures 1 shows the resulting temperature and iron abundances. Arimoto et al. (1995) have estimated the mean stellar iron abundances of about 50 ellipticals by precisely taking into account the observed metallicity gradients, among which four galaxies NGC 4374, NGC 4406, NGC 4472, and NGC 4636 are common with our *ASCA* analysis; the iron abundances of hot gas in these galaxies are 0.1, 0.2, 0.3, and 0.3 (in solar units), respectively. In contrast, the mean stellar iron abundances of these galaxies are 0.9, 0.7, 1.4, and 0.7 (in solar units), respectively. *There is clearly a macro-*

R. Bender and R. L. Davies (eds.), New Light on Galaxy Evolution, 139–142.
© 1996 *IAU. Printed in the Netherlands.*

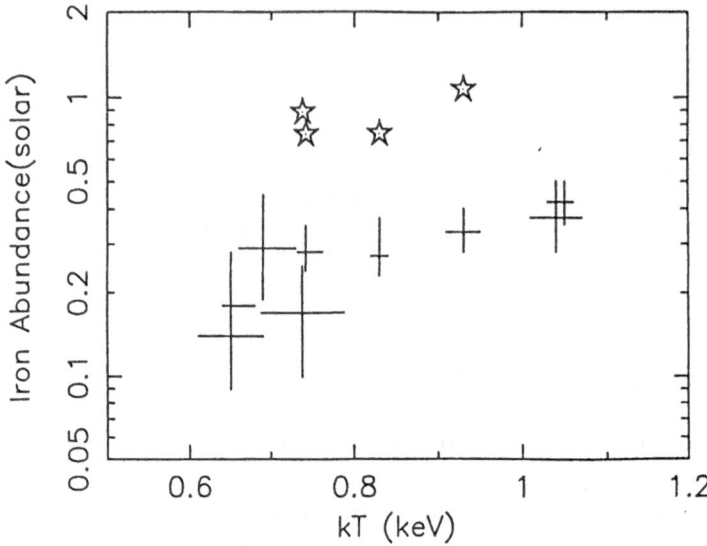

Figure 1. Iron abundance of hot ISM of elliptical galaxies. Stars indicate the mean stellar iron abundances derived by Arimoto et al. (1995).

scopic discrepancy between the expected abundance and what is indicated by the X-ray observation of elliptical galaxies with ASCA.

3. Astrophysical Implications of A Low Iron in The Hot ISM

We believe that the *optical* and the X-ray abundances cannot be easily reconciled, and therefore the existence of this macroscopic discrepancy opens three main options: 1) the optical abundances are seriously in error, 2) both optical and X-ray abundances are correct, but the ISM iron is somehow hidden to X-ray observations, and 3) there is a problem with the interpretation of the Iron-L lines.

· The low iron abundances reported require the SNIa's iron enrichment at the present time to be vanishingly small, and the average iron abundances of the stellar component to be in error by a large factor.

The drastic reduction in the rate of SNIa's has major implications for the interpretation of the X-ray properties and evolution of elliptical galaxies. With $\vartheta_{SN} \ll 1/4$ the SN heating of the ISM becomes virtually negligible for the dynamics of the gas flows in these galaxies at the present time. In absence of alternative internal sources of heat, the ISM of virtually all elliptical galaxies would now be in a cooling flow regime. *Correspondingly, all elliptical galaxies would be very bright in X-rays, at variance with the observations that indicate a large disperion in X-ray luminosity for given optical luminosity* (Canizares, Fabbiano, & Trinchieri 1987; Donnelly, Faber, &

O'Connell 1990; Ciotti et al. 1991; Fabbiano, Kim, & Trinchieri 1992).

We have implicitly assumed that all the iron ejected by stars and SNIa's is now in the gas phase of the ISM. While there is little doubt that at least a fraction of the iron is injected in the form of solid particles, the question is as to whether iron remains in a solid phase for a long enough period of time so to account for the iron discrepancy. The time required to evaporate 90% of iron particles in a 10^7 K plasma is $\sim 10^5/n_e$ yr cm^{-3} (Itoh 1989). Electron densities in the ISM of ellipticals range from $\sim 10^{-1}$ near to center to $\sim 10^{-3}$ several effective radii away, and therefore evaporation times range from 10^6 to 10^8 yr. This compares to a flow time of a few Gyr, and *we conclude that only a tiny fraction of iron could be hidden in dust particles.*

Local thermal instabilities causing a fraction of the gas to drop out of the flow locally have often been invoked as the ultimate depository for both cluster and galaxy inflows (e.g., Fabian et al. 1986). Together with this hypothesis comes the additional conjecture that such instabilities would lead to the formation of low mass, safely unobservable objects such as lower main sequence stars, brown dwarfs, or jupiters. This scenario has one testable prediction. Local thermal instabilities – if they exist – should be more efficient in dropping SNIa products out of the flow near the center where the cooling time is short, compared to the outer regions. Thus, iron should be preferentially depleted near the center, and a positive abundance gradient should develop. Actual observations seem in case to indicate the opposite gradient, with iron being more enhanced near the center (Mushotzky et al. 1994). *We conclude that neither hiding iron into jupiters appears to be a viable solution to the discrepancy.*

At the temperature that are typical of the hot ISM of elliptical galaxies ($\lesssim 1$ keV), the iron abundance is derived from the strength of a few hundreds lines originated by electron transitions down to the L shell in incompletely ionized iron ions, typically Fe XVIII-XXI, hence called Iron-L.

Spectra of ellipticals are dominated by the iron-L blends, by which both temperature and iron abundance are determined. However, is the interpretation of the iron-L lines correct? Atomic data such as ionization, recombination, and excitation rate coefficients are still controversial. New calculation of spectra of plasma (Liedahl et al 1994), using new Fe data, show that the total power of the iron-L lines is several tens of % different from Meka model (Mewe, Gronenshild & van den Oord 1985; Kaastra & Mewe 1993). Ambiguity of the iron-L blend is very large for the total power and line ratios, thus ambiguity of the resulting abundance is of no doubt significant. *We conclude that the reliability of iron-L line diagnostic tools that are currently used in conjunction with X-ray observations is in question.*

Acknowledgements

I thank to the collaborators of this work, K.Matsushita, Y.Ishimaru, T.Ohashi, and A.Renzini for allowing me to present the paper in this conference before publication.

References

Arimoto, N., K.Matsushita, Y.Ishimaru, T.Ohashi, & A.Renzini, 1995, in preparation
Canizares, C.R., Fabbiano, G., & Trinchieri, G. 1987, ApJ, 312, 503
Ciotti, L., D'Ercole, A., Pellegrini, S., & Renzini, A. 1991, ApJ, 376, 380
Donnelly, R.H., Faber, S.M., & O'Connell, R.M. 1990, ApJ, 354, 52
Fabbiano, G., Kim, D.-W., & Trinchieri, G. 1992, ApJS, 80, 531
Fabian, A.C., Thomas, P.A., Fall, S.M., and White, R.E.,III 1986, MNRAS, 221, 1049
Itoh, H. 1989, PASJ, 41, 853
Liedal, D., Osterheld, A., Mewe, R., & Kaastra, J., 1994, preprint
Mewe, R., Gronenshild, E.H.B., & van den Oord, H.J. 1985, A&AS, 62, 197
Kaastra & Mewe, 1993, Legacy Vol 3, 16
Mushotzky, R. et al. 1994, ApJ, in press
Raymond, J.C., & Smith, B.W. 1977, ApJS 35, 419
Renzini, A., Ciotti, L., D'Ercole, A., & Pellegrini, S. 1993, ApJ, 419, 52

Discussion

S.M.Faber:

We have not tried to estimate real [Fe/H] abundances for giant ellipticals because our models are clearly wrong due to non-solar Mg/Fe ratios. But if we ignore that problem, we would get *significantly* reduced [Fe/H], maybe down by a factor of 3 below solar for the whole galaxy. This would not surprise me.

N.Arimoto:

I am not sure if giant ellipticals do really have non-slar Mg/Fe ratios, although it is suggested by analyses of metallic absorption lines.

L.Vigroux:

If the wind takes place during the very early phase of galaxy formation, a dilution might take place with the remnant of the primordial gas which have been with the origin of the galaxies. This has already been used to explain the low metal abundance ($\sim 0.5[Fe/H]_\odot$) in clusters of galaxies. A clue to this problem is to use mass instead of abundance ratio. Dr.Renzini, in the previous talk, has shown that the mass of iron normalized to galaxy luminosity is constant from group to clusters of galaxies. How the elliptical galaxies of your sample compare with clusters in this ratio?

N.Arimoto:

Although it would be of great interest to derive the mass of iron normalized to galaxy luminosity, it is not yet possible to do, because total masses of hot X-ray gas cannot be accurately estimated by *ASCA*.

THE COLD INTERSTELLAR MEDIUM

An HI View of Spiral Galaxies

RENZO SANCISI
Kapteyn Astronomical Institute
University of Groningen
The Netherlands

Abstract. An HI view of spiral galaxies is presented. In the first part the standard picture of isolated, normal spiral galaxies is briefly reviewed. In the second part attention is drawn to all those phenomena, such as tidal interactions, accretion and mergers, that depend on the galaxy environment and seem to have played a significant role, in addition to the internal metabolism, in the galaxy evolution.

1. Introduction

In recent years there has been considerable progress in our knowledge of the neutral hydrogen properties of galaxies. High spatial and velocity resolution studies of nearby spirals have provided detailed information on scales of 0.1 to 1 kpc about the filamentary and diffuse structure and the temperature of the interstellar medium (Braun 1996), the HI holes and high velocity gas and, in general, on the star formation regions and the spiral arm structure (Deul and den Hartog 1990, Kamphuis 1993). The radial distribution and extent and the kinematics of the HI disks have been studied with synthesis observations of an increasing number of objects (e.g. Broeils and van Woerden 1994, Rhee and van Albada 1996). HI observations of interacting galaxies, groups and clusters have shown the presence of tails, bridges and other extended tidal structures around and between galaxies, and the dramatic effects of gas stripping like in some of the Virgo galaxies (Cayatte et al. 1990). Also, the global HI properties and their relationship with other physical properties along the Hubble sequence have been reviewed again recently (Roberts and Haynes 1994).

From this wealth of information it is important now to recognize those aspects that are significant for galaxy formation and evolution. We can

R. Bender and R. L. Davies (eds.), New Light on Galaxy Evolution, 143–150.

distinguish between 'internal' phenomena, related to processes in the disk
(e.g. star formation), and 'external' effects due to interaction with the environment. A tentative classification can be based on HI morphology and
kinematics, in which the 'normal' characteristics of well-behaved, isolated
systems are identified and the unusual, 'peculiar' aspects of the HI structure
and kinematics can be recognized.

In the first part of this review we present the essential HI features of
normal, isolated disk galaxies: the spiral arm structure and holes, the radial
extent, the disk-halo connection and the general shape of the HI disk. In the
second part we tentatively identify those peculiar aspects of the HI density
distribution and kinematics which may reveal past or present events, such
as interactions with the environment, presumably playing an important role
in the galaxy evolution.

2. Isolated 'Normal' Galaxies

2.1. SPIRAL STRUCTURE, HI HOLES AND HIGH-VELOCITY GAS

The distribution of neutral hydrogen in disk galaxies follows closely the spiral structure outlined by the young stellar population and the HII regions.
In the outer parts, beyond the optical image, the extended HI layer shows
clearly the continuation of the spiral pattern seen in the inner parts. This
is best shown by the nearby spiral galaxies M 101, NGC 628, NGC 3198
and NGC 6946.

The velocity dispersion of the HI layer, which had been reported in the
past as being approximately constant and close to 10 km/s at all radii,
appears in more recent work on nearby face-on galaxies to be decreasing
monotonically from about 10–13 km/s in the optically bright inner regions
to 6–8 km/s in the very outer parts (Kamphuis 1993).

When the HI distribution is mapped with sufficiently high angular resolution, as in M 31, M 33 (Deul and den Hartog 1990) and in other, large
nearby spirals, like M 101 and NGC 6946 (Kamphuis 1993), it shows the
presence of a large number of holes of various sizes, from about 100 pc to
a few kpc. In general, the holes are seen in the region of the optical spiral
arms and only a few are found in the far outer parts outside the bright optical image. The smaller, presumably younger ones correlate in position with
the HII regions and OB associations. It is likely, therefore, that they are
produced by collective stellar winds and supernovae. This is also supported
by the detection of high velocity gas associated with some of them, as in
the prototype, large hole with expanding shell found in M 101 (Kamphuis
et al. 1991). The velocities are usually less than 100 km/s. In NGC 6946, a
spiral galaxy rich in giant HII regions and HI holes, there is evidence of a
widespread high velocity gas component (Kamphuis and Sancisi 1993). In

conclusion, holes and high velocity gas seem to characterize the HI layers of spiral galaxies in the optical region, to be related to the star formation process and to be at the origin of the disk-halo gas circulation discussed below.

2.2. RADIAL DISTRIBUTION AND EXTENT

The HI radial distribution is known for about 100 field galaxies (see e.g. surveys by Broeils and van Woerden 1994 and by Rhee and van Albada 1996) and for a somewhat smaller number of systems in nearby clusters (Cayatte et al. 1994, Verheyen 1996). The radial surface density profile, in general, is roughly flat or slowly decreasing in the region of the optical disk, and shows a rapid, approximately exponential drop-off in the outer parts. The break occurs at a characteristic radius, which usually marks the edge of the bright optical disk and is often located close to the De Vaucouleurs or the Holmberg radius. In some objects, especially of early type, a hole or a depression is seen in the central region. The HI surface density in the dense, optical part averages about 10 $M_\odot/pc^2$. This value remains remarkably constant for galaxies of different mass and luminosity. Only low surface brightness galaxies, recently surveyed in HI by de Blok et al. (1996), have systematically lower (factor 3) HI surface densities. The HI radial extent is usually larger than the optical radius. For their sample of 50 isolated galaxies Broeils and van Woerden find an average ratio of HI to optical diameters, D_{HI} (defined at 1 $M_\odot/pc^2$) to D_{25}, of 1.8±0.4. There is a strong correlation between these HI and optical diameters. The HI disk seems to cut off sharply at a density level of about 10^{19} atoms cm^{-2} or ~0.1 $M_\odot/pc^2$, as shown by the deep 21-cm line observations of the spiral galaxy NGC 3198 by van Gorkom et al (1993). This may mean either that the gas layer ends there or that it becomes ionized.

The properties of the HI density distribution depend to some extent on galaxy morphology and on the environment. In the sample discussed by Broeils (1992) the HI surface density is very weakly correlated with morphological type, but there is no evidence of a correlation of the HI-to-optical diameter ratio with type or luminosity. The effects of dense, group and cluster, environments are described below. An investigation of galaxies in voids, recently carried out by Szomoru (1994), has shown that galaxies in very low density environments, like voids, do not differ significantly in their HI properties from field galaxies. The existence of objects made up of only neutral hydrogen, without optical counterparts, has not been proven yet. The only cases known to date of large intergalactic HI clouds are those found near interacting galaxies.

2.3. WARPS

The outskirts of galaxy disks are usually not flat, but show a large-scale, integral-sign shape distortion in the vertical direction. This phenomenon is known as warping. Warps are clearly visible in galaxies seen edge-on, or are inferred from the velocity fields of systems viewed at lower inclination angles. They seem to be quite common: according to Bosma (1991) the fraction of warped HI disks is of order 50% at least. One of the most prominent cases is that of NGC 4013 (Bottema 1995). Some edge-on galaxies, like NGC 4565, show also an optical warp; it is not at all clear, however, how frequently this occurs and whether the bend is in the old stellar disk. The systematic properties of warps have recently been investigated by Briggs (1990). He finds that the warps develop at the edge of the optical disk, and that their lines of nodes are not straight but curve so as to form a leading spiral. On the observational side there is now a clear need for better statistics on the frequency of occurrence of warps, for a more rigorous study of the possible influence of the environment, and for a more detailed analysis of the warp properties and the relationship with the optical counterpart when this is present.

2.4. DISK-HALO CONNECTION

The best information on the vertical structure of the HI layer in spiral galaxies, on the disk-halo connection and on the gas circulation between disk and halo comes from recent observations of nearby edge-on and face-on systems. Such observations are complementary and both necessary. The edge-on view allows direct determination of the vertical density distribution of the HI, but not of its vertical motion, whereas the face-on view offers the velocity information needed to understand the gas circulation upward and downward between disk and halo. The observations of giant face-on spirals like M 101 and NGC 6946, already mentioned above, and of the edge-on NGC 891 (Rupen 1991, Swaters et al. 1996) provide this kind of information.

NGC 891 shows HI wings extending up to about 5 kpc on both sides of the disk (Swaters et al. 1996). This HI gas at large distances from the plane, already detected (but with much lower S/N ratio) in earlier Westerbork observations, had been interpreted as being part of a flaring outer layer of NGC 891 (Sancisi and Allen 1979) and not as halo gas above the bright inner stellar disk. New, better sensitivity, HI observations and more sophisticated analysis with three-dimensional modelling have led to the conclusion that at least part of this gas must be located in the halo regions of the galaxy directly above and below the bright stellar disk. At the same time it has become clear that the standard assumption of cylindrical rotation is not valid and that the halo HI must have a somewhat lower (about 25 km/s) rotation velocity than the gas in the plane.

2.5. LOPSIDEDNESS

Large-scale asymmetries in the density distribution and in the kinematics of the neutral hydrogen disk are observed often in spiral galaxies. Since the first studies (Baldwin et al. 1980) based on a small number of galaxies much new evidence has accumulated. The frequency of asymmetries among spiral galaxies has been recently estimated (Richter and Sancisi 1994) from a large sample of global HI profiles of field galaxies. More than 50 percent of the systems examined show strong or mild asymmetries. Large deviations from axial symmetry seem, therefore, to be the rule rather than the exception. Examination of the HI maps and velocity fields has shown that the lopsidedness is present in general in both the density distribution and kinematics. In several such cases the rotation curves are clearly asymmetric: on one side of the galaxy they rise more slowly and reach the flat part at larger radii than on the other side. It should be noted that these are isolated, not tidally interacting systems. Although the asymmetric pattern is most clearly seen in the HI data, there is often evidence of asymmetry also in the distribution of blue light and in some cases studied recently also in the near infrared. All these facts suggest that the phenomenon of lopsidedness in spiral galaxies is quite common, structural for the disk, and long-lived.

3. Interacting and Peculiar Systems

There is clear evidence from HI observations that a number of galaxies are undergoing tidal interactions. It is useful to distinguish between two types of interaction: major and minor ones. The "major" one involves systems of comparable masses, usually produces large tidal effects and may lead to destruction of disks and to mergers with, as possible end-product, elliptical galaxies. The "minor" one takes place between a galaxy and one or more satellites or companions of smaller mass (mass ratio usually less than 0.1). This leads to gas accretion, build-up of disks and may cause local star formation and starbursts. There are also systems which, despite their being isolated, show peculiar HI properties typical of interacting systems. These may have had some recent encounter and now be in an advanced stage of

3.1. MAJOR INTERACTIONS. TIDAL TRAUMAS

Several cases are known of multiple systems with three or more members showing heavily perturbed HI images: in addition to the gas seen associated with the individual members, there are cloud complexes, tails, bridges or ring-like structures in the regions around them. Examples for this type of interaction are the M81-M82-NGC3077 (Yun et al. 1994), and NGC4631-4656-4627 (Rand 1994) groups. In all these systems it is the peculiar gas picture that unmistakably points at the ongoing strong tidal interaction.

The galaxy pairs present a simpler but similar picture of long tails and bridges. M51 and its companion form a well-known example. The VLA observations of Rots et al. (1990) have shown a highly disturbed picture. The most striking feature is a 90 kpc long HI tail without optical counterpart, connected loosely to the outer disk of M 51.

A number of such systems, characterized by optical bridges and long tails, were presented by Toomre (1977) as a possible sequence of ongoing galaxy mergers. Five of these have been recently imaged in HI with the VLA by Hibbard and van Gorkom (1995). These observations seem to indicate some trends along the merging sequence. In the early stages, large amounts of HI are still present within the galaxy disks. In the final stages there is little or no HI within the remnant bodies, and tidal material is seen falling back towards the remnant; the HI is almost completely concentrated in tidal tails often more extended than the optical parts. One of the key questions in such interactions of two disk systems is whether the end product is an elliptical galaxy or whether there are, depending on the kind of impact, possibilities for a disk to survive. An interesting case in this connection is that of NGC 3310 (Arp 217), a disk galaxy which has been recently found to have two extended, well developed HI tails (Mulder and Sancisi 1996) and may, therefore, be an advanced stage in this category of systems. And yet this system has been able to preserve its disk structure.

3.2. MINOR INTERACTIONS. GAS ACCRETION

Several galaxies have dwarf companions and when mapped in HI show clear indications of present tidal interactions. The prototype is NGC 3359. The HI map of this galaxy clearly shows a small companion with a long tail pointing back to NGC 3359 and almost connecting with its extended HI layer (Sancisi et al. 1990). The companion is a hydrogen-rich object with a very faint optical counterpart. Its HI mass is 10^8 $M_\odot$, only 2% of the HI mass, and 0.1% of the total mass of NGC 3359. Its head-tail structure indicates that it is probably being tidally disrupted. Very similar situations are found in the edge-on galaxy NGC 4565 and other spiral galaxies. The HI masses of these companions are much smaller, less than 10%, than those of the main galaxy. The picture emerging from these observations is that of the capture of gas-rich dwarfs by a massive system followed by tidal disruption and accretion, while the damage suffered by the main galaxy is small.

Cases of a probably more advanced stage in the interaction-accretion process have also been observed. The giant nearby spiral galaxy M 101 may be in such a stage. The HI complex of about 2×10^8 M$_\odot$ moving with velocities of up to 150 km/s with respect to the disk and the corresponding large cavity in the HI layer (Van der Hulst and Sancisi 1988) are interpreted as being due to the collision with a dwarf companion which has gone through the HI layer of M 101. The high velocity gas will eventually rain down back onto the M 101 disk. There are more systems, like M 101, which do not have any obvious bright companions and yet when mapped in HI display peculiar features which are reminiscent of those seen associated with strongly interacting systems. They may represent cases of past interactions and be at present in an advanced stage of accretion in which the victim is not visible any longer. Interesting examples are those of NGC 1023, Mkn 348 and NGC 628.

Similar interactions with dwarf companion galaxies and accretion phenomena are also found in a number of elliptical and S0 systems. We mention here only some of the most recently studied systems: NGC 4472 and UGC 7636 (McNamara et al. 1994), NGC 3656 (Balcells and Sancisi 1996), NGC 5128 (Cen A) (Schiminovich et al. 1994) and NGC 2865 (Schiminovich et al. 1995).

All these cases form circumstantial evidence that even in the present epoch there is episodic infall of gas onto galaxies. How often do interactions take place? and how important are they for the formation of elliptical galaxies or for the build-up of disks, star bursts and galaxy evolution in general? A rough estimate is that probably 25 to perhaps 50 percent of the galaxies show some signs of present or recent interaction.

3.3. CLUSTERS

In a cluster environment tidal interactions between galaxies and ram pressure due to the hot intergalactic gas affect the distribution of neutral hydrogen in the individual galaxies. In the central part of the Virgo cluster, in the region of the hot X-ray gas, spiral galaxies appear to have been stripped of part of their HI, especially in their outer parts (Cayatte et al. 1990, Cayatte et al. 1994). Some HI disks are even truncated inside the optical disks. The spirals in the Hydra cluster do not show any HI deficiency or stripping in spite of the X-ray emitting hot gas (Mc Mahon 1992). Similarly in Ursa Major, but perhaps less surprisingly since there is no X-ray source, no

HI deficiency or stripping are observed (Verheijen 1996). These effects depend on the location of the galaxies in the cluster and the different results probably reflect the different stages of cluster formation and evolution.

References

Balcells, M., Sancisi, R., 1996, AJ., in press.
Baldwin, J.E., Lynden-Bell, D., Sancisi, R., 1980, MNRAS, 193, 313.
Bosma, A., 1991, in Warped disks and inclined rings around galaxies, eds. S. Casertano, P. Sackett and F. Briggs, Cambridge University Press, Cambridge, p. 181.
Bottema, R., 1995, A&A, 295, 605.
Braun, R. 1996, ApJ., Submitted.
Briggs, F.H., 1990, ApJ., 352, 15.
Broeils, A.H., 1992, PhD Thesis, University of Groningen.
Broeils, A.H., and van Woerden, H., 1994 A&A Suppl. 107, 129.
Cayatte, V., van Gorkom, J.H., Balkowski, C. and Kotanyi, C., 1990, AJ., 100, 604.
Cayatte, V., Kotanyi, C., Balkowski, C. and van Gorkom, J.H., 1994, AJ. 107, 1003. de Blok, W.J.G., McGaugh, S.S. and van der Hulst, J.M., 1996, MNRAS, in press.
Deul, E.R., and den Hartog, R.H., 1990, A&A, 229, 362.
Hibbard, J.E., and van Gorkom, J.H., 1995, AJ., Submitted.
Kamphuis, J.J., 1993, PhD Thesis, University of Groningen.
Kamphuis J., and Sancisi, R., 1993, A&A, 273, L31.
Kamphuis, J., Sancisi, R., and van der Hulst, J.M., 1991, A&A, 244, L29.
Mc Mahon, P.M., Richter, O.-G., van Gorkom, J.H. and Ferguson, H.C. 1992, AJ, 103, 399.
McNamara, B.R., Sancisi, R., Henning, P.A., and Junor, W., 1994, AJ., 108, 844.
Mulder, P., Sancisi, R., 1996, in preparation.
Rand, R.J., 1994, A&A, 285, 833.
Rhee, M.-H., and van Albada, T.S., 1996, A&A Suppl., in press.
Richter, O.-G., Sancisi, R., 1994, A&A, 290, L9.
Roberts, M.S., and Haynes, M.P., 1994, Ann. Rev. Ast. Ap., 32, 115.
Rots, A.H., 1978, AJ, 83, 219.
Rots, A.H., Bosma, A., van der Hulst, J.M., Athanassoula, E., Crane, P.C., 1990, AJ, 100, 387.
Rupen, M.P., 1991, AJ. 102, 48.
Sancisi, R., 1992, in Physics of Nearby Galaxies: Nature or Nurture?, eds. T.X. Thuan, C. Balkowski and J.T.T. Van, Editions Frontieres, p.31.
Sancisi, R., Allen, R.J., 1979, A&A, 74, 73.
Sancisi, R., Broeils, A.H., Kamphuis, J., van der Hulst, J.M., 1990, in Dynamics and Interactions of Galaxies, ed. R. Wielen, Springer Verlag, p. 304.
Schiminovich, D., van Gorkom, J.H., van der Hulst, J.M., and Kasow, S., 1994, ApJ. 423, L101.
Schiminovich, D., van Gorkom, J.H., van der Hulst, J.M., and Malin, D.F., 1995, ApJ., 444, L77.
Swaters, R.A., Sancisi, R., van der Hulst, J.M., 1996, Preprint.
Szomoru, A., 1994, PhD Thesis, University of Groningen.
Toomre, A., 1977, in The Evolution of Galaxies and Stellar Populations, ed. B.M. Tinsley and R.B. Larson (New Haven: Yale Univ.), p.401.
van der Hulst, J.M., and Sancisi, R., 1988, AJ. 95, 1354.
van Gorkom, J.H., van Albada, T.S., Cornwell, T.J., and Sancisi, 1993, in preparation.
Verheijen, M., 1996, PhD Thesis, University of Groningen.
Yun, M.S., Ho, P.T.P., Lo, K.Y., 1994, Nature, 372, 530.

DETECTING CO AT HIGH REDSHIFT

LINDA J. TACCONI

Max-Planck-Institut für extraterrestrische Physik
Postfach 1603
D-85748 Garching
Germany

1. Introduction

Searches for molecular line emission from high redshift galaxies have become one of the recent highlights in millimeter astronomy, largely because detection of this emission enables one to study the potential for star formation in galaxies at epochs close to galaxy formation. Such information is crucial to models of galaxy evolution. Thus far, most of the searches have been to try to detect any of the rotational lines of CO, although many authors have also inferred the presence of molecular gas through detections of cold dust in the submillimeter region of the spectrum. In addition to providing information about the physical properties of the molecular gas in distant galaxies (when more than one transition or isotope is detected), the CO lines can be used to place stringent constrints on the dynamical masses of these systems. Moreover, since millimeter data has spectral resolutions of typically a few tens of km/s, one can pin down the redshift of the host galaxy with extremely high precision. One of the driving forces in most of the searches for CO emission at high redshift is the fact that molecular gas is known to be an important constituent in the low redshift counterparts to the types of objects that one expects to find at high redshifts, the Ultraluminous Infrared Galaxies (ULIRGs), (*e.g.* Mirabel and Sanders 1985; Sanders *et al.* 1986), powerful radio galaxies (*e.g.* Mazzarella *et al.* 1993), and nearby quasars (*e.g.* Barvainis *et al.* 1989), for example.

Here I review some of the results from the two most distant objects yet detected in CO, and list some other attempts to find CO at $z>2$. I apologize to the many authors whose work I have not included in this paper: limited space and a rapidly evolving field precludes completeness.

151

R. Bender and R. L. Davies (eds.), New Light on Galaxy Evolution, 151–154.

2. IRAS F10214+4724 at z=2.23

Many observers first turned their millimeter telescopes to high z objects in
search of molecular gas after the discovery of the extremely IR luminous
galaxy IRAS F10214+4724 at a redshift of 2.23 by Rowan-Robinson *et
al.* (1991). With the first detection of CO from this object by Brown and
vandenBout (1991, 1992), F10214 has a redshift which is nearly a factor of
10 greater than any previously detected CO sources. At $L_{IR} \sim 10^{14}$ $L_{\odot}$, it
is also the most IR luminous object in the sky. In the past few years, there
have been many studies in addition to the above references to reconfirm
the CO detections, to observe many line transitions of CO in order to get
estimates of the molecular gas properties, and to observe the emission at
high spatial resolution with millimeter interferometers to determine the
source structure (Solomon, Downes and Radford 1992; Kawabe *et al.* 1992;
Sakamoto *et al.* 1992; Radford, Brown and VandenBout (1993); Downes,
Solomon and Radford 1995). These observations show that this unusual
galaxy is very rich in molecular gas with $M(H_2) \sim M_{dyn} \sim 10^{11} h^{-1}$ $M_{\odot}$.
Here M_{dyn} is the estimated dynamical mass of the galaxy from the CO
linewidth. CO line emission ratios indicate molecular gas which on average
is both warm ($T_{kin} \sim 50$ K) and dense ($n(H_2) \sim 5 \times 10^3$ cm^{-3}, conditions
which are similar to those found in massive star forming regions in our
galaxy (Solomon, Downes and Radford 1992).

3. The Cloverleaf Quasar at z=2.56

The second strong CO emitter at z>2 is the Cloverleaf (H1413+117), a
gravitationally lensed quasar so named because the optical image is split
into four spots by the lens (Magain *et al.* 1988). It is the only broad absorption line quasar known to be lensed, and was chosen as a likely target
for a CO detection by the fact that it was found to have submm emission
due to dust (Barvainis, Antonucci and Coleman 1992). The original CO
J=3→2 detection was made by Barvainis *et al.* (1994) with the IRAM interferometer, where they detected a peak flux of 23 mJy. The detection was
immediately confirmed by the authors in several other lines of CO as well as
CI with single dish observations. Assuming that there is no amplification of
the CO emission by the lens, the Cloverleaf is very similar in its overall CO
and FIR properties to F10214+4724. That is, $M_{gas} \geq M_{dyn}$ and L$\sim 10^{14}$
$L_{\odot}$. The CO emission is bright in many transitions, and the J=4→3 line is
brighter than the J=3→2 line, likely indicative of gas which is both warm
and dense. Detailed modeling of the molecular lines in the Cloverleaf is now
underway using all available information (Barvainis *et al.* 1995).

4. What is Unique about F10214 and The Cloverleaf?

Although many observers have made long and deep searches to detect mmolecular gas from galaxies at z>2 (*e.g.* Evans *et al.* 1995; Röttgering *et al.* 1995) only F10214+4724 and the Cloverleaf have confirmed CO detections (although there are a few cases where authors are awaiting reconfirmation of tentative detections, for example see the comment at the end of this paper). The question which obviously arises from this is whether F10214 and the Cloverleaf are intrinsically unique objects in the early universe or whether they share some other common property which makes them relatively easy to detect. One possibility is that these two objects are very luminous members of the class of long sought after primeval galaxies, since in both cases the gas mass dominates the dynamical mass. This scenario is not likely given the fact that both sources contain large amounts of cold dust.

A more widely accepted picture now is that both sources are gravitationally lensed by a foreground object. Of course, the Cloverleaf is one of the most well known of the gravitational lens candidates, but based on the initial detection of CO, it was not known whether the CO emission was amplified by the lens. High resolution observations of CO at both the Owens Valley Interferometer and the IRAM interferometer show that the CO source is extended on scales of $\sim 1''$, and could well indicate that there is CO emission corresponding to the 4 optical images of the quasar. These data are as yet unpublished. There is also very strong evidence that F10214 is a gravitational lens system, however (Matthews *et al.* 1994; Graham and Liu 1995; Broadhurst and Lehár 1995; Serjeant *et al.* 1995). Gravitational lens models (*e.g.* Broadhurst and Lehár 1995) can explain the multi-component source structures seen in recent sub-arcsecond resolution near infrared images (*e.g.* Graham and Liu 1995). In this picture the symmetric arc seen in the NIR images is the lensed IRAS source, which is centered on the weaker compact intervening lensing galaxy. The NIR images also show evidence for a fainter counter image to the north of the compact component (Graham and Liu 1995). Furthermore, the recent CO images of Downes, Solomon and Radford (1995) are indicative of CO emission which is coincident with the NIR arc. In both the Cloverleaf and F10214, the CO distributions suggest, therefore, that the molecular emission is amplified by the lens. If this is the case, then the molecular mass estimates and IR luminosities have been overestimated. Revised mass estimates (*e.g.* Downes, Solomon and Radford 1995) show that the gas masses are no longer greater than the dynamical masses, and that these objects are likely the very distant counterparts to the nearby ULIRGS.

The success at studying the molecular gas properties of the Cloverleaf quasar and F10214, and the lack of strong detections in other sources seems to indicate that, with the current detection capabilities, observers need the assistance of gravitational lenses to detect molecular line emission from galaxies at high z. Careful studies of gravititional lens candidates in CO lines and the submm continuum are therefore needed to further investigate properties of molecular clouds in the early universe. Such studies will be made easier in the near future with the addition of more antennas to the current millimeter interferometers and the arrival of new large single dish telescopes.

References

Barvainis, R., Alloin, D., and Antonucci, R. 1989, $Ap.J.$, **337**, L69.

Barvainis, R., Antonucci, R., and Coleman, P. 1992, $Ap.J.$, **399**, L19.

Barvainis, R., Tacconi, L., Antonucci, R., Alloin, D., and Coleman, P. 1994, $Nature$, **371**, 586.

Barvainis, R. $et\ al.$ 1995, in preparation.

Broadhurst, T. and Lehár, J. 1995, $Ap.J.$, **450**, L41.

Brown, R.L., and VandenBout, P.A. 1991, $A.J.$, **102**, 1956.

Brown, R.L., and VandenBout, P.A. 1992, $Ap.J.$, **397**, L19.

Downes, D., Solomon, P.M., and Radford, S.J.E. 1995, $Ap.J.$, submitted.

Graham, J.R. and Liu, M.C. 1995, .J., **449**, L29.

Kawabe, R., Sakamoto, K., Ishizuki, S., and Ishiguro, M. 1992, $Ap.J.$, **397**, L23.

Magain, P., Surdej, J., Swings, J.-P., Borgeest, U., Kayser, R., Kuhr, H., Refsdal, S., and Remy, M. 1988, $Nature$, **334**, 325.

Mazzarella, J.M., Graham, J.R., Sanders, D.B., and Djorgovski, S. 1993, $Ap.J.$, **409**, 170.

Radford, S.J.E., Brown, R.L., and Vanden Bout, P.A. 1993, $A&A$, **271**, L21.

Rowan-Robinson, M., $et\ al.$ 1991, $Nature$, **351**, 719.

Sakamoto, K., Ishizuki, S., Kawabe, R., and Ishiguro, M. 1992, $Ap.J.$, 397, L27.

Sanders, D.B., and Mirabel, I.F. 1985, $Ap.J.$, **298**, L31.

Sanders, D.B., Scoville, N.Z., Young, J.S., Soifer, B.T., Schloerb, F.P., Rice, W.L., and Danielson, G.E. 1986, $Ap.J.$, **305**, L45.

Solomon, P.M., Downes, D., and Radford, S.J.E. 1992, $Ap.J.$, **398**, L29.

D. Clements: There are an increasing number of high redshift objects being detected in the submm continuum, so prospects for further CO detections are good. The submm continuum acts as a signpost to where to point your CO spectrometer.

R. Windhorst: It is perhaps worth pointing out that Yamada $et\ al.$ (1995, A.J. in press) detected the high redshift weak radio galaxy 53w002 at z=2.390 in CO with the Nobeyama telescope. Perhaps Toru, who is in the audience wants to comment on this. The interpretation of this result is possibly complicated because 53w002 is surrounded by at least 3 other confirmed objects at z=2.40 (see poster 144 with the deep HST images of Windhorst $et\ al.$.

DARK MATTER HALOS AROUND GALAXIES

DARK MATTER HALOS

R.P. SAGLIA
Institut für Astronomie und Astrophysik
Scheinerstraße 1, D-81679 München, Germany

Abstract. The observational evidence for dark halos around galaxies is shortly reviewed. New and old techniques and results constraining the mass, the distribution, the shape and the nature of dark halos are discussed.

1. Introduction

The casuality problem (why is the Cosmic Background Radiation so homogeneous, with temperature fluctuations $\Delta T/T \approx 10^{-5}$, Bennett *et al.* 1994?) and the fine-tuning of the density of the Universe (the present density Ω_0 in units of the critical density should be near zero, if its initial value is different from 1, Peebles 1993) force astronomers to believe in the inflation scenario (Guth, 1981), which produces a $\Omega = 1$, flat Universe. At the same time, the constraints coming from the primordial nucleosynthesis of light elements require that at most 10% (Mathews *et al.*, 1993) of this density can be in the form of baryons. Astronomers are therefore desperate for finding evidence of this remaining 90% dark component of the Universe.

This paper shortly summarizes the observational facts pointing to dark halos around galaxies and the properties of these halos. Comprehensive reviews of the subject are given by Carr (1994), Ashman (1992), Trimble (1987). The paper is organized as follows. Section 2 reviews the observational evidence for dark halos around galaxies, in the Milky Way (2.1), in spirals (2.2), in dwarfs (2.3), in ellipticals (2.4), in binary systems, galaxies with satellites or groups of galaxies (2.5), closing with the critical point of view of the "non-believers" (2.6). Sect. 3 summarizes the global properties of galaxy dark halos as derived from the observations: their tridimensional shape (3.1), their structure parameters and correlations (3.2), the constraints on their nature (3.3). Section 4 summarizes the conclusions and

R. Bender and R. L. Davies (eds.), New Light on Galaxy Evolution, 157–165.
© *1996 IAU. Printed in the Netherlands.*

the future prospects.

2. Observational Evidence

2.1. THE MILKY WAY

Disk Dark Matter. The first attempt to find dark matter in galaxies is due to Oort (1932, 1960). Following his idea and using a sample of F dwarfs and K giants, Bahcall (1984) finds that the combined Poisson $\partial^2\Phi(z)/\partial z^2 = 4\pi G\rho$ and Boltzmann $\sigma_{zi}^2\partial\rho_i(z)/\partial z = -\rho_i(z)\partial\Phi(z)/\partial z$ equations required that at least 50% of the material in the disk has to be dark, confirming Oort's earlier result. Here $\Phi(z)$ is the gravitational potential in the solar neighborhood at a distance z orthogonal from the disk and ρ the disk density, while σ_{zi} and ρ_i are the velocity dispersions and the densities of stars of a given type i. However, more recent work (see Flynn & Fuchs 1994 and references therein) tends to find a total dynamical surface-mass density in agreement with the visible surface density, without need for a dark component associated to the disk.

Halo Dark Matter. A dark halo is in any case needed, when an analysis similar to what done for external galaxies (see 2.2 and 2.5) is performed. Kochaneck (1995a) considers the Galaxy's (flat) rotation curve and the constraints coming from the local escape velocity of the stars, the motions of its satellites and the Local Group timing model. He derives a mass inside 50 kpc between 3.3 and $6.1\times10^{11}M_\odot$ and a half-mass scale length of the halo between 110 and 300 kpc.

Baryonic Dark Matter. The most recent and exiting studies of the Milky Way dark halo come from the microlensing experiments. If the halo of the Galaxy is made of massive compact objects (MACHOs), by monitoring $\approx 10^6$ stars in the LMC, the SMC or the galactic bulge one should detect several achromatic luminosity enhancement events every year, caused by gravitational microlensing. The characteristic temporal dependence of the microlensing luminosity curve allows one to discriminate against variable stars and to constraint the masses of the microlensing objects. Comparing the rates of detections towards the different lines of sight, information about the MACHO halo shape can be gained. Following the ideas of Paczyński (1986), a number of collaborations started to search and find microlensing events: MACHO (Alcock *et al.*, 1993), EROS (Aubourg *et al.*, 1993), OGLE (Udalski *et al.*, 1994). The rates derived up to now in the LMC direction rule out a dark halo made completely of MACHOs with masses between 8×10^{-5} and $0.3M_\odot$ but allow for 20% of the (spherical) halo to be made of MACHOs in the mass range $0.01 - 0.1M_\odot$ (Alcock *et al.*, 1995). Many more events than expected are instead derived towards the galactic bulge. This could imply that a new component of the Galaxy has been detected,

such as a heavy disk or a bar, or that the galactic halo is more disk-like
than spherical (Alcock *et al.*, 1995). Finally, note that star counts in deep
HST images rule out that a significant contribution to the mass of the halo
comes from faint, hydrogen-burning stars (Bahcall *et al.*, 1994).

2.2. SPIRAL GALAXIES

During the last 25 years the extended "flat" HI rotation curves of spiral galaxies have provided the best evidence for dark halos around extragalactic objects. Spiral galaxies are two-dimensional, rotationally supported
and gas-rich stellar systems. The exponential radial decline of their surface brightness distribution should generate a keplerian ($\propto r^{-1/2}$) rotation
curve already at a few exponential scale lengths h from the center. Rotational velocities are instead constant out to $\geq 11h$ (NGC 3198, van Albada
et al. 1985), thus requiring the presence of a probably spherical (see 3.1),
isothermal-like (with density $\propto r^{-2}$) halo, with mass $\propto r$. Mass models constructed to fit the measured rotation curves try to minimize the amount of
dark matter needed ("maximum-disk" solution) and produce the so-called
"conspiracy problem" (van Albada & Sancisi, 1986). The expected declining velocities of the disk component must be exactly compensated by the
rising halo curves to give the nearly featureless flat data. At the same time,
the constraint coming from the Tully-Fisher relation (Tully & Fisher, 1977),
linking the maximum rotational velocity to the luminosity of the galaxy,
must be satisfied. Finally, a proper mass decomposition should also take
into account the dynamical constraints coming from the observed presence
of spiral structure (Athanassoula *et al.*, 1987). Nowadays HI and optical
rotation curves are available for more than a thousend galaxies (see 3.2).
As a general result, the amount of dark mass needed to fit the rotation
curves equals the luminous mass M_L inside the optical limits R_{opt} of the
galaxies, and is $\approx 4M_L$ inside $2R_{opt}$, giving a typical total mass to blue
luminosity ratio in the range $10 - 30 M_\odot / L_\odot$.

2.3. DWARF GALAXIES

Dwarf galaxies have low luminosity ($M_B > -17$) and therefore a low baryonic content, offering the optimal "laboratory" to study the properties of
dark matter halos. Rotation curves of gas-rich dwarf irregulars, analysed as
described in 2.2, give large dark to luminous mass ratios (≈ 10) and imply
large dark matter central densities (DDO 154, Carignan & Freeman 1988;
see also this book). Masses of (gas-free) dwarf spheroidals can be estimated
from their luminosity core radii r_C and their central stellar velocity dispersion σ_0, measured from repeated observations of radial velocities of single
stars (Mateo, 1994). High dark matter central densities $\rho_0 = 9\sigma_0^2/(4\pi G r_C^2)$

and mass-to-light ratios ($\approx 100 M_\odot / L_\odot$) are also derived here. These results rule out that the dark matter halos of dwarf galaxies are made of massive neutrinos. In this case, the present ν phase-space density should be less or equal than the one at the decoupling, implying either implausibly high neutrino masses or very short dynamical friction decay times (Gerhard & Spergel, 1992a).

2.4. ELLIPTICAL GALAXIES

Until recently, no clear evidence for dark matter in ellipticals was available. These stellar systems are tridimensional, pressure supported and without cold gas, so that the dynamical information comes only from the analysis of their (difficult to measure) absorption line spectra. In addition, the interpretation of the so derived (projected) velocity and velocity dispersion profiles $\sigma(r)$ was affected by a degeneracy problem: not knowing the intrinsic distribution of stellar orbits, a flat $\sigma(r)$ profile (the analogous of the flat rotation curves of spirals) could have been interpreted as evidence for dark halos in the presence of radial or isotropic orbits, or as evidence for tangential orbits and constant M/L ratio. Only in a couple of clear-cut cases (i.g. NGC 7144, Saglia *et al.* 1993) an increasing σ profile rules out the tangential case. The analysis of the line of sight velocity distributions (LOSVD) in addition to $\sigma(r)$ offers now a general tool to break this degeneracy: tangential orbits produce flat-top LOSVDs, while radial orbits produce triangular shape LOSVDs (Bender *et al.*, 1994). Jeske *et al.* (1995, also this book) apply this technique to the E0 galaxy NGC 6703 to constrain its gravitational potential and stellar orbital structure. A number of ellipticals (Saglia *et al.*, 1995) are being analysed following this approach. Using similar data (Carollo *et al.* 1995) de Zeeuw and Carollo (this book) find evidence for dark halos in ellipticals using a stellar orbit superposition algorithm. Promising constraints on the extent of dark halos around ellipticals should come from the analysis of the radial velocity distributions of planetary nebulae (Arnaboldi *et al.*, 1994) and globular clusters (Grillmair *et al.*, 1994), although only with large amounts (> 500) of velocities the degeneracy due to the unknown orbital structure (analogous to what discussed above) can be disentangled.

Bright ellipticals have usually an X-ray corona, generated by thermal bremsstrahlung of a hot, massive gaseous component, in hydrostatic equilibrium in the gravitational potential of the galaxies (Fabricant *et al.*, 1980). The gas density $\rho_g(r)$ and temperature $T_g(r)$ profiles can therefore be used to constrain the mass of the system as $M(r) \propto r T_g [d \log \rho_g / d \log r + d \log T_g / d \log r]$. This approach allows one to probe the gravitational potential in the outer regions of ellipticals ($\leq 8 R_e$), but suffers some severe uncertainties. Only recently temperature profiles have started to become available (Mushotzky *et al.*, 1994) and it is difficult to estimate the role of the external pressure (Bertin *et al.*, 1993). Evidence for dark halos around single ellipticals has also been gained when a gaseous ring or disk is present (Franx *et al.*, 1994), or a gravitational lense is modeled (Kochanek, 1995b). Gravitational lensing statistics (Maoz & Rix, 1993) requires also the presence of dark halos around early-type galaxies.

2.5. BINARIES, SATELLITES, GROUPS

The kinematics of binary galaxies, galaxies with companions or groups of galaxies give the opportunity to probe the typical total mass and scale-length of dark matter halos. Projected separations r_P and relative velocities Δv can be combined and averaged to estimate masses M using the Virial Theorem: $M(3 - 2 < e^2 >) = 32 < r_P \Delta v^2 > /(\pi G)$ (Zaritsky & White, 1994), where $< e^2 >$ is the mean of the square ellipticity of the orbits. Mass estimates of single systems are however very uncertain, due to the problems caused by interlopers, the unknown orbital structure and the small number statistics. More stringent limits can be obtained, if models for halo formation are used to estimate these problems. Zaritsky & White (1994) analyse a sample of 69 satellites using this approach and find a typical halo mass inside 200 kpc of $1.5 - 2.6 \times 10^{12} M_\odot$ ($H_0 = 75$).

2.6. THE OTHER POINT OF VIEW

In the previous sections the observed dynamical discrepancies have been interpreted by invoking the presence of dark halos, using the newtonian laws. But also alternative interpretations have been proposed. Valentijn (1990) suggests that the luminous mass of the disks of spirals had been underestimated due to the presence of dust. More recent investigations (Peletier *et al.*, 1995) seem to rule out this possibility. Lequeux (1994), Pfenniger *et al.* (1994) and Gerhard & Spergel (1995, see also this book) argue that molecular hydrogen may have been strongly underestimated, especially in the outer regions of spiral galaxies. Finally, and more radically, Milgrom (1983) suggests MOND (modified newtonian dynamics) as the solution for the observed mass discrepancies. Accelerations in the low acceleration regime typical of the external regions of galaxy should tend to a (cosmic) constant, thus producing flat rotation curves. However, the dynamics of dwarfs (Gerhard & Spergel, 1992b) and the existence of declining rotation curves (Persic *et al.*, 1995) present strong challanges to this model.

3. Properties of Dark Halos

3.1. SHAPES

Polar ring galaxies, disk galaxies with a gaseous ring orthogonal to the plane of the disk, offer the possibility to measure the velocity field in two perpendicular planes, and therefore to probe the gravitational potential and the dark halo shape. The analysis of the data has to involve the self-gravity contribution of the polar ring and in the best case studied, NGC 4650A, indicates a rather flattened (E6) halo (Sackett *et al.*, 1994a). A similar result (E5-E7) is found using a completely different technique, which compares the isophotes of optical and X-ray surface brightnesses of the elliptical NGC 720 (Buote & Canizares, 1994). However, a rather round halo (E0.1-1.6) is needed to explain the twisted disk of NGC 4753 (Steiman-Cameron *et al.*, 1992). Useful constrains should come from the analysis of the flaring of HI disks (Olling, 1995): very flattened halos should produce very little flaring and, therefore, could be ruled out. Inferring halo shapes from the modeling of warps may prove more dangerous (Dubinski & Kujken *et al.*, 1995).

Constraints on the axisymmetry of dark halos are much more stringent. The scatter that deviations from axisymmetry would induce on the Tully-Fisher relation (Franx & de Zeeuw, 1992), the expansion of the HI velocity fields in harmonics (Franx *et al.*, 1994), and the azimuthal structure of stellar disks in the infrared band (Rix & Zaritsky, 1995) all imply that deviations from axisymmetry of the dark halos of galaxies are in the range E0-E0.6. Dissipationless dark matter N-body simulations generate instead rather strongly triaxial dark halos (Katz & Gunn, 1991), which however may turn more axisymmetric when gas dynamics is included.

3.2. GLOBAL CORRELATIONS

The large data-base of rotation curves nowadays available (Persic *et al.*, 1995) allows one to correct the simple picture of "flat rotation curves" of 2.2 and to investigate possible correlations of dark matter properties with the properties of galaxies. A luminosity dependence of the rotational velocities seems now well established, with low-luminosity galaxies showing rising rotation curves and high-luminosity galaxies having flat or declining curves (Persic *et al.* 1995, Casertano & van Gorkom 1991). Equivalently, one finds that the dark to luminous mass ratio inside R_{opt} scales with galaxy luminosity as $M_D/M_L(R_{opt}) \propto L_B^{-0.6}$.

A further step can be made by computing the core radii r_C, the central densities ρ_0 and velocity dispersions $\sigma_0 \approx V_{max}/\sqrt{2}$ of the dark halos found around dwarfs and late-type spirals, where the influence of the baryonic component should be minimal (see 2.2 and 2.3). Kormendy (1990) finds that dark matter halos in low luminosity galaxies are smaller, denser, and with lower velocity dispersion than those in high luminosity galaxies, with scaling laws similar to those obtained for the visible component.

3.3. NATURE

Are the dark halos around galaxies baryonic? Sackett *et al.* (1994b) detect a faint stellar halo around the edge-on galaxy NGC 5907, with a density profile able to reproduce the observed (flat) rotation curve, and suggest that faint stars could make the halos of galaxies. However, the limits coming from HST faint star counts exclude this possibility (see 2.1). The microlensing experiments suggest also that non hydrogen-burning, jupiter like compact objects cannot make all of the halo, but may be a fraction of it (see 2.1). Molecular hydrogen cloudlets (see 2.6) are still a possibility, although unlikely (Wilson & Mauersberger, 1994).

Are the dark halos around galaxies non-baryonic? Massive neutrinos cannot make the halos of dwarfs (see 2.3) and therefore are not very appealing, However, a large number of weakly interacting massive particles (WIMPS) is "on the market", from axions to the lightest particle of supersymmetric theories (Primack *et al.*, 1988). Although yet undetected in the laboratory, WIMPS remain the best candidate constituents of dark matter halos, because of the successes of Cold Dark Matter type models (see Frenk, this book) in explaining the large scale structure of the Universe.

4. Conclusions

The existence of dark halos around spiral galaxies is well established, thanks to 25 years of measuring rotation curves. The next years of research should give us the same degree of confidence for the case of elliptical galaxies and assess the intrinsic shapes of dark halos, with gravitational lensing playing an important role. The steadily growing dynamical data-set should soon allow one to study the statistical properties of dark halos (for example, their mass function), and, possibly, their evolution with redshift (see Franx and Rix, this book). Hopefully, particle physicists will in the mean time detect the particles that constitute the dark matter.

Acknowledgments Financial support by the Deutsche Forschungsgemeinschaft under SFB 375 is gratefully acknowledged.

References

Alcock, C., *et al.* 1993, Nature **365**, 621
Alcock, C., *et al.* 1995, preprint

Arnaboldi, M., Freeman, K.C., Hui, X., Capaccioli, M., Ford, H. 1994, The Messenger **76**, 40

Ashman, K.M. 1992, PASP **104**, 1109

Aubourg, E. *et al.* 1993, Nature **365**, 623

Athanassoula, E., Bosma, A., Papaioannou, S. 1987, A&A **179**, 23

Bahcall, J.N. 1984, ApJ **276**, 169

Bahcall, J.N., Flynn, C., Gould, A., Kirhakos, S. 1994, ApJ **435**, L51

Bennett, C.L., *et al.* 1994, ApJ **436**, 423

Bender, R., Saglia, R.P., Gerhard, O. 1994, MNRAS **269**, 785

Bertin, G., Pignatelli, E., Saglia, R.P. 1993, A&A **271**, 381

Buote, D.A., Canizares, C.R. 1994, ApJ **427**, 86

Carignan, C., Freeman, K.C. 1988, ApJ **332**, L33

Carollo, C.M., de Zeeuw, P.T., van der Marel, R.P., Danziger, I.J., Qian, E.E. 1995, ApJ **441**, L25

Carr, B. 1994, ARAA **32**, 531

Casertano, S., van Gorkom, J.H 1991, AJ **101**, 1231

Dubinski, J., Kuijken, K. 1995, ApJ **442**, 492

Fabricant, D., Lecar, M., Gorenstein, P. 1980, ApJ **241**, 552

Flynn, C., Fuchs, B. 1994, MNRAS **270**, 471

Franx, M., de Zeeuw, T. 1992, ApJ **392**, L47

Franx, M., van Gorkom, J.H., de Zeeuw, T. 1994, ApJ **436**, 642

Gerhard, O.E., Spergel, D.N. 1992a, ApJ **397**, 38

Gerhard, O.E., Spergel, D.N. 1992b, ApJ **397**, 38

Gerhard, O.E., Spergel, D.N. 1995, ApJL, in press

Grillmair, C.J., Freeman, K.C., Bicknell, G.V., Carter, D., Couch, W.J., Sommer-Larsen, J., Taylor, K. 1994, ApJ **422**, L9

Guth, A. 1981, Phys. Rev. D**23**, 347

Jeske, G., Gerhard, O., Saglia, R.P., Bender, R. 1995, MNRAS submitted

Katz, N., Gunn, J.E. 1991, ApJ **368**, 325

Kochanek, C.S. 1995a, preprint

Kochanek, C.S. 1995b, ApJ **445**, 559

Kormendy, J. 1990, ASP **10**, 33

Lequeux, J., 1994, A&A **287**, 368

Maoz, D., Rix, H-W. 1993, ApJ **416**, 425

Mateo M. 1994, ESO/OHP Workshop on Dwarf Galaxies, B. Bingelli, G. Meylan Eds., ESO, Garching, p. 209

Mathews, G.J., Schramm D.N., Meyer, B.S. 1993, ApJ **404**, 476

Mushotzky, R.F., Loewenstein, M., Awaki, H., Makishima, K., Matsuchita, K., Matsumoto, H. 1994, ApJ **436**, L79

Milgrom, M. 1983, ApJ **260**, 365

Olling R.P. 1995, AJ in press

Oort, J.H. 1932, Bull. Astron. Inst. Netherlands **6**, 249

Oort, J.H. 1960, Bull. Astron. Inst. Netherlands **15**, 45

Paczyński, B. 1986, ApJ **304**, 1

Peebles, P.J.E. 1993 *Principles of Physical Cosmology*, Princeton University Press, Princeton

Peletier, R.F., Valentijn, E.A., Moorwood, A.F.M., Freudling, W., Knapen, J.H., Beckman, J.E. 1995, A&A **300**, L1

Persic, M., Salucci, P., Stel, F. 1995, MNRAS submitted

Pfenniger, D., Combes, F., Martinet, L. 1994, A&A **285**, 79

Primack, J.R., Seckel, D., Sadoulet, B. 1988, Ann. Rev. Nucl. Part. Sci. **B38**, 751

Rix, H-W., Zaritsky, D. 1995, ApJ **447**, 82

Sackett, P.D., Rix, H.-W., Jarvis, B.J., Freeman, K.C. 1994a, ApJ **436**, 629

Sackett, P.D., Morrison, H.L., Harding, P., Boroson, T.A. 1994b, Nature **370**, 441

Saglia, R.P., *et al.* 1993, ApJ **403**, 567

Saglia, R.P., Bender, R., Jeske, G., Gerhard, O. 1995, in preparation
Steiman-Cameron, T.Y., Kormendy, J., Durisen, R.H. 1992, AJ **104**, 1339
Trimble, V. 1987, ARAA **25**, 425
Tully, R.B., Fisher, J.R. 1977, A&A **54**, 661.
Udalski, A., Szymański, M., Kaluzny, J., Kubjak, M., Mateo, M., Krzemiński, W. 1994, ApJ **426**, L69
Valentijn, E.A., 1990, Nature **346**, 153
van Albada, T.S., Bahcall, J., Begeman, K., Sancisi, R. 1985, ApJ **295**, 305
van Albada, T.S., Sancisi, R. 1986 Phil. Trans. Roy. Soc. London A **320**, 447
Wilson T.L., Mauersberger R., 1994, A&A **282**, L41
Zaritsky, D., White, S.D.M 1994, ApJ **435**, 599

Discussion

Fabbiano: Is there an explanation for the luminosity dependence of the rotation curve of spirals? I find it especially difficult to understand if the DM is non-baryonic.

Saglia: This is indeed a problem of CDM-like models. Navarro (this book) discusses the difficulties emerging when such a luminosity dependence is modeled.

Peletier: You say that measurements of the local mass density by Bahcall et al. imply that halos cannot be made out of faint stars. What about the possible detection by Morrison *et al.* (1994, AJ **108**, 1191) and the modeling of Sackett *et al.* (1994b) of an optical halo in NGC 5907?

Saglia: see discussion in Sect. 3.3.

Faber: can you comment on evidence from X-rays around elliptical galaxies?

Saglia: see discussion in Sect. 2.4.

Dejonghe: I would like to remind you of the case of NGC 7144 (Saglia *et al.*, 1993), which has a rising velocity dispersion profile up to 2 R_e. I do not know of any dynamical model, irrespective of dynamical structure and/or presence of LOSVDs, that can reproduce this without dark matter.

Saglia: I agree with you. However, this is a *single* case. The method described in 2.4 allows us to study the problem of dark matter in *all* ellipticals.

Dehnen: From the theories of early density perturbations (in good agreement with COBE) we expect the (dark) matter to have power on all scales, especially on those smaller than can be resolved by current simulations of structure formation in the universe. This implies that dark matter halos must also have cusps. Is there any evidence in this direction?

Saglia: Dark halos of CDM-like models without gas dynamics *do not* reproduce the density profiles of observed dark halos, see discussion by Burkert and Navarro in this book.

Roberto Saglia and Claire Halliday

BARYONIC DARK HALOS: A MODEL WITH MACHOS AND COLD GAS GLOBULES

ORTWIN GERHARD[1] AND JOSEPH SILK[2]
[1] *Astronomisches Institut, Univ. Basel, Switzerland*
[2] *Dpt. of Astronomy, Univ. of California, Berkeley, U.S.A.*

Abstract. The dark matter in the halos of galaxies may well be baryonic, and much of the mass within them could be in the form of clusters of substellar objects within which are embedded cold gas globules. Such halos might play an active role in galaxy formation and evolution.

1. Introduction: The Case for Baryonic Halos

A number of arguments point towards the dominance of baryonic dark matter in galaxy halos. Some of the strongest arguments have been known for many years: (i) Cosmic nucleosynthesis predicts considerably more baryons than are seen in luminous form (Kolb & Turner 1990). (ii) While the visible mass in spiral galaxies fails to explain the *amplitudes* of the observed rotation curves, their *form* is often well-predicted from scaling up the distribution of gas mass (Bosma 1981). With the discovery of MACHOs, we now have evidence for *some* of the halo dark matter being baryonic (Alcock *et al.* 1993, Aubourg *et al.* 1993). Since this is the only observational evidence to date on the nature of any dark matter, it is natural to ask whether the entire mass in galaxy halos could be baryonic.

2. The Old Model: Gas Clouds Only

In this talk a brief account is given of a model of Gerhard & Silk (1995; hereafter GS) which asserts that a substantial fraction of halo dark matter may be in the form of dense, cold gas clouds. In its original form, as submitted to Nature in 1993, the model considered these gas clouds as isolated objects. We found that collisional dissipation and evaporation can be avoided if the clouds have sufficiently high column density and correspondingly low area

167

R. Bender and R. L. Davies (eds.), New Light on Galaxy Evolution, 167–170.
© *1996 IAU. Printed in the Netherlands.*

filling factor, and for the same reasons detection in emission or absorption is very difficult. However, as emphasized by several colleagues, the theoretical problem of preventing the clouds from collapsing and forming stars is more serious than in the disk ISM because of the long time-scales involved. The nearby gas might moreover have been seen in γ-rays via cosmic ray interactions and ensuing π^0-production, and in UV-illumination when these clouds traversed a nearby HII-region.

3. The New Model: Gas Globules Stabilized by Surrounding MACHO Clusters

In the meantime, MACHOs of mass $\sim 0.01\,M_\odot$ had been discovered, and been found to contribute an appreciable fraction (Alcock et al. 1995) of the Milky Way's halo mass. This implies that at early times some halo gas did fragment and collapse, and the question arises as to whether all the remaining gas collapsed onto the disk or whether most of it remained in the halo as dense cold clouds. In fact, since such low mass MACHOs do not inject as much energy into their surroundings as do normal main sequence stars, the gas might remain near the MACHOs. Further, we realized that, if the MACHOs are clustered by analogy with the formation of most low-mass stars, they can exert a stabilizing influence on any remaining embedded gas by removing some of the self-gravity response. We have considered this in the context of a simple model in which both the MACHO cluster and the gas cloud are spherical and do not rotate, and in which the equation of state is polytropic. One then finds that $\sim 10 - 50\%$ of the MACHO cluster mass can be embedded as a stable sphere of gas. For more general configurations, up to of order half the total mass could be stabilized. This result has led to a modified model in which the halo is assumed to be made of clusters of low-mass objects within which cold gas clouds are embedded.

4. Physical Parameters of the Gas Clouds

The thermal state of these clouds is uncertain, both because of unknown cosmic ray heating rates in the outer halo, and because modern cooling calculations with revised molecular excitation rates (Neufeld et al. 1995) are not available for the low abundances and cosmic ray ionization rates expected in the outer halo. However, at the high densities implied by the model ($n \sim 10^7\,cm^{-3}$), the cooling rate per molecule appears to decrease with density. If the temperature remains above the CMB value, this will lead to a polytropic rather than an isothermal equation of state.

A critical condition for gas to survive in the halo is that the area filling factor of the clouds is sufficiently small, $f \sim 0.01$ in the outer halo. Given the rotation curve, this sets a lower limit to the cloud column density,

$N \sim 10^{23} \, \text{cm}^{-2}$ if collisions are to be avoided in the outer halo, and larger for clouds further in. Together with a temperature of, say, $T \simeq 10 \, \text{K}$, the condition that the clouds be partially self-gravitating determines their mass and radius. Typical values are $M_c \lesssim 1 \, \text{M}_\odot$ and $L \lesssim 0.02 \, \text{pc}$.

5. Parameters of the MACHO Clusters

The masses of individual fragments are limited by the opacity argument to $m \gtrsim 10^{-3} \, \text{M}_\odot$ (Rees 1976), and the MACHO experiments favour masses somewhat larger. So that the gas content of the halo is non-negligible, the MACHO cluster mass should not exceed a small multiple of the mass of its embedded gas cloud. The requirement that the cluster should not evaporate in a Hubble time (Moore & Silk 1995) is then marginally satisfied, requiring some fraction of cloud support in macroscopic motions. It also sets an upper limit to the cloud column density. Because the constraints on cloud collisions require high column densities, whereas the constraints on cluster evaporation favour low column densities, the new model now leads to fairly well-defined typical parameters: cluster masses of $\sim 10 \, \text{M}_\odot$, MACHO masses of $0.01 \, \text{M}_\odot$, and cold gas content up to of order 50%, with typical column density $N \sim 10^{23} \, \text{cm}^{-2}$ and temperature $T \sim 10 \, \text{K}$.

6. Observational Constraints

With these parameters, one furthermore predicts that the dissipational evolution is a strong function of radius, in the sense that inner halos should by now be largely depleted of gas, while outer halos should still contain substantial gas fractions. This also greatly reduces the observational problems that remained in the original model. While at present, observations cannot constrain our hypothesis, nevertheless a number of observational techniques can be improved to search for the postulated cloud-clusters. In particular, more sensitive FIR, gamma ray, local and high redshift mm-line observations, and further analysis of the microlensing experiments will provide tighter constraints on the cloud parameters and might falsify the model.

7. Implications

If indeed a significant fraction of the mass of galactic halos is or once was in the form of cold gas, then the radial dependence of the cloud survival rate has cosmogonical implications for galaxy formation. The halos of galaxies would then play a much more active role in the build-up of galactic disks and their chemical evolution. Moreover, these processes would then naturally occur fastest in massive galaxies, with much of the mass in dwarf systems remaining in dark halos even today.

References

Alcock, C. A., *et al.* , 1993, Nature **365**, 621
Alcock, C. A., *et al.* , 1995, ApJ , in press
Aubourg, E., *et al.* , 1993, Nature **365**, 623
Bosma, A., 1981, AJ **86**, 1825
Gerhard, O.E., Silk, J., 1995, ApJ , submitted
Kolb, E.W., Turner, M.S., 1990, *The Early Universe* (Addison Wesley, New York).
Moore, B., Silk, J., 1995, ApJ **442**, L5
Neufeld D.A., Lepp S., Melnick G.J., 1995, preprint.
Rees, M.J., 1976, MNRAS **176**, 483

Discussion

RIX: If you have 0.01 of sky covered with dense, presumably optically opaque, molecular clouds, would you not expect a distinct time-dependent absorption signature in the MACHO and OGLE surveys? If so, what is the time-scale for such time variations?

GERHARD: The time-scale is $\sim 80\,$yr for our standard parameters, so it should be detectable statistically. We have not worked the details out yet.

MOORE: The dynamics of the Magellanic Stream (Moore & Davis 1994) constrain the total mass of diffuse gas within 50 kpc to be $\sim 0.5\%$ of the total halo mass - consistent with ROSAT observations. Is it possible to confine 99.5% of the halo mass in such small cold gas clouds, without winds, evaporation etc. liberating a significant fraction of gas to a diffuse component?

GERHARD: The evaporation rate is about $10^{-11}\,\mathrm{yr}^{-1}$ for a hot gas density of $10^{-4}\,\mathrm{cm}^{-3}$; collisions yield a comparable rate of $\gtrsim (100 t_{\mathrm{dyn}})^{-1}$. This is about consistent with 1% of the halo gas in diffuse form if the relevant cooling time is $10^9\,$yr.

BOSMA: There are galaxies which have flat rotation curves, yet with a light distribution completely different from a standard exponential disk (Malin-1 "cousins"). If the dark halo participates in the evolution of the stellar populations, how do you explain such large differences between such galaxies and the more "standard" ones?

GERHARD: We have not tried sofar, but these halos could be unusually diffuse or they could have formed late.

INTERACTION OF DISKS AND DARK HALOES OF DWARF SPIRALS

B. FUCHS, V. FRIESE, H. REFFERT, R. WIELEN

Astronomisches Rechen - Institut Heidelberg
Mönchhofstr. 12 - 14, 69120 Heidelberg, Germany

1. Introduction

In a number of dwarf spiral galaxies the HI-emission has been studied with sufficient resolution to derive the rotation curves of the galaxies. These show that the disks of dwarf spirals are imbedded in extended haloes of dark matter, quite similar to the disks of giant spiral galaxies.

The presence of such massive, extended haloes affects the dynamics of the disks of the galaxies. We concentrate on two aspects here: First we discuss the effects of large amounts of dark matter on the spiral structure of the disks. Unfortunately the velocity dispersion of the stars in dwarf spirals is not known, but we have to use the vertical scale heights of the disks to derive estimates of the velocity dispersions. Once the velocity dispersions are known Toomre's Q stability parameter and 'X = 2' criterion (Toomre 1964, 1981, Athanassoula et al. 1987) can be evaluated.

Second, it has been suggested that the main constituents of the dark haloes of giant spiral galaxies might be very compact objects such as black holes with masses of the order of a few 10^6 $\mathcal{M}_\odot$ (Lacey and Ostriker 1985). This scenario has been put forward among others in order to explain the observed increase of the velocity dispersion of the stars in the disk of the Galaxy (Wielen 1977). We have discussed the scattering of stars by massive black holes penetrating the disk in some detail elsewhere and found many attractive features, fitting nearly ideally the locally or globally observed kinematics of the stars in the Galaxy (Wielen et al. 1992). Thus it is interesting to carry this conjecture over to the haloes of dwarf spiral galaxies, especially since the disks of these galaxies are much less massive and thus more fragile than in giant spirals.

R. Bender and R. L. Davies (eds.), New Light on Galaxy Evolution, 171–174.

2. Spiral Structure

We have analyzed in detail the disks of a set of 11 dwarf spirals listed in table 1 for which rotation curves are available in the literature, mainly by Carignan and collaborators (cf. Côté et al. 1991 and references therein). Dwarf spirals have no bulge component (Freeman 1987). So a mass model comprising an exponential disk and a dark halo component described by a pseudo isothermal density law has been fitted to the rotation curve of each galaxy. We discuss usually a best χ^2-fit and a 'maximum disk' fit. Next, the vertical velocity dipersions of the stars are estimated from the vertical scale height of each disk using the vertical hydrostatic equilibrium condition. The vertical velocity dispersions are then converted to radial velocity dispersions by adopting the same axial ratio of the velocity ellipsoid as in the Milky Way. Since even the vertical scale heights of the galaxies in table 1 are not known we estimate the vertical scale heights from the radial scale lengths. For this purpose we have performed a statistical flattening analysis of faint dwarf spirals in the ESO-Uppsala catalogue in order to derive the intrinsic ratio of vertical to radial exponential scale lengths z_0/h. From a sample of 115 galaxies with $M_B \geq -18$ and Hubble type Sc or later we find $z_0/h = 0.2 \pm 0.05$ corresponding to an intrinsic flattening $q_0 = 0.14$.

In this way we have modelled individually the disk of each galaxy. There is a clear distinction between the dynamical states of galactic disks which develop ordered spiral structure or which appear irregular. NGC 300 is a typical example of the first class. The stability parameter is in the range where swing amplification is effective. The expected number of spiral arms, estimated using the 'X = 2' criterion, indicates a two-armed pattern in the inner parts of the disk which breaks up in filaments in the outer parts, exactly as observed. NGC 55 is a typical irregular galaxy. The values of the stability parameter are so large that any spiral structure is suppressed. The transition from Hubble types Sc, Sd to Sm, Im is correlated, as can be seen from table 1, to the ratio of disk to dark halo mass within one Holmberg radius. Although our dynamical models rely on rather crude estimates of the stellar velocity dispersions and ambiguous decompositions of the rotation curves our conclusions appear to be rather robust, because we could interpret the Hubble type of each galaxy consistently.

3. Disk Heating

Assuming now that the stochastic heating of the disks is due to the black holes of the dark haloes, we can deduce from the velocity dispersions of the stars typical black hole masses. Wielen (1977) has determined empirically the diffusion coefficient, which describes the increase of the velocity dispersion of stars in the solar neighbourhood. Lacey and Ostriker (1985) have

TABLE 1.

	type	M_B	$M_d^\dagger$	$M_h(r < R_{Holm})$	M_d/M_h	m_h
		mag	$10^8 \mathcal{M}_\odot$	$10^8 \mathcal{M}_\odot$		$10^6 \mathcal{M}_\odot$
UGC 2259	Scd	-16.5	31 (41)	30 (15)	1 (3)	2 (40)
NGC 247	Sd	-18.0	87 (102)	121 (79)	0.7 (1.3)	12 (81)
NGC 300	Sd	-17.9	35 (52)	67 (41)	0.5 (1.3)	1.6 (7)
NGC 1560	Sd	-16.4	14 (17)	29 (28)	0.5 (0.6)	1.2 (1.9)
NGC 7793	Sd	-18.3	35 (–)	106 (–)	0.3 (–)	0.4 (–)
DDO 168	IBm	-15.2	2 (4)	6 (4)	0.3 (1)	0.05 (0.17)
DDO 170	Irr	-15.2	1.4 (2.7)	9.4 (7.7)	0.2 (0.4)	0.06 (0.2)
NGC 5585	Sd	-17.5	7.4 (14)	108 (99)	0.07 (0.14)	0.04 (0.1)
DDO 154	Im	-13.8	– (0.28)*	– (2.9)	– (0.1)	—
NGC 55	Sm	-18.6	3 (14)*	170 (159)	0.02 (0.09)	– (0.07)
NGC 3109	Sm	-16.8	– (1.8)*	– (56)	– (0.03)	—

$\dagger$Maximum disk estimates given in parantheses. *HI disk mass not included.

calculated the diffusion coefficient theoretically and shown that it is proportional to the space density times the individual mass of the black holes. This implies a mass of $3 \cdot 10^6$ $\mathcal{M}_\odot$ of a typical black hole in the Galaxy. Quite a similar value of $2 \cdot 10^6$ $\mathcal{M}_\odot$ can be derived for NGC 3198, another giant Sc galaxy ($M_B = -19.4$), using the the disk and dark halo parameters of Bottema (1988) and van Albada et al. (1985). The typical black hole masses, which we find for the Sd galaxies are of the same order of magnitude as in giant spirals. The velocity dispersions of the disk stars of the irregular galaxies, however, are hardly larger than the turbulent velocity dispersion of the interstellar gas, so that the disks of these galaxies cannot have experienced much stochastic heating. Consequently, the masses of the hypothetical black holes in these galaxies turn out to be low, of the order of 10^4 to 10^5 $\mathcal{M}_\odot$. This was noted before by Rix and Lake (1993) and Fuchs and Wielen (1993). Such low masses contradict the supposed universal nature of the black holes and cast severe doubts on the scenario of dark haloes of galaxies made of massive black holes.

Furthermore, if the dark haloes of the irregular galaxies were made up of black holes with 'standard' masses of a few 10^6 $\mathcal{M}_\odot$, the two–body–relaxation time of such a system would be much shorter than a Hubble time. In a galaxy like DDO 154 the entire dark halo would consist of about only 1000 black holes. We have run several N–body simulations which show that the ensemble of black holes evolves very fast within a Hubble time, developing a concentrated core of a few objects and a diluted halo expanded to 100 kpc scale.

References

van Albada, T.S., Bahcall, J.N., Begeman, K., & Sancisi, R. 1985, ApJ 295, 305
Athanassoula, E., Bosma, A., & Papaioaunou, S. 1987, A&A 179, 23
Bottema, R. 1988, A&A 197, 105
Côté, S., Carignan, C., & Sancisi, R. 1991, AJ 102, 904
Freeman, K. C. 1987, IAU Symp. No. 117, p.119
Fuchs, B., & Wielen, R. 1993, AG Abstract Ser. 8, 145
Lacey, C., & Ostriker, J. P. 1985, ApJ 299, 633
Rix, H. W., & Lake, G. 1993, ApJ 417, L1
Toomre, A. 1964, ApJ 139, 1214
Toomre, A. 1981, in The Structure and Evolution of Normal Galaxies, eds. S. M. Fall & D. Lynden-Bell (Cambridge: Cambridge Univ. Press) p.111
Wielen, R. 1977, A&A 60, 263
Wielen, R., & Fuchs, B. 1983, in Kinematics, Dynamics and Structure of the Milky Way, ed W. L. H. Shuter (Dordrecht: Reidel) p.81
Wielen, R., Dettbarn, C., Fuchs, B., Jahreiß, H., & Radons, G. 1992, IAU Symp. No. 149, p.81

Discussion

Kormendy: Your point that the dynamical timescale for a DM halo made of $100 \cdot 10^6 \, \mathcal{M}_\odot$ BHs is short, is well taken and can be strengthened still more by considering the lowest luminosity dwarf spheroidal galaxies. They have halo masses that are so small that only a few – 10 BHs would be required. Then the dynamical timescale for halo evolution and galaxy evaporation is extremely short – only a modest number of crossing times. I have felt for some years that this is one of the strongest arguments against the idea that DM halos are made of $10^6 \, \mathcal{M}_\odot$ BHs.

Moore: Moore (1994, *ApJ* **413**, L93) and Moore & Silk (1995, *ApJ* **442**, L5) use the existence of low luminosity globular clusters and the invariance with position of the globular cluster luminosity function, to constrain the mass of possible dark matter candidates in galactic halos. Our limits are of the order of 10^4 to $10^5 \, \mathcal{M}_\odot$.

McGaugh: $M_H(r < R_*)/M_{disk}$ is strongly correlated with L. At fixed L, it is strongly correlated with surface brightness. Understanding these correlations is fundamental to the problem of galaxy formatiom and evolution in dark matter halos. Is there any reason to expect these correlations in either baryonic or non–baryonic dark matter scenarios?

Fuchs: One might add that in irregular galaxies the dark haloes tend to be more concentrated and have higher central densities than in Sd galaxies. However, these correlations do not seem to be related to the present day dynamics of the galaxies but to the formation processes of the galaxies.

THE STRUCTURE OF DARK MATTER HALOS IN DWARF GALAXIES

A. BURKERT
Max-Planck-Institut für Astronomie
Königstuhl 17,69117 Heidelberg, Germany

1. Introduction

Some dwarf galaxies have HI rotation curves that are completely dominated by a surrounding dark matter (DM) halo (e.g. Carignan & Freeman 1988). These objects represent ideal candidates for an investigation of the density structure of low-mass DM halos as the uncertainties resulting from the subtraction of the visible component are small, even in the innermost regions. Flores & Primack (1994) and Moore (1994) compared the observed DM rotation curves with the profiles, predicted from cosmological cold dark matter (CDM) calculations. They found an interesting discrepancy: whereas the calculations lead to a DM density distribution which diverges as $\rho \sim r^{-1}$ in the inner parts, the observed rotation curves indicate shallow DM cores which can be described by an isothermal density profile with finite central density.

2. The universal density profiles of DM halos

More recently, Burkert (1995) investigated the DM rotation curves in detail and found that they are self-similar, as expected from scale free cosmological models. Figure 1 shows the DM mass distribution as derived from the HI-rotation curves of those dwarf spiral galaxies which are known to be completely DM dominated. All four profiles indeed follow the same universal mass relation. The mass distribution as predicted from cosmological CDM simulations (Navarro et al. 1995) leads to too much mass at small radii as a result of the central density cusp. Even the frequently used modified isothermal profile does not provide a good fit to the observations in the outer regions due to the linear divergence of mass with radius. The solid

175

R. Bender and R. L. Davies (eds.), New Light on Galaxy Evolution, 175–178.

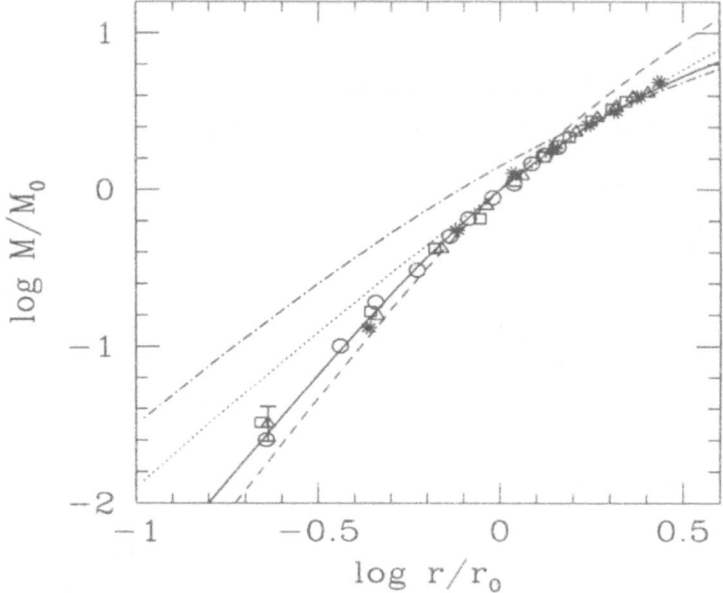

Figure 1. Dark matter mass profiles are shown for the following dwarf spiral galaxies (Burkert, 1995): DDO154 (open triangle), DDO105 (open square), NGC3109 (open circle) and DDO170 (starred). The errorbar at the innermost triangle represents the observational uncertainty in the inner region. The isothermal fit with core radius r_0 is shown as dashed curve, the solid line shows the revised profile, given in the text. The dotted and dot-dashed curve show the mass profiles as predicted from CDM calculations with formation redshifts of $z = 0.6$ and $z = 1.5$, respectively.

line shows a density distribution which fits the observations very nicely over the whole observed radius range:

$$\rho_{DM}(r) = \frac{\rho_0 r_0^3}{(r + r_0)(r^2 + r_0^2)} \tag{1}$$

where ρ_0 and r_0 are free parameters which represent the central DM density and a scale radius, respectively.

3. Scale relations for dark matter halos

According to equation 1 the density profiles of dark matter halos are completely described by two parameters: ρ_0 and r_0. It is important to investigate whether there exists a relation between these parameters which would provide information on the primordial fluctuation spectrum from which these

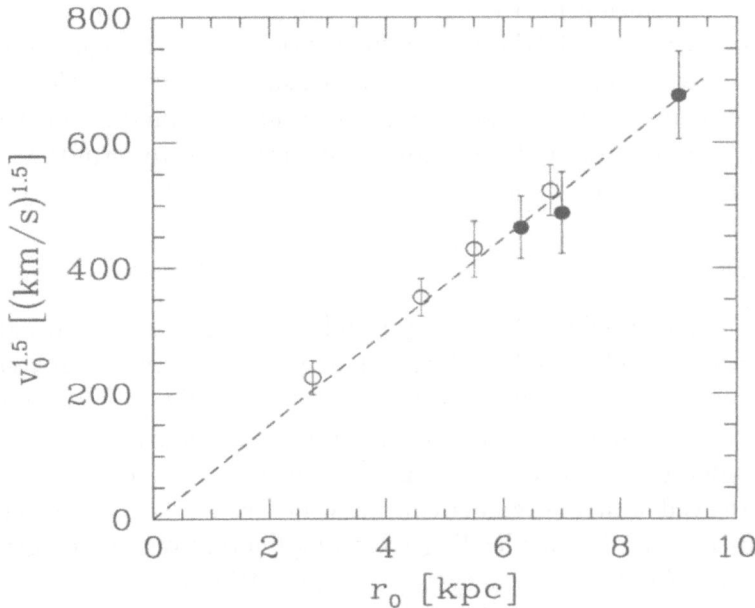

Figure 2. The scaling relation between the rotational velocity v_0 measured at r_0 is shown for the DM halos, investigated by Burkert (1995). Open circles represent the four DDO galaxies which have been used also in figure 1. The filled circles show three additional galaxies: NGC55, NGC300 and NGC1560. The dashed line is a fit through the data points.

objects formed. Instead of ρ_0 which cannot be observed directly, figure 2 shows the rotational velocity v_0 of the DM rotation curves at r_0 as a function of r_0. We find a strong linear correlation between r_0 and $v_0^{1.5}$. Using equation 1 and assuming spherical symmetry one can derive the following scaling relations for low-mass DM halos:

$$v_0 = 17.7 \left(\frac{r_0}{kpc}\right)^{2/3} \frac{km}{s}$$

$$M_0 = 7.2 \times 10^7 \left(\frac{r_0}{kpc}\right)^{7/3} M_\odot \qquad (2)$$

$$\rho_0 = 2.7 \times 10^{-2} \left(\frac{r_0}{kpc}\right)^{-2/3} \frac{M_\odot}{pc^3}$$

where M_0 is the total DM mass inside r_0.

These scaling relations indicate that the observed DM halos represent a one-parameter family with the density profiles being completely determined by the scale length r_0. As Burkert (1995) demonstrates, there exists a cosmological explanation for the equations 2, if one assumes that all halos formed from primordial CDM density fluctuations with the same initial amplitude. Otherwise one would expect a much larger spread of the data points shown in figure 2. It is however puzzling that only fluctuations with a certain fixed primordial amplitude should have managed to form the DM halos around dwarf spiral galaxies.

4. Discussion

Cosmological models predict DM density profiles with shapes that disagree with the observations. This might indicate that secular dynamical processes affected the central halo regions, leading to shallower DM profiles at the end. Suppose, for example, that the dwarf spirals had a much larger baryonic mass originally. In the case of gas it is likely that these low-mass systems experienced a strong galactic wind which could have removed a substantial gas fraction by this leading to strong fluctuations in the gravitational potential. At the end, the systems became DM dominated even in the inner regions and the gravitational fluctuations could have affected the inner DM density structure, leading to shallow central density profiles. It is however unlikely in such a scenario, that the energetic processes which lead to the galactic wind are so fine tuned that the resulting potential changes always lead to the same, self-similar dark matter density profiles, representing a one-parameter family as described by the equations 1 and 2. Even if the DM halos initially were a one-parameter family, one would expect that mass loss introduces a second independent parameter which reflects the efficiency with which the baryonic processes affect the inner DM halo. The discrepancy between the observed and theoretically predicted DM profiles and the existence of scaling relations are more likely directly coupled with the formation history of DM halos and therefore might indicate that some important physical features which are related to the nature and origin of dark matter are still missing in cosmological models.

References

Burkert, A. 1995, *Astrophys. J.*, **447**, L25
Carignan, C. and Freeman, K.C. 1988, *Astrophys. J.*, **332**, L33
Flores, R.A. and Primack, J.R. 1994, *Astrophys. J.*, **427**,L1
Moore, B. 1994, *Nature*, **370**, 629
Navarro, J.F., Frenk, C.S. and White, S.D.M. 1995, *MNRAS*, **275**, 56

MERGERS IN THE LOCAL UNIVERSE

MERGERS AND THE FORMATION OF MASSIVE ELLIPTICALS

RALF BENDER
Universitäts-Sternwarte
Scheinerstr. 1, D-81679 München, Germany
email: bender@usm.uni-muenchen.de

1. Introduction and Abstract

Many nearby luminous elliptical galaxies exhibit counter-rotating cores, minor axis rotation or peculiar velocity fields. These phenomena require that merging and accretion events played a major role in the formation of ellipticals and, equally important, that stars dominated significantly over gas in the merger progenitors. Low redshift analogues to the last stages in the formation of massive ellipticals can be observed in ultraluminous IRAS mergers. However, this does not imply that ellipticals generally formed in spiral-spiral mergers. Mergers can also occur as a consequence of inhomogenous collapse or hierarchical bottom-up structure formation and it is actually more likely that the progenitors of present-day ellipticals did not resemble present-day spirals.

Although *some* ellipticals most certainly were and still are formed at low redshift, the *general* formation epoch of massive ellipticals is likely to be found at redshifts > 1. The evidence for this is threefold: (a) the homogenous stellar populations of ellipticals, most notably the tight correlation between colors/absorption linestrengths and velocity dispersion, (b) the high overabundance of light elements relative to iron and (c) the very weak evolution of colors and linestrengths of ellipticals with redshift.

2. Peculiar cores in ellipticals

Peculiar cores in ellipticals are characterized by angular momentum vectors opposite or perpendicular to those of the main bodies of the galaxies. It is evident that this kind of structure could not have been formed if the galaxy had been assembled from mainly gaseous constituents because efficient angular momentum exchange would have lined up the angular momentum vectors of inner and outer parts. In other words, the merging constituents

R. Bender and R. L. Davies (eds.), New Light on Galaxy Evolution, 181–190.
© 1996 *IAU. Printed in the Netherlands.*

did already consist, to at least a significant fraction, of stars. Despite of kinematic decoupling, peculiar cores are in general stable, long-lived phenomena and so, unlike shells and ripples in the outer parts of galaxies, represent a long-term memory of the formation history of an object.

About 1/3 of nearby *luminous* ellipticals show peculiar core kinematics (Bender 1990). This fraction is, e.g., also found in the Virgo cluster: of the nine brightest Virgo ellipticals ($M_B < -20.5$, $H_o = 50\,km/s/Mpc$) three show kinematically decoupled cores (NGC 4365, 4406, 4472). Because of projection statistics it then follows that **more than 50% of all luminous ellipticals contain kinematically decoupled cores**. In most of them, the analysis of the line-of-sight velocity distributions shows that the peculiar cores are likely to be rapidly spinning thick disks or torus-like components (Franx and Illingworth 1988, Bender 1988, 1990, Rix and White 1992, Surma and Bender 1995). They formed dissipatively and must have involved substantial star formation (because of high v/σ and metallicities higher than in the main bodies), see Bender and Surma (1992), Davies et al. (1993), Surma and Bender (1995). These components have masses in the range $10^9...10^{10}\,M_\odot$ and radii of up to about one kiloparsec.

Most of the ellipticals with peculiar cores belong to the class of boxy ellipticals or ellipticals with irregular isophotes. There is only one elliptical known so far (NGC 1700, Franx et al. 1989) which has a counter-rotating core and *significantly disky* isophotes. Boxy ellipticals are in general more luminous than disky ellipticals (e.g., Bender et al. 1993b). Therefore the natural interpretation of these findings is that mergers that lead to counter-rotating peculiar cores will usually form luminous, mainly boxy, ellipticals. Of course, it is to be expected that pre-existing disks cannot survive major mergers (e.g. Quinn, Hernquist and Fullagar 1993) and, indeed, recent simulations by Steinmetz (1995) also show that dissipationless merging produces preferentially boxy isophotes outside the core. Consequently, the observed correlation between peculiar cores and boxy isophotes of the main bodies is very plausible.

A note of caution: The peculiar core kinematics in some ellipticals can possibly be explained in ways different from the above scenario. E.g., the decoupled core could be due to streaming in a triaxial figure, obliquely projected (Binney 1985, Franx et al. 1991, Statler 1994), or due to *dissipationless* merging with a small compact elliptical or a bulge dominated S0 (Kormendy 1984, Balcells and Quinn 1990). It seems clear however that these latter scenarios cannot account for the formation of the majority of ellipticals with peculiar cores. The reasons are that rotation amplitudes in the cores are in general too high and that core metallicities are enhanced with respect to the main body (Bender and Surma 1992). Similarly unlikely is the accretion of gas-rich irregulars because they simply do not contain enough gas to form a massive central component. Typically, a much larger amount of gas is needed, like the one found in massive spirals.

Finally, it is also noteworthy that ellipticals with peculiar cores are found in all environments, in rich clusters as well as in small groups; examples for the latter are NGC 5322 (Bender 1988), IC 1459 (Franx and Illingworth 1988), or NGC 3608 (Jedrzejewski and Schechter 1988).

3. The analogy between IRAS mergers and the formation of massive elliptical galaxies

A plausible formation scenario for ellipticals with kinematically decoupled cores can be sketched by inspecting the properties of ultraluminous IRAS galaxies and by N-body simulations of merging spirals.

Hernquist and Barnes (1991), Barnes and Hernquist (1996), and Barnes, this conference showed in their simulations that ellipticals with counter-rotating cores can originate in spiral-spiral mergers. The model spirals consisted of dark matter, stars and gas in the usual mix. While the stars undergo a process of violent relaxation during merging and form a smooth $r^{1/4}$ main body, the cold gas is efficiently transported to the center of the merger where it settles into a rapidly rotating thick disk or torus. This result of the merger simulations is consistent with the observations of the molecular gas distributions in IRAS mergers, e.g. NGC 520, Arp 220 or NGC 7252 (e.g. Sanders et al. 1988, 1991; Schweizer 1990; Kormendy and Sanders 1992). Both simulations and observations show that the molecular gas tori can be kinematically decoupled from the main bodies of the galaxies, i.e. their angular momentum vectors can be opposite or perpendicular to the ones of the main bodies. The molecular gas masses observed in the centers of luminous IRAS mergers are very similar to those of the counter-rotating cores ($10^9 - 10^{10}\,M_\odot$, Sanders et al. 1991); the same is true with respect to the radii which are typically of the order of a few hundred parsecs or smaller.

The high concentration of molecular gas in the center of the merger leads to violent star formation and forms a rotationally flattened central stellar component. This component is likely to be very metal-rich because the molecular clouds were pre-enriched and also because the IMF may be top-heavy in mergers (e.g., Wright et al. 1988, Bernlöhr 1993). In some of the IRAS mergers we can observe this process just now (e.g. NGC 520, Arp 220). Once the IRAS mergers have aged by about 5 Gyrs, the relics of the central starbursts are likely to resemble the decoupled cores observed in ellipticals today, both with respect to kinematics and metallicity (Bender and Surma 1992). In some mergers, newly formed stars may not only be found in the center but also at larger radii. These stars are due to star formation triggered in the early phases of the merging (Fritze-von-Alvensleben and Gerhard 1994) and may contribute to a smooth overall appearance of the line-strengths gradients after several Gigayears.

These considerations show that a *qualitative* understanding of the formation of ellipticals via *dissipative merging* can be reached in consistency with observations of present day mergers and N-body simulations. However, the plausible analogy between IRAS mergers and the formation of ellipticals does of course neither imply that ellipticals must have formed via merging of spirals nor that they formed late (i.e. at low z) in general. It is equally possible that *both* ellipticals and spheroids formed at higher redshifts by (hierarchical) processes involving *both* merging-induced violent relaxation and dissipation. The relative amount of dissipation varied as a function of luminosity and other protogalactic parameters (like density and environment) and determined the degree of anisotropy of the final object. More luminous ellipticals may on average have assembled from more evolved progenitors (in which most of the baryonic matter had already been transformed into stars) and, thus, not only velocity anisotropy but also kinematic de-coupling between core and main body may have been produced in these objects. The important parameter determining whether the final object would show peculiar kinematics and features in the line-strength gradients is the *ratio between star formation timescale and the timescale over which violent mergings occured*. For a more detailed discussion of these points see Bender and Surma (1992), Bender, Burstein and Faber (1992, 1994).

As the discussion of the next paragraphs will show, it is indeed indicated that most luminous ellipticals formed the bulk of their stars at relatively high redshifts and on rather short time scales.

4. The mean ages and star formation histories of massive elliptical galaxies

Despite the large variety of structural properties, the stellar populations of elliptical galaxies are to first order surprisingly homogenous. Colors and line-strengths are one-to-one correlated to such an extent that it makes sense to discuss their stellar populations in wholistic terms (e.g. Sandage and Visvanathan 1978, Burstein et al. 1988; Peletier 1989; Faber et al. 1992). The stellar populations of elliptical galaxies are very tightly related to their central velocity dispersions (σ) (e.g., Dressler et al. 1987, Bender, Burstein and Faber 1993a). It is important to note that there is no difference between ellipticals that appear to have 'normal' kinematics and ellipticals with kinematically decoupled cores.

From existing stellar population synthesis models (e.g., O'Connell 1986;

Bruzual and Charlot 1993; Worthey 1994) one can estimate the combined scatter in age and metallicity from the observed scatter in the Mg−σ relation. Bender, Burstein and Faber (1993) found for *luminous ellipticals* that the scatter in age and/or metallicity at a fixed σ must be smaller than 15%. This implies that, **for a given σ, luminous ellipticals cannot have formed continuosly over the Hubble time** (this is consistent with a recent analysis of Schweizer and Seitzer 1992).[1]

Similar constraints on the range of age and/or metallicity at a fixed σ were reached independently for Coma and Virgo cluster luminous ellipticals by Ellis and collaborators (e.g., Ellis 1992) on the basis of the (V-K)−σ correlation, and by Renzini and Ciotti (1993) and Ciotti et al. (this conference) on the basis of the small scatter in M/L perpendicular to the fundamental plane.

Further and independent information about the star formation history of ellipticals can be derived from their element abundance ratios. For *luminous* ellipticals, Peletier (1989), Faber et al. (1992) and Davies et al. (1993) found that Mg is overabundant relative to Fe[2]. Over a larger luminosity range, [Mg/Fe] seems to be correlated with velocity dispersion: faint ellipticals have [Mg/Fe]$\approx$ 0 while luminous ellipticals reach [Mg/Fe]$\approx$ 0.4 (Gonzalez 1993, Fisher, Franx and Illingworth 1995). Furthermore, Paquet (1994) could show that, in luminous ellipticals, other light elements like Na and CN are overabundant relative to Fe as well. Evidently, the enrichment of *massive* (high velocity dispersion) ellipticals was in general dominated by Supernovae II, as these are the only significant source of light elements. Supernovae Ia, for comparison, only provide iron peak elements, e.g., Truran and Thielemann (1986). Consequently, the enrichment history of luminous ellipticals differed significantly from the one of the Solar Neighborhood, see Faber et al. (1992), Matteucci and Greggio (1986), Truran and Burkert (1995).

Ellipticals with peculiar cores show the same overabundance in light elements as luminous ellipticals on average. Within the galaxies, the [Mg/Fe] overabundance is in general radially constant up to at least their effective radii (Davies, Sadler and Peletier 1993, Surma and Bender 1995, Paquet 1994). So far, we found in a sample of five peculiar core ellipticals only one object which had solar element ratios in the core and outer parts (NGC

---

[1] Low luminosity ellipticals show larger scatter around the mean Mg−σ relation and therefore may have a larger age spread. In fact, Gonzalez (1993) showed that low-luminosity Es ($M_T \approx -18$) seem to be systematically younger than giant Es ($M_T \approx -21$), see Faber et al. (1995) and Worthey, this conference. Note that this trend runs opposite to the one expected in a cold-dark-matter model (Kauffman et al. 1993).

[2] This relies mostly on measurements inside the half-light radius of the galaxies.

5322). Again, it is indicated that there is no distiction between 'normal' luminous ellipticals and ellipticals with kinematically decoupled cores.

The prevalence of Supernovae II and in turn the light element over-abundance in massive ellipticals can be due to the following effects: (a) a star formation time scale smaller than 10^9 years (SNI explode in significant numbers only after about 1 Gyr after star formation started, e.g. Truran and Burkert 1995), (b) a top heavy initial mass function, (c) a reduced frequency of binary stars (leading to fewer SNI events), (d) selective mass loss mechanisms that resulted in a more efficient loss of SNI elements. Option (c) and (d) are rather unlikely because one expects the binary frequency to be determined by the local process of star formation rather than by global galaxy properties and because selective mass loss processes are likely to work, if at all, only in low mass galaxies (see Gilmore and Wyse 1991). However, also option (b) does not work well, if the overabundance in ellipticals approaches [Mg/Fe]$\approx$ 0.4 dex. Both from the abundance analysis of SNII events as from the abundance pattern of Galactic halo stars (e.g. Fuhrmann, Axer and Gehren 1995), it is very likely that the [Mg/Fe] overabundance is produced by SNII alone without any significant input of Fe from SNI. Turning the argument around, this means that in most luminous ellipticals ([Mg/Fe]$\approx$ 0.3) one can allow for a only rather modest enrichment by SNIa events. The consequence is that **star formation time scales for the bulk of the stars in luminous ellipticals most likely were shorter than roughly 2Gyr.** A moderately top-heavy IMF and *significant* star formation extending over more than 2Gyr are unlikely to solve the overabundance problem because after 2Gyr the Fe enrichment via SNI would start to reduce [Mg/Fe] below the observed value. Note that these considerations do not only apply to the cores of ellipticals but for the bulk of their stars, since the [Mg/Fe] overabundane is similar at all radii (see above).

An important further conclusion can be drawn from these findings: since most present day spirals have gas-to-star ratios smaller than 0.2 and most stars show solar element ratios, **merging of objects similar to present-day spirals cannot produce objects similar to most present-day ellipticals.** However, some ellipticals (e.g. NGC 5322) have [Mg/Fe]$\approx$ 0 and could be late merger products.

5. Ages of ellipticals from their color and line-strength evolution with redshift

In order to constrain formation ages of ellipticals, Aragon-Salamanca et al. (1993) investigated the evolution of V-K colors of Brightest Cluster Members (BCM) up to $z = 0.9$. From the rather small color evolution they concluded that ellipticals have mostly formed at $z > 5$. Although selection effects and the strong dependence of colors on metallicity are major caveats in this analysis, the small color evolution is indeed an important indication for high ages of the bulk of the stars in luminous ellipticals.

Relatively high redshifts of formation are also indicated by the very small and almost non-measurable redshift evolution of the bright end of the galaxy luminosity function (Lilly, this conference), by the Tolman-test (Dickinson, this conference), and by the evolution of the mass-to-light ratios of ellipticals with redshift (Franx, this conference).

Yet another method to constrain the formation ages of ellipticals can be based on the evolution of the $Mg-\sigma$ relation with redshift (Bender, Ziegler and Bruzual 1996, Ziegler and Bender, this conference). Relative to the latter three methods, which measure luminosity evolution and depend on the slope of stellar initial mass function, the $Mg-\sigma$ method is virtually independent from the IMF and, in addition, relatively insensitive to selection effects. This method is likely to represent the least ambiguous test of the redshift evolution of elliptical galaxies.

Bender, Ziegler and Bruzual (1996) have measured the $Mg-\sigma$ relation for a sample of brightest cluster ellipticals at redshifts around 0.4 (see also the poster paper to this conference by Ziegler and Bender). There is clear evidence for the evolution of the stellar populations in ellipticals with redshift. At any given velocity dispersion, the strongest Mg absorption found at $z = 0.4$ is significantly weaker than at $z = 0$, typically by about $\Delta Mg_b \approx 0.4$ Å. Translating this difference into relative age differences using Worthey's (1994) models implies that the bulk of the stars in luminous ellipticals has indeed formed at $z > 2$ (note that this does *not* exclude mergers with *minor* star formation to happen at lower redshift).

This result is roughly consistent with the predicted mean ages of elliptical galaxies in a cold dark matter universe (Kauffmann 1995). However, the number of data points is still too small to allow a discrimation between the standard ($\Omega = 1$) cold dark matter model and a low density CDM model.

6. Conclusions

Most of the arguments given in the previous sections are still rather qualitative. Nevertheless, the following conclusions can be reached with reasonable confidence:

- Elliptical Galaxies formed from merging of massive progenitor objects that consisted partly of stars and partly of gas.
- The bulk of the stars in the majority of massive cluster ellipticals formed at redshifts above two.

- The star formation time scale for the bulk of the stars in massive ellipticals was most likely shorter than about two Gigayears.
- The high [Mg/Fe] overabundance of massive ellipticals excludes that they have formed from objects similar to present-day spirals.

These conclusions do not contradict the hypothesis that present-day dissipative mergers can form ellipticals. However, these late ellipticals are but a minority among the overall population of ellipticals. The above conclusions neither rule out the possibility of *minor* accretion or merging events taking plaece and leading to the E+A phenomenon. However, these minor events are unlikely to add a large fraction of mass to the already existing underlying old stellar population in ellipticals.

Acknowledgements: I thank David Burstein, Gustavo Bruzual, Sandy Faber, Ulrich Hopp, Guinevere Kauffmann, Alvio Renzini, Roberto Saglia and Bodo Ziegler for interesting discussions and comments.

References

Aragon-Salamanca, A., Ellis, R.S., Couch, W.J., Carter, D.: 1993, *Mon. Not. R. astr. Soc.* **262**, 764

Balcells, M., Quinn, P.: 1990, *Astrophys. J.* **361**, 381

Barnes, J., Hernquist, L.: 1996, *Astrophys. J.* , in press

Bender, R.: 1988, *Astr. Astrophys.* **202**, L5

Bender, R.: 1990, in *Dynamics and Interactions of Galaxies*, ed. R. Wielen, Springer Verlag Heidelberg, p.232

Bender, R., Surma, P.: 1992, *Astr. Astrophys.* **258**, 250

Bender, R., Burstein, D., Faber, S.M.: 1992, *Astrophys. J.* **399**, 462, BBF1

Bender, R., Burstein, D., Faber, S.M.: 1993a, *Astrophys. J.* **411**, 153, BBF2

Bender, R., *et al.*: 1993b, in *Structure,Dynamics and Chemical Evolution of Elliptical Galaxies*, ESO/EIPC workshop, eds. J. Danziger et al., European Southern Observatory, München

Bender, R., Burstein, D., Faber, S.M.: 1994, in *Panchromatic View of Galaxies*, eds. G Hensler et al., Editions Frontieres, Gif-sur-Yvette, p.99

Bender, R., Ziegler, B., Bruzual, G.: 1996, *Astrophys. J.* , in press

Bernlöhr, K.: 1993, *Astr. Astrophys.* **270**, 20

Binney, J.: 1985, *Mon. Not. R. astr. Soc.* **212**, 767

Bruzual, G, Charlot, S.: 1993, *Astrophys. J.* **405**, 538

Burstein,D., Davies, R.L., Dressler, A., Faber, S.M., Stone, R.P.S., Lynden-Bell, D., Terlevich, R., Wegner, G.: 1988, in *Towards Understanding Galaxies at Large Redshift*, eds. R.G. Kron & A. Renzini, Kluwer Dordrecht, p.17

Davies, R.L., Sadler, E., Peletier, R.: 1993, *Mon. Not. R. astr. Soc.* **262**, 650

Dressler, A., Lynden-Bell, D., Burstein, D., Davies, R.L., Faber, S.M., D., Terlevich, R., Wegner, G.: 1987, *Astrophys. J.* **313**, 42

Ellis, R.: 1992, in *The Stellar Populations of Galaxies*, IAU Symp. 149, eds. B. Barbuy & A. Renzini, Kluwer Dordrecht

Faber, S.M., Worthey, G., Gonzalez, J.: 1992, in *The Stellar Populations in Galaxies*, IAU Symp. 149, eds. B. Barbuy & A. Renzini, Kluwer Dordrecht

Faber, S.M., Trager, S., Gonzalez, J., Worthey, G.: 1995, in *Stellar Populations*, IAU Symp. 164, eds. P.C. van der Kruit & G. Gilmore, Kluwer Dordrecht

Fisher, D., Franx, M., Illingworth, G.: 1995, *Astrophys. J.* , in press

Franx, M., Illingworth, G.: 1988, *Astrophys. J.* **327**, L55
Franx, M., Illingworth, G., Heckman, T.: 1989, *Astrophys. J.* **344**, 613
Franx, M., Illingworth, G., de Zeeuw, T.: 1991, *Astrophys. J.* **383**, 112
Fritze-von Alvensleben, U., Gerhard, O.: 1994,*Astr. Astrophys.* **285**, 775
Fuhrmann, K., Axer, Gehren, T.: 1995, *Astr. Astrophys.* , in press
Gilmore, G., Wyse, R.F.G.: 1991, *Astrophys. J.* **367**, L55
Gonzalez, J.: 1993, PhD thesis, University of California at Santa Cruz
Hernquist, L., Barnes, J.E.: 1991, *Nature* **354**, 210
Jedrzejewski, R., Schechter, P.: 1988, *Astrophys. J.* **330**, L87
Kauffmann, G., White, S.D.M, Guiderdoni, B.: 1993, *Mon. Not. R. astr. Soc.* **264**, 201
Kauffmann, G.: 1995, *Mon. Not. R. astr. Soc.* , in press
Kormendy, J.: 1984, *Astrophys. J.* **287**, 577
Kormendy, J., Sanders, D.B.: 1992, *Astrophys. J.* **390**, L53
Matteucci, F., Greggio, L.: 1986, *Astr. Astrophys.* **154**, 279
O'Connell, R.: 1986 in *Stellar Populations*, eds. C. Norman et al. Cambridge University Press
Paquet, A.: 1994, PhD thesis, University of Heidelberg
Peletier, R.: 1989, PhD thesis, University of Groningen
Quinn, P.J., Hernquist, L., Fullagar, D.P.: 1993, *Astrophys. J.* **403**, 74
Renzini, A., Ciotti, L.: 1993, *Astrophys. J.* **416**, L49
Rix, H.-W., White, S.D.M.: 1992, *Mon. Not. R. astr. Soc.* **254**, 389
Sandage, A., Visvanathan, N.: 1978, *Astrophys. J.* **223**, 707
Sanders, D.B., Scoville, N.Z., Sargent, A.I., Soifer,B.T.: 1988, *Astrophys. J.* **324**, L55
Sanders, D.B., Scoville, N.Z., Soifer, B.T.: 1991, *Astrophys. J.* **370**, 158
Schweizer, F.: 1990, in *Dynamics and Interactions of Galaxies*, ed. R. Wielen, Springer Verlag Heidelberg, p.232
Schweizer, F., Seitzer, P.: 1992, *Astron. J.* **104**, 1039
Statler, T.S.: 1994, *Astrophys. J.* **425**, 500
Steinmetz, M.: 1995, in 'Galaxies in the Young Universe', ed. H. Hippelein, Springer Verlag, Heidelberg
Surma, P., Bender, R.: 1995, *Astr. Astrophys.* **298**, 405
Truran, J., Thielemann, F.: 1986 in *Stellar Populations*, eds. C. Norman et al. Cambridge University Press
Truran, J., Burkert, A.: 1995, in it Panchromatic View of Galaxies, eds. G Hensler et al., Editions Frontieres, Gif-sur-Yvette, p.389
Worthey, G.: 1994, *Astrophys. J.-Suppl. Ser.* **95**, 107
Wright, G.S., Joseph, R.D., Robertson, N.A., James, P.A., Meikle, W.P.S.: 1988, *Mon. Not. R. astr. Soc.* **233**, 1

DISCUSSION:

Dickinson: A Tolman-type analysis of ellipticals in clusters at $z \approx 0.4$ suggests somewhat less B-band luminosity evolution than you measure with the Mg$-\sigma$ relation. Do you know that the galaxies you observed at $z \approx 0.4$ are bona fide ellipticals?

Bender: Most of the objects we observed are in Abell 370 and have HST morphologies. For the other clusters we are currently taking HST images. However, I believe this does not matter very much, because at any given velocity dispersion, even the strongest lined ellipticals at $z \approx 0.4$ have weaker Mg absorption than present day ellipticals. Also note that the evolution as measured by a Tolman-type analysis is IMF dependent while the evolution as derived from Mg$-\sigma$ is not (see Section 5).

Djorgovski: Is there any systematic difference between field and cluster ellipticals in the Mg vs. Fe diagram?

Bender: Frankly, I do not know. But I guess, the effect cannot be very large, because otherwise it would be noticable in the $Mg-\sigma$ and/or $Fe-\sigma$ relation which have been checked for environmental dependencies.

Fritze-von Alvensleben: Damped $Ly\alpha$ systems at $z = 2...4$ contain 10^11 $M_\odot$ of gas, i.e., they contain already the mass of stars + gas in present day spirals. So, early mergers of massive spirals are possible, they will be accompanied by very strong starbursts and could from ellipticals.

Bender: No doubt. But if these early spirals are similar to present-day spirals in gas-to-star ratio and abundance pattern (i.e. $[Mg/Fe]\approx 0$), then they cannot be the progenitors of most luminous ellipticals (because these have $[Mg/Fe]\approx 0.4$). On the other hand, if you allow them to contain a much higher gas fraction than 0.2, then of course, it may work out. But in this case, I would not call these high z objects genuine spirals.

Fritze-von Alvensleben: The increase in $[Mg/Fe]$ in a starburst is a direct function of the burst strength. For strong bursts in gas-rich spirals the models I calculated with Ortwin Gerhard seem to indicate that it should well be possible to reach $[Mg/Fe]$ up to 0.5.

Bender: I agree. But you will reach $[Mg/Fe]$ of 0.5 only in the newly formed stars. Outside the core of the merger remnant these will constitute only a small fraction of the total stellar mass and, therefore, after a few Gigayears, the spectrum of the remnant's main body will not show a very high $[Mg/Fe]$ anymore.

DYNAMICAL EFFECTS OF MERGERS

J.E. BARNES
Institute for Astronomy, University of Hawai'i
2680 Woodlawn Drive, Honolulu, HI 96822, USA

1. Introduction

The bridges and tails of interacting galaxies were elegantly explained when
Toomre & Toomre (1972) showed that such features arise from tides acting
on disk galaxies. Emboldened by this success, Toomre & Toomre proposed
that certain twin-tailed fuzzballs without obvious interaction partners were
in fact the merged relics of interacting pairs, and that such relics eventually
become elliptical galaxies. Observations of twin-tailed systems found evidence for both their tidal origins and their elliptical destinies (*e.g.* Schweizer
1986), while self-consistent numerical simulations substantiated theoretical
predictions of rapid orbital decay and the elliptical-like outcome of violent relaxation during merging (*e.g.* Barnes 1988, and references therein).
But while our basic picture of merging seems solid, the precise role of this
process in galaxy evolution is not so clear. Dynamical studies provide some
insight into this issue; at present we can distinguish three levels of reliability
in the numerical work: solid results, good bets, and hopeful guesses.

2. Solid Results

Confidence in numerical models may be grounded in the realization that
self-consistent N-body simulation is essentially a Monte-Carlo technique for
solving the coupled collisionless Boltzmann and Poisson equations (White
1982, Hernquist & Barnes 1990). As with other Monte-Carlo methods, N-
body simulations yield results with errors scaling as $N^{-1/2}$; the steady
progress of computers enables experimenters to perform increasingly subtle
calculations.

Tidal models of interacting systems can often be built by combining "off-
the-shelf" components. Fig. 1 shows a self-consistent lookalike of Arp 252
produced by a parabolic encounter of two equal-mass galaxies, each con-

R. Bender and R. L. Davies (eds.), New Light on Galaxy Evolution, 191–198.
© 1996 IAU. Printed in the Netherlands.

sisting of a central bulge, an exponential disk, and a dark halo comprising 80% of the mass. Even without velocity data, such modeling can yield interesting insights; for example, to match the phase of the bar in the lower disk, the bulge must not be so compact and massive as to dominate the rotation curve at small radii. And as Stockton (1974) showed by checking Toomre & Toomre's (1972) reconstruction of The Mice, models matching the appearance of interacting systems can make qualitative predictions about velocities.

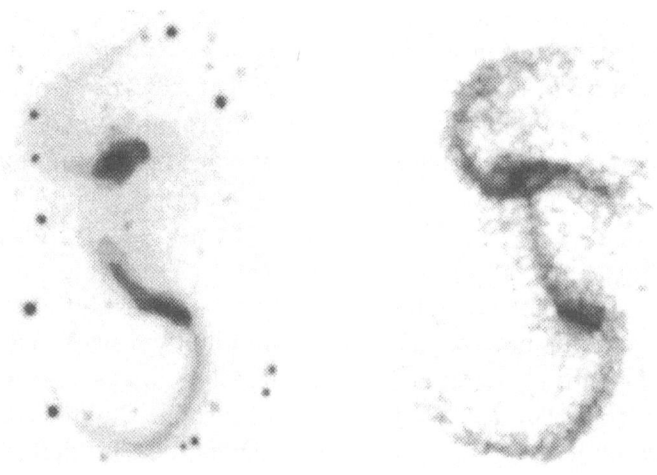

Figure 1. R-band image of Arp 252, courtesy Bill Keel (left) and a grey-scale representation of an *N*-body lookalike (right). The dark halos are not shown.

Strong encounters like this one, extrapolated into the future, merge after only one or two additional orbits. Tidal interactions between dark halos are largely to blame for such rapid orbit decays (White 1978); halos efficiently absorb orbital angular momentum from the luminous material (Barnes 1988, 1992). Indeed, close-passing disks can merge well before their tidal tails disperse, creating "twin-tailed" relics like NGC 7252 (Schweizer 1982, Hibbard *et al.* 1994), as a self-consistent model for that very galaxy illustrates (Hibbard & Mihos 1995).

Self-consistent simulations of equal-mass disk galaxy encounters have repeatedly produced merger remnants with some characteristics of normal ellipticals (Farouki & Shapiro 1982, Negroponte & White 1983, Gerhard 1983a, Barnes 1988, Hernquist 1993a). Violent relaxation due to orbit scattering by time-dependent gravitational fields tends to erase existing structures (Lynden-Bell 1967); in merger simulations, the generic result is a pressure-supported ellipsoidal hulk. And here as in other studies of violent relaxation (May & van Albada 1984), these hulks have cores with radii determined by the peak central phase-space densities of the initial galaxies, grafted onto luminosity profiles resembling a de Vaucouleurs (1948) law at larger radii.

The gravitational field of a merger generally settles down before particle orbits are fully randomized, so traces of progenitor structure *do* survive the process. Radial abundance and color gradients are likely to be only moderately reduced by dissipationless mergers since initial and final binding energies are fairly well-correlated (White 1980, Barnes 1988). More detailed simulations have shown that kinematic signatures of the initial disks can also survive merging (Gerhard 1983b, Barnes 1992, Hernquist 1993a). For example, these simulated merger remnants often exhibit large misalignments between their spin and minor axes, due to streaming of particles on major-axis tube orbits (Levison 1987); on the other hand, most elliptical galaxies seem to have smaller misalignments (Franx, Illingworth, & de Zeeuw 1991).

3. Good Bets

While gas-dynamical modeling does not yet inspire as much confidence as collisionless N-body simulation, calculations combining N-body and Smoothed Particle Hydrodynamic (SPH) methods are probably on the right track. SPH converges to the standard equations of motion for a compressible fluid in the $N \to \infty$ limit (*e.g.* Monaghan 1992, and references therein), although existing adaptive implementations do not get the gas thermodynamics quite right (Hernquist 1993b, but see Nelson & Papaloizou 1994). In any case, the actual dynamics of the ISM are vastly more complex than those of any foreseeable numerical model. But over kpc scales, rapid cooling and momentum conservation appear to be the key ingredients required for numerical simulations; SPH and "sticky-particle" codes both produce reasonable first approximations in this regime.

One outcome seen consistently in gas-dynamical models is rapid inflows of dissipative material in tidally disturbed disks (Noguchi 1988, 1991, Hernquist 1989, Combes, Dupraz, & Gerin 1990, Barnes & Hernquist 1991, 1996). In a disk with of *any* kind of non-axisymmetric structure, gravitational torques transfer angular momentum from dissipative to collisionless components (*e.g.* Simkin, Su, & Schwarz 1980, Carlberg & Freedman 1985, von Linden *et al.*, these proceedings). The structures produced in tidally-perturbed disks are quite effective at creating such torques, and the inflows which result may well explain the kpc-scale central starbursts seen in interacting disk systems (Bushouse 1987).

Indeed, gravitational torques and dynamical friction are so remarkably efficient at transporting spin and orbital angular momentum in dissipative

merger simulations that about half the entire gas inventory is promptly
concentrated within a central blob of radius comparable to the model's
spatial resolution (Negroponte & White 1983, Barnes & Hernquist 1991,
Noguchi 1991; also Fig. 2). Violent relaxation alone can't explain how so
much gas is repeatedly scattered *inward*; a general although rather qualitative explanation might be that the inward scattering of gas maximizes
the dissipation rate and consequently the production of entropy (Barnes &
Hernquist 1996).

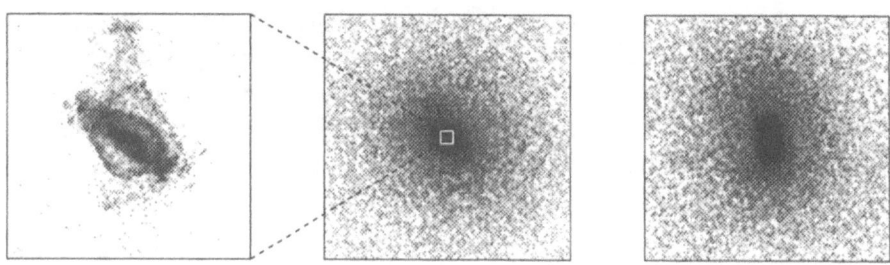

Figure 2.   Nuclear gas cloud (left) in simulated merger remnant (middle), and a pure
N-body version of the same calculation (right).

In terms of their masses and radii, these simulated gas blobs are comparable to the central clouds of molecular material detected in merger remnants like as NGC 520 (Sanders *et al.* 1988) and Arp 220 (Scoville *et al.*
1991). Although the coincidence of radii may be partly due to modeling
limitations, mergers do seem capable of assembling massive central clouds
on the short time-scales needed to fuel ultra-luminous starbursts (Larson
1987).

Gas driven to the center of a remnant by strongly time-dependent gravitational torques may become kinematically decoupled from the galaxy it
resides in. Out of four *N*-body/SPH simulations of equal-mass encounters,
one produced a striking example of such kinematic decoupling in the form
of a pair of counter-rotating gas disks (Hernquist & Barnes 1991, Barnes &
Hernquist 1996). Similar counter-rotating disks are found in the prototypical merger remnant NGC 7252 (Schweizer 1982), and the idea that such
structures can evolve into kinematically-decoupled cores is supported by
observations of steep metalicity gradients in such cores (Bender & Surma
1992, Davies, Sadler & Peletier 1993).

Detailed analysis of a few cases indicates that the central gas clouds
formed in simulated remnants can markedly affect the shape of the *stellar*
distributions of these systems, reducing overall flattening and favoring more
oblate structures (Barnes & Hernquist 1996; see also the right-hand panel
of Fig. 2). A similar result has been found in cosmological simulations (Katz

& Gunn 1991, Urdy 1993). This effect may be due to destabilization of box orbits in the deep potential wells produced by the central gas (Dubinski 1994). If this explanation is right, there should be a correlation between the initial central concentrations of the galaxies involved and the oblateness of the remnants they produce which could be studied in pure N-body simulations.

Gas which returns to the merger remnant at later times has too much angular momentum to reach the center; this material forms a warped disk extending to several effective radii (*e.g.* Barnes & Hernquist 1986). Similar disks are also seen in early-type galaxies like NGC 4753 (Steiman-Cameron, Kormendy & Durisen 1992). Star formation in well-settled gas disks could produce the vestigial *stellar* disks found in some elliptical galaxies (*e.g.* Bender 1988, Scorza & Bender, these proceedings).

4. Hopeful Guesses

Finally, there are some topics which are as yet mostly matters of conjecture. While it is not clear what role merging plays in galaxy evolution, it would be very puzzling if merger remnants were shown to evolve into objects categorically unlike real galaxies!

Star formation is at the core of many unsolved problems concerning effects of mergers. To begin with, there is the worry that star formation or subsequent supernovae might deplete or disperse the interstellar material before it can form clouds like the one shown in Fig. 2. Simulations are not a very effective way to address this worry, since star formation can only be crudely included; it may be more productive to use observations of such gas clouds in luminous IR galaxies (Scoville *et al.* 1991) to constrain the disruptive effects of star formation. A more subtle concern involves the *timing* of starbursts; here calculations show that the rate at which high-density gas accumulates is sensitive to the encounter parameters (Barnes & Hernquist 1996) and to the internal structure of the galaxies involved (Mihos & Hernquist 1994a). At present it is not clear if most merging galaxies go through brief but very intense starbursts, or if only mergers with rather special initial conditions attain high rates of star formation (Hibbard 1995).

A related worry concerns the distribution of the stars produced in a merger-induced starburst. Numerical simulations with star-formation rules have produced remnants with pronounced central luminosity spikes (Mihos & Hernquist 1994b). Such spikes are not evident in real elliptical galaxies; to be consistent with the observations, nuclear starburst populations should join smoothly onto the bodies of galaxies, without terribly obvious discontinuities in luminosity, color, or abundance. It is quite possible that feedback mechanisms not yet included in the simulations may be involved in matching nuclear starburst populations onto the galaxies they inhabit; the need for similar feedback mechanisms has long been recognized in hierarchical models for galaxy formation (White & Rees 1978; see Kaufmann, these proceedings).

The focus on mergers of equal-mass galaxies in existing simulations is partly due to computational limitations; unequal-mass encounters take longer to merge and require larger N to reveal more subtle effects. Shown in Fig. 3 is the result of a close retrograde passage of two disk galaxies with a 3:1 mass ratio; the larger disk has survived the merger – albeit with considerable heating – because of its relatively weak response to its retrograde companion. In a small sample of such 3:1 mergers, disk-like kinematics persist in about half of the remnants produced. This suggests that disk galaxies may sometimes accrete small companions without suffering too much heating. Yet mass ratios of 10:1 or more, which are probably more relevant for typical disk galaxies (Toth & Ostriker 1992), are still largely beyond the reach of systematic numerical studies (Walker, Mihos & Hernquist 1995).

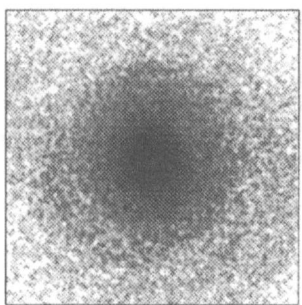 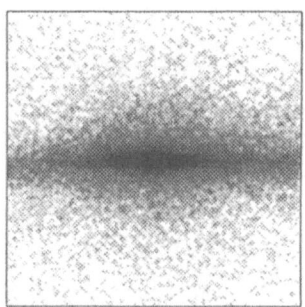

Figure 3. Face-on (left) and edge-on (right) views of an S0-like merger remnant produced by a 3:1 retrograde encounter.

Another topic yet scarcely touched concerns multiple mergers in groups and small clusters. It seems reasonable that such mergers should tend to smear or average out the kinematic signatures of individual disks, and this expectation is supported by the few simulations which have been performed (Barnes 1989, Weil & Hernquist 1994). It also seems likely that dissipationless mergers would disrupt the kinematically decoupled cores seen in many elliptical galaxies, although simulations have not yet been done to check this point. If multiple mergers do play an important role in the formation of cluster ellipticals, decoupled cores should be rare in such galaxies.

I thank Lars Hernquist for permission to discuss our unpublished work, and Frank van den Bosch and Susanne von Linden for help with the slides and video for this talk. The Pittsburgh Supercomputing Center and the Maui High Performance Computing Center provided computing time for this research. I acknowledge partial support from NASA grant NAG 5-2836.

References

Barnes, J.E. 1988. *Ap.J.* **331**, 699.
Barnes, J.E. 1989. *Nature* **338**, 123.
Barnes, J.E. 1992. *Ap.J.* **393**, 484.
Barnes, J.E. & Hernquist, L. 1991. *Ap.J.* **370**, L65.
Barnes, J.E. & Hernquist, L. 1996. *Ap.J.*, in press.
Bender, R. 1988. *A&A* **202**, L5.
Bender, R. & Surma, P. 1992. *A&A* **258**, 250.
Bushouse, H.A. 1987. *Ap.J.* **320**, 49.
Carlberg, R. & Freedman, W.L. 1985. *Ap.J.* **298**, 486.
Combes, F., Dupraz, C., & Gerin, M. 1990. In "Dynamics and Interactions of Galaxies", ed. R.Wielen (Springer: Berlin), p. 205.
Davies, R.L., Sadler, E.M., & Peletier, R.F. 1993. *MNRAS* **262**, 650.
de Vaucouleurs, G. 1948. *Ann. d'Astrophys.* **11**, 247.
Dubinski, J. 1994. *Ap.J.* **431**, 617.
Farouki, R.T. & Shapiro, S. 1982. *Ap.J.* **259**, 103.
Franx, M., Illingworth, G.D., & de Zeeuw, T. 1991. *Ap.J.* **383**, 112.
Gerhard, O.E. 1983a. *MNRAS* **202**, 1159.
Gerhard, O.E. 1983b. *MNRAS* **203**, 19P.
Hernquist, L. 1989. *Nature* **340**, 687.
Hernquist, L. 1993a. *Ap.J.* **409**, 548.
Hernquist, L. 1993b. *Ap.J.* **404**, 717.
Hernquist, L. & Barnes, J. 1990. *Ap.J.* **349**, 562.
Hernquist, L. & Barnes, J. 1991. *Nature* **354**, 210.
Hibbard, J.E. 1995. *PhD Thesis*, Columbia University.
Hibbard, J.E. *et al.* 1994. *A.J.* **107**, 67.
Hibbard, J.E. & Mihos, J.C. 1995. *A.J.* **110**, 140.
Katz, N. & Gunn, J.E. 1991. *Ap.J.* **377**, 365.
Larson, R.B. 1987. In "Starbursts and Galaxy Evolution", eds. T.X. Thuan, T. Montmerle, J.T.T. Van, (Editions Frontieres: Gif sur Yvette), p. 467.
Levison, H. 1987. *Ap.J.* **320**, L93.
Lynden-Bell, D. 1967. *MNRAS* **136**, 101.
May, A. & van Albada, T.S. 1984. *MNRAS* **209**, 15.
Mihos, J.C. & Hernquist, L. 1994a. *Ap.J.* **431**, L9.
Mihos, J.C. & Hernquist, L. 1994b. *Ap.J.* **437**, L47.
Monaghan, J.J. 1992. *ARA&A* **30**, 543.
Negroponte, J. & White, S.D.M. 1983. *MNRAS* **205**, 1009.
Nelson, R.P. & Papaloizou, J.C.B. 1994. *MNRAS* **270**, 1.
Noguchi, M. 1988. *A&A* **203**, 259.
Noguchi, M. 1991. *MNRAS* **251**, 360.
Sanders, D.B. *et al.* 1988. *Ap.J.* **324**, L55.
Schweizer, F. 1982. *Ap.J.* **252**, 455.
Schweizer, F. 1986. *Science* **231**, 227.
Scoville, N.Z. *et al.* 1991. *Ap.J.* **366**, L5.
Simkin, S.M., Su, H.J. & Schwarz, M.P. 1980. *Ap.J.* **237**, 404.
Steiman-Cameron, T.Y., Kormendy, J. & Durisen, R.H. 1992. *A.J.* **104**, 1339.
Stockton, A. 1974. *Ap.J.* **187**, 219.
Toomre, A. & Toomre, J. 1972. *Ap.J.* **179**, 623.

Toth, G. & Ostriker, J.P. 1992, *Ap.J.* **389**, 5.

Urdy, S. 1993. *A&A* **268**, 35.

Weil, M.L. & Hernquist, L. 1994. *Ap.J.* **431**, L79.

White, S.D.M. 1978. *MNRAS* **184**, 185.

White, S.D.M. 1980. *MNRAS* **191**, 1P.

White, S.D.M. 1982. In "Morphology and Dynamics of Galaxies", eds. L. Martinet & M. Mayor (Geneva Observatory: Sauverny), p. 291.

White, S.D.M. & Rees, M.J. 1978. *MNRAS* **183**, 341.

Discussion

Khachikian: I have a strong point of view: not all unusual, complex galaxies, especially those with complex nuclei, are results of merging. It is necessary to be careful. As an example you showed NGC 520 which is known as an Irr galaxy of the M 82 type and was studied in detail by many authors.

Barnes: I too would resist the tendency to label all galaxies with complex or unusual nuclear regions as mergers. But in NGC 520 the diagnosis is based *not* on the nuclear regions, but on outer features – including what seem to be tidal tails. It is very hard to explain such nonequilibrium features in an *old* stellar population without invoking a large-scale time-dependent gravitational field. Since NGC 520 has no nearby companions which could produce such a field, I believe this galaxy is merging.

Khachikian: I would like to ask a question more philosophical than astrophysical: do you first assume that a given galaxy is a merger, and then study it, or do you first study it and then conclude that it is a merged galaxy?

Barnes: If one studies a galaxy and finds that it has features characteristic of a merger, then merging is a reasonable *hypothesis* – to be accepted or rejected on the basis of further study.

Windhorst: Two questions: 1. What would change in your simulations if you added a 10^9 M$_\odot$ point source (AGN) to the center of each galaxy before their approach? 2. In ultradeep HST images, various groups have seen "tadpole"-like objects. Cowie *et al.* (1995 preprint) call them "chain-galaxies" and suggest they may be spirals in formation. To me they seem to be more dynamically disturbed objects in groups or perhaps distant clusters. What can they be dynamically? Can they form normal spirals?

Barnes: 1. I think the simulations may still be too crude to reliably show the dynamical effects of even 10^9 M$_\odot$ black holes. 2. I doubt that chain-galaxies can collapse to form normal disks – too much scope for angular momentum transport.

THE BUTCHER-OEMLER EFFECT IN NEARBY CLUSTERS

R.M. SHARPLES
University of Durham
Department of Physics, South Road, Durham, UK.

1. Introduction

Probably the most striking evidence for galaxy evolution at recent epochs has been the discovery of a rapid change in the nature of galaxy populations in clusters over the redshift range z=0 to z=0.5. The 'classical' Butcher-Oemler effect (Butcher & Oemler 1978, 1984) used photomteric studies to reveal an unexpected increase in the fraction of *blue* galaxies in the cores of distant (z~ 0.4) rich concentrated clusters when compared with nearby (z< 0.05) clusters of similarly high richness and central concentration. An alternative view, based on spectroscopic studies (Dressler & Gunn 1982; Couch & Sharples 1987), manifests itself as an increase in the fraction of *active* galaxies which show signs of recent star formation and/or nuclear activity. Some of these galaxies are indeed blue but some (e.g. in Cl0016+16, Dressler & Gunn 1992) are red. Although it is the very **absence** of blue galaxies in nearby clusters which defines the classical Butcher-Oemler effect, comparable spectroscopic studies of nearby cluster populations with the appropriate completeness and high signal-to-noise required for population (as opposed to dynamical) studies have only recently been undertaken. In at least one case these have revealed unexpected similarities to the spectroscopic signatures which appear so prevalent at higher redshifts.

2. Environmental Effects on E/S0 Galaxies in Nearby Clusters

Early-type (E/S0) galaxies are conventionally viewed as old stellar systems which formed the bulk of their stars in the first 1-2 Gyrs after their formation epoch. However, there is now a growing body of evidence to suggest that recent star formation has occured in at least *some* nearby E/S0 galaxies (O'Connell 1980; Rose 1985, Davies 1995). Bower *et al.* (1990) used gravity-sensitive spectral indices in the blue spectral region to show that star formation in early-type galaxies in the cores of nearby rich clusters was

R. Bender and R. L. Davies (eds.), New Light on Galaxy Evolution, 199–202.

truncated at an earlier epoch (on average) than similar galaxies in the field. A important handle on the question of whether such differences are primarily the result of nature (formation mechanism) or nurture (environmental effects) can be obtained by studying the early-type galaxy population in the outer parts of rich clusters where the environment is intermediate between the dense core regions and that of the field.

3. Star Formation in Early-Type Galaxies in the Coma Cluster

Some years ago we began such a study of the Coma cluster with the goal of studying distribution of E/S0 galaxies in the SrII/Hδ/FeI plane (Rose 1985) at a projected distance of R$\sim 1\,h^{-1}$ Mpc from the cluster centre. Since no deep wide-area morphological classification surveys were available, the sample was selected primarily by isolating early-type galaxies using the colour-magnitude sequence and subsequently rejecting any objects which showed evidence of spiral structure. Spectra have now been obtained for 184 galaxies in 3 fields.

3.1. SPATIAL DISTRIBUTION

The surprising result to emerge from the first two fields analysed by Caldwell *et al.* (1993) was that a significant fraction (20/125) of the objects studied had abnormal spectra for early-type galaxies, exhibiting either emission-lines or strong Balmer absorption lines similar to the so-called 'post-starburst' galaxies (Dressler & Gunn 1982) in the distant Butcher-Oemler clusters. The great majority of these unusual spectra (15/20) occured in galaxies located in the SW field but, since only 2 of the 3 fields could be observed on this first run due to weather, it was unclear whether this effect was a real radial gradient in the galaxy population of the cluster or due to some special conditions in the SW field. We have now obtained spectra for galaxies in the NE field and the fraction of abnormal spectra is again low (similar to that for the central field) thus confirming that the effect cannot be primarily due to a radial gradient (see Fig. 1)

The first indications of what might give rise to the peculiar nature of the SW field came with the publication of the ROSAT wide-field X-ray map which showed a secondary peak of X-ray emission, probably associated with a group of galaxies surrounding the cD galaxy NGC 4839 which lies near the centre of the SW field. This association fits nicely with a picture in which the star formation activity in the Coma galaxies (and by implication the activity seen in higher redshift clusters) was triggered by the interaction of a bound subclump of galaxies with the main cluster. More detailed analysis of the velocity distribution and the X-ray morphology (Burns *et al.* 1994) in fact favours a model in which the subclump has already passed through

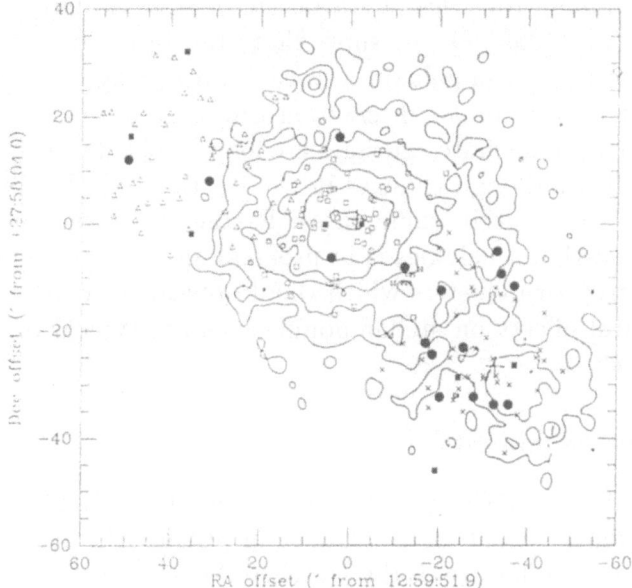

Figure 1. Spatial distribution of Coma E/S0 galaxies overlaid on a ROSAT X-ray map from Briel *et al* (1992). Spectra have been obtained in three 40 arcmin diameter fields using the Hydra fibre spectrograph at KPNO. Normal spectrum galaxies are denoted by open symbols, abnormal galaxies by filled symbols.

the cluster core where the effects of the dense intra-cluster medium (Evrard 1991) or frequent galaxy interactions (Moore 1995) may have triggered the recent star formation activity.

3.2. SPECTRAL ANALYSIS

The majority of the abnormal spectra in the SW clump exhibit strong Balmer lines in addition to the usual metallic features (CaII, G band, Mgb) expected for an old stellar population. Their spectra are reminiscent of the 'E+A' galaxies found in abundance in the cores of distant Butcher-Oemler clusters but with generally redder colours. The most direct analogues are the class of 'Hδ-strong' galaxies identified in three z=0.31 clusters by Couch & Sharples (1987).

Because the galaxies are relatively nearby, the Coma sample gives us an opportunity to study in unprecedented detail an environmental effect which is probably closely related to the Butcher-Oemler phenomenon. Two of the most obvious questions are when did the star-formation activity which produces the strong Balmer-line objects take place, and was this activity confined to the nuclear regions of the galaxies or was it a global phenomenon ? Leonardi & Rose (1995) present a new technique for accurately determining the ages of starbursts in post-starburst galaxies which Caldwell *et al.* (1995) use to show that star-formation bursts in a small subsample of the Coma galaxies ceased 1-1.5 Gyr ago, consistent with the dynamical estimates of the interaction timescale from Burns *et al.* (1994). The remnant intermediate age stars are found to be distributed over a large range in radius although changes in the fraction of light coming from this population is evident in some cases. Further work on the nature of the Coma post-starburst galaxies is in progress along with comparable studies of other nearby rich clusters which should reveal the extent to which these environmental effects on stellar populations are typical of clusters at the present epoch.

4. Acknowledgements

I am grateful to my colleagues Jim Rose, Nelson Caldwell and Richard Ellis for permission to use data in advance of publication and for many useful discussions on the nature of the Butcher-Oemler effect.

References

Bower, R.G., Ellis, R.S., Rose, J.A. and Sharples, R.M. (1990) *AJ*, **Vol. no. 99**, p. 530
Briel, U.G, Henry, J.P. and Bohringer, H. (1992) *A&A*, **Vol. no. 259**, p. L31
Burns, J.O., Roettiger, K., Ledlow, M. and Klypin, A. (1994) *ApJ*, **Vol. no. 427**, p. L87
Butcher, H.R. and Oemler, A. (1978) *ApJ*, **Vol. no. 219**, p. 18
Butcher, H.R. and Oemler, A. (1984) *ApJ*, **Vol. no. 285**, p. 426
Caldwell, C.N., Rose, J.A., Franx, M., and Leonardi, A. (1995) *Preprint*
Caldwell, C.N., Rose, J.A., Sharples, R.M., Ellis, R.S., and Bower, R.G. (1994) *AJ*, **Vol. no. 106**, p. 473
Couch, W.J. and Sharples, R.M. (1987) *MNRAS*, **Vol. no. 229**, p. 42
Davies, R.L. (1995) in, *New light on Galaxy Evolution*, IAU **171**, p. XXX.
Dressler, A. and Gunn, J.E. (1982) *ApJ*, **Vol. no. 263**, p. 533
Dressler, A. and Gunn, J.E. (1992) *ApJSupp*, **Vol. no. 78**, p. 1
Evrard, A.E. (1991) *MNRAS*, **Vol. no. 248**, p. 8P
Leonardi, A. and Rose, J.A. (1995) *Preprint*
Moore, B. (1995) in, *New light on Galaxy Evolution*, IAU **171**, p. XXX.
O'Connell, R. (1980) *ApJ*, **Vol. no. 236**, p. 430
Rose, J.A. (1985) *AJ*, **Vol. no. 90**, p. 1927

DISCUSSION

LIU: What is the fraction of abnormal galaxies in the SW clump and in Coma as a whole, and how does this compare with the active galaxy fraction in more distant B-O clusters ?

SHARPLES: In the SW clump it is 15/64 or 23%. Over all three fields this drops to 14%. In the distant clusters it is $\sim$ 20% based on either the blue fraction or the number of post-starburst spectra.

DYNMAMICAL EFFECTS ON GALAXIES IN CLUSTERS

BEN MOORE, NEAL KATZ AND GEORGE LAKE

Department of Astronomy
University of Washington
Seattle, WA 98195, USA

1. INTRODUCTION

For nearly 20 years, we've known that clusters at $z \gtrsim 0.3$ have a substantial population of "blue galaxies" seen only as fuzzy blobs in ground based images (Butcher & Oemler 1978, 1984). Hubble Space Telescope (HST) images reveal that these "fuzzy blue blobs" are low luminosity, often disturbed, spiral galaxies "Sp" (Dressler *et al* 1994a,b, Couch *et al* 1994). Today, rich galaxy clusters are dominated by elliptical "E" and lenticular "S0" galaxies (Dressler 1980), mostly low luminosity dwarfs.

A successful model for the "Butcher-Oemler effect" has three requirements: 1) a mechanism for creating disturbed galaxies with enhanced star formation, 2) a cosmological context that explains why the mechanism operates most efficiently at $z \sim 0.4$, and 3) an identification of the remnants of the distorted blue galaxies in clusters today. Kauffman (1995) showed that hierarchical clustering models produce an enhancement at $z \sim 0.4$ given a mechanism that operates when a spiral first enters a cluster, handling the second requirement. Several mechanisms have been examined qualitatively: mergers (Icke 1985, Miller 1988), compression of gas in the high pressure cluster environment (Dressler & Gunn 1983, Evrard 1991) and tidal compression by the cluster (Byrd and Valtonen 1990, Valluri 1993). Each of these mechanisms can produce starbursts, but none address morphological evolution and identify remnants. By analyzing their HST images, Oemler *et al* (1995) conclude that merging is implausible as the blue galaxy fraction is large and the merging probability is low. They observed disturbed spirals throughout the cluster, whereas both ram pressure stripping and global tides will only operate efficiently near the cluster's center.

203

R. Bender and R. L. Davies (eds.), New Light on Galaxy Evolution, 203–206.
© *1996 IAU. Printed in the Netherlands.*

2. GALAXY HARASSMENT AND CLUSTER EVOLUTION

Although direct mergers are extremely rare, every galaxy experiences a high speed close encounter with a bright galaxy once per Gyr. The masses of bright galaxies in clusters will determine the havoc wreaked by these encounters. Galaxies in the field have massive dark halos, but there was speculation that these were stripped from individual galaxies within clusters (c.f. White and Rees 1978). Recently, Moore, Katz and Lake (1995) examined all of the physical processes that strip mass from galaxies within clusters. All galaxies are tidally limited by the potential field of the cluster. Over the 5 Gyr life of a cluster of galaxies, they find that bright galaxies retain most of the mass within their tidal radius (defined at the pericenter of their orbit), the rest being liberated by fast encounters with other bright galaxies. Hence, an L_* galaxy in a rich cluster with a pericenter of 300 kpc will have a total mass of $\sim 4 \times 10^{11} M_\odot$. A rapid encounter with such a galaxy causes a tidal compression of the stellar and dark components. While these encounters are the cause of the stripping of mass from the dark halo (considered self-consistently throughout this work), their effect on small galactic disks is even more dramatic. The differential impulse violently redistributes the orderly motions of gas into non-circular intersecting orbits that promote star formation.

As clusters form, the gas rich disks of newly infalling spirals experience fast fly-by collisions when they enter the dense cluster environment. They are strongly perturbed resulting in disturbed morphologies and rapid bursts of star formation. To distinguish this from other collisional effects such as galaxy mergers and galaxy cannibalism, we refer to this process as "galaxy harassment". We use numerical simulations to follow the evolution of a small bulgeless spiral galaxy orbiting a dense cluster modeled on Coma. Our simulations use smoothed particle hydrodynamics (TREESPH, c.f. Hernquist & Katz 1989) to evolve the gas component of the disk at resolutions of 100 - 500 pc. Further details are found in MKL95.

We simulated galaxies on circular and elliptical orbits in smooth cluster potentials before examining the effects of harassment. The disk galaxy shows little evolution over 5 Gyrs when placed on a 450 kpc circular orbit in a smooth cluster potential. The disk becomes bar unstable after the first pericentric passage (c.f. Byrd and Valtonen 1990). Thereafter, each time the galaxy passes through pericenter, the halo loses a small fraction of its mass but stars and gas remain bound. The most dramatic evolution occurs when we include the fast fly-by collisions of other galaxies. In these simulations, the galaxy has strong encounters throughout its evolution. The first encounter leads to a pronounced bar instability, but continued heating of the disk by perturbers results in a variety of new effects: morphological transformation, strong flow of gas into the center of the galaxy and creation of stellar/gaseous debris arcs.

One observational puzzle has been the ubiquity of disturbed galaxies with no sign of current interaction (Dressler *et al* 1994a). This feature is clearly seen in our simulated images. Over the course of 3 Gyr, the closest approach of any encounter is more than 30 kpc away. Since the relative velocity of strong encounters is $\sim 1,500$ km s^{-1}, and the velocity impulse internal to the galaxy is only $\lesssim 50$ km s^{-1}, the perturbing galaxy moves ~ 100 kpc by the time that disk's response is visible. The bulk of the evolution is driven by $\lesssim$ five strong encounters with galaxies brighter than $\sim L_*$. As a result, the evolution is chaotic: whereas one fragile disk galaxy can avoid strong encounters for a few Gyr, another may be completely destroyed.

After several strong enounters, the loss of angular momentum to their own dark halos and the perturbing galaxies, combined with impulsive heating, leads to a prolate figure supported equally by random motions and rotation. The gas sinks to the very center of the galaxy and the stellar distribution is heated to the extent that it closely resembles a dwarf elliptical, although some retain very thick stellar disks and would be classed as dwarf lenticulars. Encounters cease to create sharp distortions and fail to remove any more material from the compact remnant.

Below L_*, two distinct classes of elliptical galaxies are observed. Low luminosity Es with high central surface brightness are a rare extension to the sequence of bright ellipticals; the archetype is M32. The most numerous galaxies in clusters are in a second class of dwarf ellipticals, also known as dwarf spheroidals (dE/dSph) - at least 3 magnitudes below L_*. Their exponential surface brightness profiles resemble those of spirals, as does the correlation of their low central surface brightnesses with total luminosity (Kormendy 1985).

The final stellar systems have a large degree of rotational support, surface density profiles and shapes that are in good agreement with observations (Ferguson & Binggeli 1994 and references therein). The observed stellar populations of dE galaxies implies recent star formation activity that can easily be understood in our model as a result of recent encounters with cluster galaxies. Harassed Sd spiral galaxies undergo a remarkable transformation between morphological classes without any merging taking place. Their dynamical states can account for all of the dissimilarities between dwarf elliptical and normal elliptical galaxies. Harassment provides the link between the dominant populations of galaxies in clusters at $z \sim 0.4$ and the present-day.

Discussion

G. Illingworth *It would be very interesting to correlate the angular momentum and energy changes in the test galaxy with the perturber relative velocity to see if the slower encounters were those causing the biggest changes.*

If the encounters are all in the impulsive regime then the global energy change is proportional to $M_{pert}^2/v_{pert}^2/b_{impact}^4$, although the response in the disk will depend on other factors.

B.M. Poggianti *Could you outline the main similarities and differences between your "harassment" simulations and those that Barnes showed before? In particular, could you distinguish observationally between harassment and merging?*

Barnes's merger simulations are between bound galactic systems, whereas rapid "fly-by" encounters drive the morphological evolution in clusters, where galaxies rarely come within 30 kpc of each other. Preliminary work indicates that the harassed remnants are prolate and have a large degree of rotational support, unlike merger remnants which are typically oblate and pressure supported.

Acknowledgments   This research was funded by NASA through the LTSA and HPCC/ESS programs.

References

Byrd G. & Valtonen M. 1990, *Ap.J.*, **350**, 80-94.
Butcher H. & Oemler A. 1978, *Ap.J.*, **219**, 18-33.
Butcher H. & Oemler A. 1984, *Ap.J.*, **285**, 426-38.
Couch W.J., Ellis R.S., Sharples R. & Smail I. 1994, *Ap.J.*, **430**, 121-38.
Dressler A. 1980, *Ap.J.*, **236**, 351-65.
Dressler A. & Gunn J.E. 1983, *Ap.J.*, **270**, 7-19.
Dressler A, Oemler A., Butcher H. & Gunn J.E. 1994a, *Ap.J.*, **430**, 107-20.
Dressler A, Oemler A., Sparks W.B. & Lucas R.A. 1994b, *Ap.J.Lett.*, **435**, L2
Evrard A.E. 1991, *M.N.R.A.S.*, **248**, 8p-10.
Ferguson H.C. & Binggeli B. 1994, *Astronomy and Astrophysics Review*, **6**, 67
Hernquist L. & Katz N. 1989, *Ap.J.Supp.*, **70**, 419-46.
Icke V. 1985, *A.A.*, **144**, 115-23.
Kauffmann G. 1995, *M.N.R.A.S.*, **274**, 153-60.
Kormendy J. 1985, *Ap.J.*, **295**, 73.
Miller R.H. 1988, *Comment. Astrophys.*, **13**, 1-11.
Moore B., Katz N. & Lake G. 1995, *Ap.J.*, in press.
Oemler A. 1995, This proceedings.
Valluri M. 1993, *Ap.J.*, **408**, 57-70.
White S.D.M. & Rees M. 1978, *M.N.R.A.S.*, **183**, 341.

HIGH REDSHIFT GALAXIES

THE EVOLUTION OF FIELD GALAXIES

S.J. LILLY
University of Toronto, Toronto, Canada

O. LE FEVRE AND F. HAMMER
Observatoire de Paris, Meudon, France

D. CRAMPTON
Dominion Astrophysical Observatory, Victoria, Canada

D.J. SCHADE AND J.D. HUDON
University of Toronto, Toronto, Canada

AND

L. TRESSE
University of Cambridge, Cambridge, UK

1. Introduction

During the late 1980's, successively deeper redshift surveys carried out with multi-object spectrographs on 4-m class telescopes produced growing evidence for evolution in the galaxy population. While some evolution ha^d been expected from analysis of the galaxy number counts, the surpris^i indication from the first deep redshift surveys was that this appear^r involve moderate luminosity galaxies lying at moderate redshifts (hurst *et al.* 1988, Colless *et al.* 1990, Cowie *et al.* 1991). However. results were suggestive, these early surveys suffered a number c^r problems that hampered their interpretation:

(a) the samples were small, especially at the faintest levels.
   weight was limited and analysis was based on crude
   of the data such as the median redshift of sample^r

(b) the typical redshifts were small ($z << 0.5$), so t'
   could only be seen against "local" populations
   quite different - indeed the local luminosity fι
   poorly defined (Loveday *et al.* 1992, Marzke ·

R. Bender and R. L. Davies (eds.), New Light on Galaxy Evolution, 209–21
© *1996 IAU. Printed in the Netherlands.*

(c) the samples were selected in the observed B-band, so that comparison with local samples was based on the poorly constrained ultraviolet properties and relative numbers of galaxies of different types.

As reviewed by Koo and Kron (1992), the evidence for there being *any* evolution was thus not completely watertight, and as recently as 1993, Koo *et al.* (1993) were able to produce unevolving models for the galaxy population which were consistent with the available B-band faint galaxy data.

In order to improve on this situation, two independent French and Canadian programs (Tresse *et al.* 1993, Lilly *et al.* 1993) merged into the Canada-France Redshift Survey (CFRS). We adopted the ambitious observational goal of securing spectroscopic identifications for 1000 I-band selected objects with $17.5 < I_{AB} < 22.5$. The I-band selection was chosen to correspond to selection at rest-wavelengths between B and V for $0.5 < z < 0.9$, and the depth was chosen to yield a median redshift $< z > \sim 0.6$, corresponding to a look-back time of 50% of the age of the Universe (for $\Omega = 1$).

1.1. THE CANADA-FRANCE REDSHIFT SURVEY

The data and methods of the Canada-France Redshift Survey (CFRS) have been described in detail elsewhere (Lilly *et al.* 1995a, Le Fevre *et al.* 1995a, Lilly *et al.* 1995b, Hammer *et al.* 1995a, Crampton *et al.* 1995). The final "statistically-complete" sample consists of 943 objects: 591 galaxies with secure redshifts $0.0 < z < 1.3$ and a median $< z > = 0.55$, 6 quasars, 200 stars and 146 objects (believed to be nearly all galaxies) that were unidentified or had insecure spectroscopic identifications.

The CFRS was designed to allow three separate approaches to be taken to studying the evolution of galaxies and large scale structure over the last 2/3 of the Hubble time. Specifically, we designed the program:

(a) To define the basic statistical properties of the galaxy *population*, such as the luminosity function $\phi(L, colour, z)$ and the spatial correlation function, $\xi(r, z)$, over the redshift range $0 < z < 1$, using samples selected in as similar a way as possible to local samples of galaxies.

(b) To produce large well-defined sub-samples of the galaxy population which could then be studied in more detail (spectroscopically, morphologically, kinematically) to reveal the most important physical processes occurring in galaxies at earlier epochs, and thus to link the evolution of *individual* galaxies to the changes seen in the *population* as a whole.

To relate faint field galaxies (the "normal" population) to objects selected in deep surveys at other wavelengths (e.g. the radio, infrared and millimeter wavebands. As an example, Hammer *et al.* (1995b) discuss he identifications of a large well-defined sample of μ-Jy radio sources.

2. Statistical descriptions of the galaxy population at high redshift

2.1. THE LUMINOSITY FUNCTION

Fig 1 shows the luminosity function $\phi(L, colour, z)$ over the interval $0 < z < 1.3$ (computed in the rest-frame B-band for $q_0 = 0.5$ and $H_0 = 50$ kms^{-1}Mpc^{-1}). The most remarkable feature of this diagram is the clear difference between the evolution of the red and blue populations of galaxies (split at the colours of the Coleman *et al.* (1980) Sbc spectral energy distribution). It should be stressed here that the luminosity function, of course, reflects changes in the *population* and not (neccessarily) changes in *individual* objects.

The luminosity function of the population of red galaxies shows virtually no change (for $\Omega = 1$) all the way back to $z \sim 1$. This suggests that the population of massive, reasonably quiescent galaxies, was to a very large degree "in place" at $z \sim 1$. Interestingly, we see no evidence for significant luminosity evolution (of order 0.8 magnitudes) that might be expected from the passive evolution of the stellar population. However, the expected passive luminosity evolution depends sensitively on the slope of the stellar initial mass function as well as the age (Tinsley and Gunn, 1976). Changes of a few tenths of a magnitude in L* of the red population back to $z \sim 0.8$ are certainly not ruled out (Lilly *et al.*, 1995c) and adoption of a lower value of q_0 would also imply greater evolution. An improved estimate of the expected passive luminosity evolution would greatly help in the interpretation of the red luminosity function.

The luminosity function of bluer galaxies shows a substantial enhancement at $z > 0.5$ in the numbers of galaxies with roughly present-day L* ($M_{AB}(B) \sim -21.0$). This evolving blue population, now clearly revealed in the luminosity function is responsible for steepening the galaxy number counts down to $B \sim 24$. The observed changes in the blue luminosity function between $z \sim 0.3$ and $z \sim 0.7$ could be due to either (a) a uniform brightening of all blue galaxies by about 1.5 magnitudes, (b) an increase in the number density of galaxies of constant luminosity by a factor of three, (c) a combination of both effects, or (d) an even more complicated scenario involving galaxies crossing the red-blue divide. Further information on the environments, morphologies, kinematics, and stellar populations of the galaxies (see below) should allow us to discriminate between these scenarios and will lead to an understanding of the physical evolution of individual galaxies as well as the statistical evolution of the population as a whole.

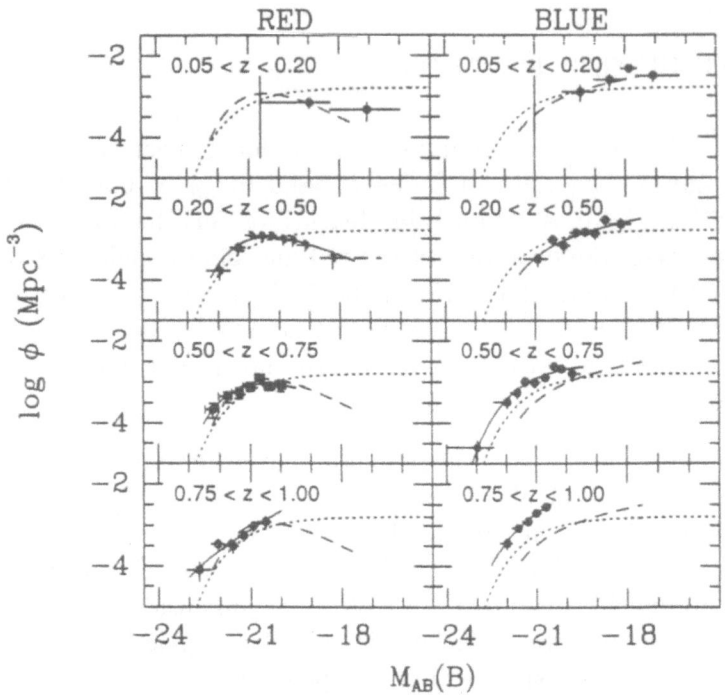

Figure 1. The CFRS luminosity function $\phi(L, colour, z)$ split by intrinsic colour and redshift (computed for $H_0 = 50$ kms^{-1}Mpc^{-1} and $q_0 = 0.5$). This shows that the luminosity function of red galaxies changes very little over the interval $0 < z < 1$, while the luminosity function of blue galaxies changes significantly at $z > 0.5$. See Lilly *et al.* (1995c) for details.

2.2. THE CLUSTERING CORRELATION FUNCTION

The evolving amplitude of the correlation function $\xi(r, z)$ reflects both the growth of large scale structure and the nature of the faint galaxy population, since the population at high redshift may be differently clustered (intrinsically) to that at low redshift. The strength of galaxy clustering at high redshift has been studied with the CFRS in two ways: (a) from the redshift catalogue itself (Le Fevre *et al.*, 1996) using the $w(r_p)$ formulism (Davis and Peebles, 1983) and (b) by using the observed overall $N(z)$ to invert the projected two-dimensional correlation function $w(\theta)$ obtained for much larger samples of galaxies (Hudon and Lilly, 1996). These two approaches give consistent results (see Fig 2).

The physical correlation length $r_0(z)$ that is measured (Fig 2) may be parameterized as

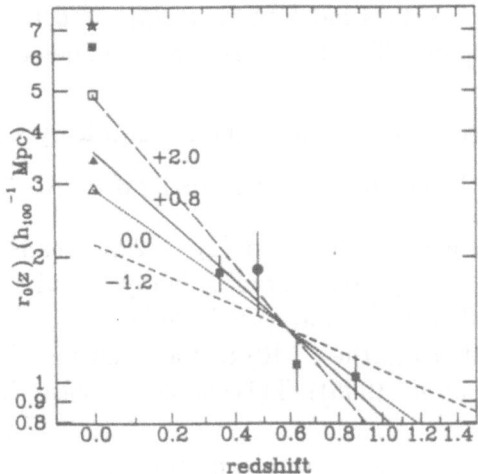

Figure 2. The scale length of the correlation function, $r_0(z)$ at high redshift as derived directly from the CFRS sample (solid squares) and from the projected $w(\theta)$ and the CFRS N(z) (solid disk) (computed for $H_0 = 100$ kms^{-1}Mpc^{-1} and $q_0 = 0.5$). The diagonal lines indicate various evolutionary models parameterized and labelled by the ϵ parameter. If the faint galaxy population is to evolve into the local population of massive galaxies seen in the CfA or IRAS samples, then evolution with $\epsilon \geq +0.8$ is required. See Le Fevre *et al* (1996) for details.

$$r_0(z) = r_0(0)(1+z)^{-(3+\epsilon)/\gamma} \qquad (1)$$

with $\gamma = 1.8$ as the slope of the power-law correlation function. If clustering does not grow (i.e. is fixed in comoving space) $\epsilon = \gamma - 3 \sim -1.2$. A scenario in which clustering is represented as bound systems in an expanding Universe has $\epsilon = 0$, while the *mass-distribution* in CDM-like hierarchical models has $\epsilon \sim +0.8$ on these scales.

If, as seems reasonable (from the luminosity function results, above, and the imaging results, below), the galaxy population at high redshift evolves into something like the local CfA sample, with $r_0 \sim 5.5h_{100}^{-1}$ Mpc (Davis and Peebles, 1983), or the 60-μm-selected IRAS sample of star-forming galaxies, with $r_0 \sim 3.7h_{100}^{-1}$ Mpc (Fisher *et al.*, 1994) then we require, for $q_0 \sim 0.5$, strong growth of clustering, with $+0.8 < \epsilon < +2.0$. A scenario involving no growth of structure would produce a local population that would be much more weakly clustered than *any* local population. For $q_0 \sim 0$, less evolution is required and evolution to an IRAS-like sample is consistent with the $\epsilon = 0$ "stable clustering" growth case.

An important result (Le Fevre *et al.*, 1996) is the demonstration that the red and blue populations at $z > 0.5$ have very similar auto-correlation (and also cross-correlation) function amplitudes. Thus the spatial segregation between blue and red galaxies seen locally (and seen at marginal significance in our own data at $z < 0.5$) has largely disappeared at $z > 0.5$. This may be seen as an analogous effect to the Butcher-Oemler effect.

3.   Towards a physical interpretation of galaxy evolution

3.1. GALAXY MORPHOLOGIES FROM HST IMAGING

The high number density of sources at $I_{AB} \sim 22.5$ means that each WFPC2 field in our redshift survey areas contains many galaxies with spectroscopically determined redshifts. To date, we have images in F450W and F814W of four WFPC2 fields from the CFRS and a resulting sample of 32 galaxies at $z > 0.5$ ((Schade et al., 1995)). This should increase sixfold within a year or so.

The 32 galaxies at $z > 0.5$ show the full range of Hubble types from ellipticals through spirals to irregulars. Spiral arms and bars are clearly visible in many galaxies and 4 of the 32 clearly look as if they are merging/interacting galaxies. As expected, the galaxies are considerably less regular on the shorter wavelength images.

Our analysis has centered on two-dimensional fitting of the galaxies using bulge+disk models and examination of the subsequent residual images (Schade et al., 1995). About 70% of the galaxies form a regular Hubble sequence of red bulges and blue disks. The remainder of the sample are what we have termed "blue nucleated galaxies" - galaxies with concentrated blue components that nearly always have large, highly asymmetric, residuals from the fitting process. While the parameters of the fits give that these are "bulge-dominated" galaxies, the large and asymmetrical residuals have led us to interpret these as either irregular galaxies with off-centered star-formation or as systems that are the result of mergers or interactions. These galaxies are presumably the "peculiar" or "irregular" class of galaxies shown to have a steep number count in the HST Medium Deep Survey by Glazebrook et al. (1995) and Driver et al. (see these proceedings).

In the 50% of galaxies that are both well-represented by bulge-plus-disk models and have $B/T < 0.5$, the disk surface brightnesses are significantly brighter (by about one magnitude) than seen in local samples, indicating a higher star-formation rate per unit area (see Fig 3).

Thus our present interpretation of the imaging data is that at least two processes are occurring to produce the changes in the blue galaxy luminosity function - the disk brightening in normal regular galaxies and the emergence of a population of relatively bright "irregular" galaxies whose nature is less clear. As our sample of galaxies with both HST imaging and spectroscopically determined redshifts grows (we should have 350 galaxies by the end of HST Cycle 5) we should be able to study statistical distribution functions such as the disk size function, the B/T distribution, the

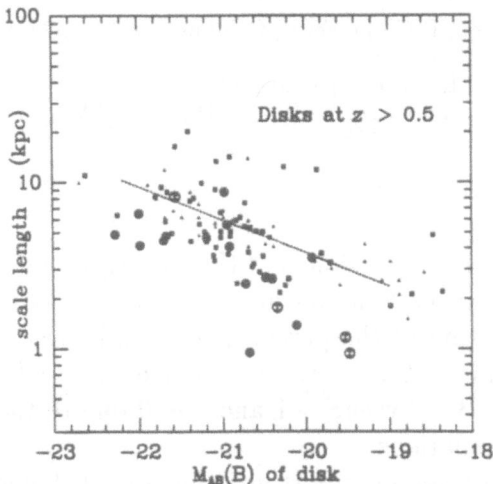

Figure 3. The disk scalelength as a function of disk luminosity for the $z > 0.5$ galaxies that have $B/T < 0.75$ (computed for $H_0 = 50$ kms^{-1}Mpc^{-1} and $q_0 = 0.5$). Small symbols are data on local galaxies from Kent (1985) and van der kruit (1987). Dashed line is the Freeman "law" $\mu(B) = 21.65$ mag arcsec^{-2}. The galaxy disks at $z > 0.5$ are of systematically higher surface brightness. See Schade *et al.* (1995) for details.

bulge luminosity function, etc.. This will enable us to constrain much better the range of allowable scenarios for the evolution of field galaxies.

References

Broadhurst, T.J., Ellis, R.S., Shanks, T., 1988, MNRAS, 235, 827
Coleman, G. Wu, C., Weedman, D., 1980, ApJS, 43, 393
Colless, M.M., Ellis, R., Taylor, K., Hook, R.N., 1990, MNRAS, 244, 408
Cowie, L.L., Songaila, A., Hu, E.M., 1991, Nature, 354, 460
Crampton, D., Le Fevre, O., Lilly, S.J., Hammer, F., 1995, ApJ, 455, in press
Davis, M., Peebles, P.J.E., 1983, ApJ, 267, 465
Fisher, B.F., Davis, M., Strauss, M., Yahil., A., Huchra, J., 1994, MNRAS, 266, 50.
Fomalont, E.B., Windhorst, R., Kristian, J.A., Kellerman, K., 1991, AJ, 102, 1258
Hammer, F., Crampton. D., Le Fevre, O., Lilly, S.J., 1995a, ApJ, 455, in press
Hammer, F., Crampton, D., Lilly, S., Le Fevre, O., Kenet, T., 1995b, MNRAS, in press
Hudon, J.D., Lilly, S.J., 1996, ApJ, in press.
Kent, S., 1985, ApJS, 59, 115
Koo, D.C., Kron, R.G., 1992, ARAA, 30, 613.
Koo, D.C., Gronwall, C., Bruzual, G., 1993, ApJ(Lett), 415, L21
Le Fevre, O., Crampton. D., Lilly, S.J., Hammer, F., Tresse, L., 1995a, ApJ, 455, in press,
Le Fevre, O., Hudon, D., Lilly, S.J., Crampton. D., Hammer, F., 1996, ApJ, in press
Lilly, S.J., 1993, 411, 501
Lilly, S.J., Le Fevre, O., Crampton. D., Hammer, F.,Tresse, L., 1995a, ApJ, 455, in press
Lilly, S.J., Hammer, F., Le Fevre, O., Crampton. D., 1995b, ApJ, 455, in press
Lilly, S.J., Tresse, L., Hammer, F., Le Fevre, O., Crampton, D., 1995c, ApJ, 455, in press
Loveday, J., Peterson, B.A., Efstathiou, G., Maddox, S.J., 1992, ApJ, 390, 338.
Marzke, R.O., Geller, M.J., Huchra, J.P., Corwin, H., 1994, ApJ, 428, 43.
Schade, D.J., Lilly, S.J., Crampton, D.C., Hammer, F., Le Fevre, O., Tresse, L.; 1995,
    ApJLett, 451, 1.

Schade, D.J., Crampton, D.C., Le Fevre, O., Hammer, O., Lilly, S.J.; 1996, MNRAS, in press.
Tinsley, B.M., Gunn, J.E., 1976, ApJ, 203, 52.
Tresse, L., Hammer, F., Le Fevre, O., Proust, D., 1993, A&A, 277, 53
van der Kruit, P., 1987, A&A, 173, 59

DISCUSSION:

Simon White: Your conclusion that the early-type galaxies have the *same* rest-frame B-luminosity function at $z \sim 1$ as at $z \sim 0$ would, if correct, imply *strong* evolution of the population. This is because if all early-type galaxies were formed at $z > 1$, we would expect a fading of at least a magnitude at rest-B between $z \sim 1$ and $z \sim 0$ due to the passive evolution of their stellar populations.

Simon Lilly: You raise an important point that I should have emphasized more in my talk - it is discussed at length in Lilly *et al.* (1995c). It is important to remember that the amount of passive evolution expected is very sensitively dependent on the poorly constrained slope of the i.m.f. in early-type stellar systems. Ideally, we will soon get some observational constraints on this from studying the fundamental plane (or projections thereof) at high redshift.

Ortwin Gerhard: From dynamical considerations, one might expect that some bulges formed by a bar instability followed by a bending instability scattering disk stars out of the plane. If this happened, then an observational signature would be a blue bulge. Do you see any objects in your sample that might plausibly be interpreted in this way.

Simon Lilly: Our sample observed with HST is still small (32 galaxies at $z > 0.5$. Of these, 70% can be decomposed into a red bulge and a blue disk. The remainder have compact blue coponents which may conceivably be "bulges". However, these galaxies are almost always associated with large and asymmetric residuals in our two-dimensional fitting algorithm, and so we have tended to interpret them so far as either irregular galaxies with off-centered dominant HII regions or as the result of mergers or interactions (see Schade et al 1995). So this question is still open, I think.

Alvio Renzini: Qualitatively, it looks like the clusters and field do the same thing. Is this true quantitatively? i.e. is the evolution of the LF of blue galaxies in the field sufficient to "explain" the B-O effect?

Simon Lilly: That has not been looked at in detail. My impression is that, at a given redshift, the B-O effect may involve more luminous blue galaxies than the field evolution. However, as you say, there are indeed many qualitative similarities. One related result is the fact that the clustering properties of intrinsically red and blue galaxies are the same at high redshift, implying that the red-blue spatial segregation at low redshift has been eliminated.

A GLIMPSE OF FIELD GALAXIES AT REDSHIFTS $Z \sim 1$ USING HST AND THE KECK TELESCOPE

D. C. KOO

UCO/Lick Obs. and Board of Astronomy and Astrophysics
University of California, Santa Cruz, CA 95064 USA

Abstract. Data from the Keck and Hubble Space Telescopes have been combined to explore the nature of very faint $I > 22$ field galaxies. At a redshift $z \sim 1$, such galaxies have luminosities similar to that of typical galaxies today. Though small, our sample of 33 redshifts already suggest that the median redshift for $I > 22$ galaxies is higher than the $z = 0.6$ expected for the "maximum merger model" of Carlberg (1995). At redshifts $z > 0.8$, mergers, interactions, and infall of minor galaxies into larger hosts appear to be common events; a wide diversity of morphological types existed; and some stellar populations were already so red that their major formation epoch occurred at redshifts $z > 2$.

1. Introduction

The nature of very faint, blue field galaxies is still a major cosmological mystery. Although deep refurbished Hubble Space Telescope (HST) images to $I \sim 24$ or fainter show a predominance of late-type or unusual galaxy morphologies in both clusters (see Oemler in these proceedings) and the field (see Driver in these proceedings), their redshifts remain largely unknown. Yet such redshifts are crucial to determine whether particular galaxies are near or far, intrinsically blue or red, bright or faint, large or small, low or high surface brightness, etc. Even the interpretation of a galaxy's morphology is dependent on its observed rest-frame wavelength and thus its redshift. With sufficient resolution and quality, spectra can also serve as probes of internal velocities and hence masses (see Illingworth in these proceedings), ages, and metal abundances of galaxies. Such a program, using the Keck 10m Telescope for the spectroscopy, is now underway as a new initiative called the Deep Extragalactic Evolutionary Probe, or DEEP (Mould 1993, Koo 1995).

R. Bender and R. L. Davies (eds.), New Light on Galaxy Evolution, 217–220.
© *1996 IAU. Printed in the Netherlands.*

Here we provide early results of a new DEEP survey that combine redshifts from the Keck Telescope with photometry, colors, and morphologies
from refurbished HST images taken by Groth *et al.* (1995). Though this
Keck redshift sample has only 33 galaxies (due to the loss of 80% of the
run to weather), the magnitudes are so faint (11 with $I < 22$, 13 with
$22 < I < 23$, and 9 with $I > 23$) and the redshifts so high (median $z \sim 0.8$),
that this survey provides a unique glimpse of the nature of faint, distant
field galaxies *of typical luminosities (L^*)* at an epoch beyond half the Hubble age. We highlight several intriguing hints that have already emerged.

2. Observations

Photometry and morphology are extracted from HST images taken by
Groth *et al.* (1995). The survey region consists of 28 overlapping WFPC2
fields with each observed in the F606W (V) filter for 2800s and F814W (I)
for 4400s. One field, however, was exposed for 7 hours in the same filters.
Our spectroscopic survey was centered on this very deep field, but also
covered 4 other flanking fields of shallower depth.

All but two bright galaxies were observed through masks cut with multiple slitlets, which allowed simultaneous exposure of ~ 25 or more targets
with the Low Resolution Imaging Spectrograph (LRIS, see Oke *et al.* 1995).
We adopted a slitwidth of 1.1 arcsecs and achieved a dispersion of 1.28Å
per px (3-4px resolution) over a spectral range of 6500Å to 9100Å.

The galaxies chosen for spectroscopy do not constitute a totally random,
magnitude-limited sample, but were instead chosen to be *representive* of a
variety of morphologies, magnitudes, and colors to a limit of $(V+I)/2 \sim 24$.
Three of the faintest emission-line redshifts were found serendipitously in
the slit of the primary target; eight galaxies, all with $I > 22$, had no or
uncertain redshifts. Also note that [OII] 3727Å is redshifted beyond our
spectral limit of ~ 9100Åfor redshifts $z > 1.4$. Thus we were gratified to
achieve an overall completeness of 80% (33/41) to a limit of $I > 24$. This is
presumably due to the high incidence of strong emission lines among very
faint galaxies.

3. Results and Discussion

Figure 1 summarizes our results in a V-I color versus redshift diagram for
our entire Keck sample, including 18 galaxies from Forbes *et al.* (1995). The
curves provide guides to the intrinsic colors of the galaxies. One striking
aspect of our data is the high concentration at redshifts $z \sim 0.8$ and 1.1,
which yields a median $z \sim 0.8$ to 1.0, regardless of the redshifts of the 9
failures . This result is inconsistent with $z \sim 0.6$ as predicted by the "maximum merger models", which otherwise fit existing brighter observations

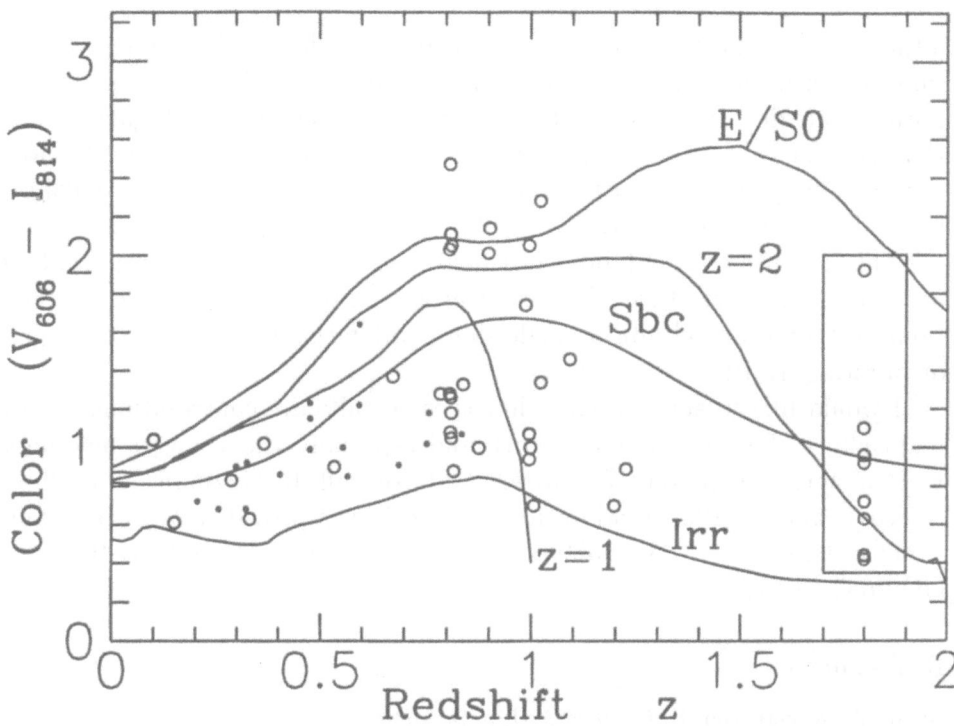

Figure 1. $V - I$ color vs redshift plot of Keck targets from this work (open circles) and Forbes *et al.* 1995 (points). Objects without redshifts are placed in the separate box to the right. Several labeled lines show the expected colors for various spectral energy distributions, including one resulting from an instantaneous burst of star formation at redshift $z = 2$ (using the models of Bruzual and Charlot 1993) that becomes almost as red as a non-evolving local elliptical or SO (E/SO) by $z < 1$; another resulting from a model burst at $z = 1$ that might be compared to the bursting dwarfs in the model of Babul and Ferguson (1995); another to the colors of a local Sbc galaxy; and the bluest one for N4449, a very actively star-forming Irr galaxy.

(Carlberg 1995). The higher median, however, matches an extrapolation of the landmark $I = 22$ Canada-France Redshift Survey (Lilly *et al.* 1995) or even some simple luminosity evolution models (e.g., Gronwall and Koo 1995).

Figure 1 shows how well the galaxies fall within the bounds seen in the colors of normal, local galaxies. We find neither unusually blue nor unusually red galaxies. Other implications from the figure are that (i) *field* galaxies do exist at high redshifts $z \gtrsim 0.7$ which have intrinsic colors comparable to that found for local ellipticals; (ii) that the most recent major star-formation event in these red galaxies presumably occurred at redshifts $z \gtrsim 2$; (iii) that the current model of Babul and Ferguson (1995) does not quite match the observed color distribution nor the presence of very blue galaxies beyond redshifts $z \sim 1$; and (iv) that intrinsically blue galaxies partake in the strong clustering seen at redshifts $z \sim 0.8$ and 1.0. Furthermore, although HST images show strong and frequent hints of mergers, interactions, other peculiar patterns, and infall of minor galaxies into larger hosts, normal galaxies are also visible. The morphologies of $z \sim 1$ galaxies are thus not confined to late-type, peculiar systems and, conversely, the late-type galaxies seen in deep HST images are not necessarily of low redshifts and thus of low-luminosity.

This glimpse of very faint $z \sim 1$ field galaxies strongly suggests that we need to invoke a mixture of physical processes to account for the observations, rather than rely on a single dominant mechanism, such as mergers or bursting dwarfs.

I would like to acknowledge that these results are contributions from a team effort of DEEP and E. Groth and especially the younger members: R. Brunner, A. Connolly, D. Forbes, C. Gronwall, R. Guzman, A. Phillips, N. Vogt, and K. Wu. Funding for this work was provided by NSF grants AST91-20005 and AST-8858203 and NASA grants AR-5801.01-94A and GO-2684.04-87A.

References

Babul, A., & Ferguson, H. C. 1995, ApJ, in press.
Bruzual, G. A., and Charlot, S. 1993, ApJ, 405, 538
Carlberg, R. G. 1995, in proceedings of Ringberg Workshop, preprint
     Sparks, W. B., & Lucas, R. A. 1994, ApJ, 435, L23
Forbes, D. A., Phillips, A. C., Koo, D. C., & Illingworth, G. D. 1995, ApJ, submitted
Gronwall, C., & Koo, D. C. 1995, ApJ, 440, L1
Groth, E. J., et al. 1995, in preparation
Koo, D. C. 1995, in "Wide Field Spectroscopy and the Distant Universe", eds. S. Maddox
     and A. Aragón-Salamanca (World Scientific: Singapore), 55
Lilly, S. J., Le Fèvre, O., Hammer, F., Crampton, D., & Tresse, L. 1995, ApJ, in press
Mould, J. 1993, in "Sky Surveys: Protostars to Protogalaxies", ed. B. T. Soifer, (ASP
     Conf. Ser., 43), 281
Oke, J. B., Cohen, J. G., Carr, M., Cromer, J., Dingizian, A., et al. 1995, PASP, 107, 375

Questions and Answers

E. Kachikian: Did you find Seyfert-type spectra among your objects?

D. Koo: Not yet, but this is not unexpected for our small sample.

M. Dickinson: Just a comment on the red galaxies – for *clusters*, at least, we are seeing very red, ordinary, boring $r^{1/4}$-law profile ellipticals out to $z = 1.2$.

D. Koo: We see these too. I was just pointing to one with an unusual profile and was *not* implying that the finding of an exponential profile was common.

S. Zepf: The unusual morphology of the red object suggests that their colors might be affected by dust. What are the spectral types of these red galaxies?

D. Koo: We have not yet examined the spectral types, but dust is visible in some cases.

U. Hopp: How many single-lined objects do you have at $z \sim 1$? What further arguments did you use to assign a redshift to a single-line spectrum?

D. Koo: Most are single lines of [OII]3727. This identification was usually based on a rising continuum with Balmer absorption to the red, sometimes on the resolution of the doublet, and on the lack of [NII] or [SII] if H_α is a possibility. Some uncertainties definitely remain.

CAUGHT IN THE ACT: THE IDENTIFICATION OF THE GALAXIES RESPONSIBLE FOR THE FAINT BLUE EXCESS

SIMON P. DRIVER AND ROGIER A. WINDHORST

Arizona State University,
Department of Physics and Astronomy,
Box 871504, Tempe, AZ85287-1504, USA

AND

RICHARD E. GRIFFITHS

Bloomberg Center for Physics and Astronomy, JHU

Abstract. We summarise recent Hubble Space Telescope results on the morphology of faint field galaxies. Our two principle results are: (1) the galaxies responsible for the faint blue excess have late-type/irregular morphology and (2) the number counts of normal galaxies, ellipticals and early-type spirals, are well fit by standard no-evolution models implying that the giant population was in place and mature by a redshift of ≥ 0.7.

1. Introduction

One of the great leaps forward in the study of faint field galaxies, and made possible only by the Wide Field and Planetary Camera 2 (WFPC2) on board the Hubble Space Telescope (HST), is the ability to discern detailed morphological information to faint magnitudes ($I \leq 24.24$). The Medium Deep Survey (a key HST project, c.f. Griffiths *et al.* 1994) has exploited this new area resulting in a number of recent morphology related publications (Casertano *et al.* 1995; Driver, Windhorst & Griffiths 1995; Driver *et al.* 1995; Glazebrook *et al.* 1995). Here we briefly summarise and coalate some of the key results from these papers, and interpret them by comparison to the predictions of some generic faint galaxy models. We also discuss the redshift dsitribution for Late-type/Irregulars which can now be directly derived based on the mounting observational evidence that the "giant" galaxies are not strongly evolving over the redshift range $0.0 < z < 1.0$ (c.f. Driver *et al.* 1995; Mutz *et al.* 1994; Lilly; Dickinson these proceedings).

221

R. Bender and R. L. Davies (eds.), New Light on Galaxy Evolution, 221–224.
© *1996 IAU. Printed in the Netherlands.*

2. Summary of the MDS morphology papers

CASERTANO *et al.* 1995 - Casertano analysed the entire WF/PC database (13,500 galaxies), using an automated technique to classify the sample into bulge dominated (E/S0), disk Dominated ($\sim$ Sabcd) or other systems (M/Irr/?). The principle results are that at faint magnitudes ($20 < I_{785LP} < 21$) the number counts are dominated by very small disk systems with a mean scale-length of ~ 0.3 arcseconds.

DRIVER, WINDHORST & GRIFFITHS 1995 - A complete magnitude limited sample ($I_{814} < 22$) of 144 field galaxies drawn from six MDS fields. Classifications are by eye into ellitpicals (E/S0), early-type spirals (Sabc) and late-type spirals/Irregulars (Sd/Irr). At $I_{814} \sim 22.0$ mag, these three populations are observed in almost equal proportions. While the E/S0's and Sabc's are well fit by "conventional" models, the Sd/Irr's have about 1 dex higher surface densisty than expected.

GLAZEBROOK *et al.* 1995 - A similar analysis to that above was made independently based on a sample of 301 galaxies drawn from 10 MDS fields over a comparable magnitude range. The findings are similar demonstrating a robustness to the observational result.

DRIVER *et al.* 1995 - The most recent survey based on a single *ultradeep* HST field (now totalling 67 orbits, c.f. Windhorst - this conference; and Windhorst & Keel 1995). The results confirmed the earlier work and find that the galaxy counts beyond $I814 = 22$ continue to become increasingly dominated by Sd/Irrs. (see colour plate by Windhorst, these proceedings).

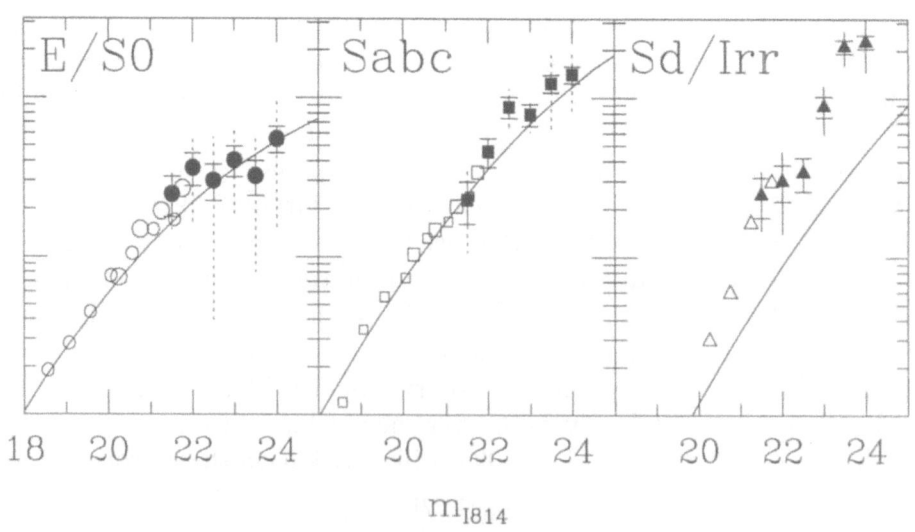

Figure 1. A compendium of morphological galaxy counts compared to the "standard" no-evolution models. Note that the models are renormalized at $b_J \sim 18$ mag, which is somewhat naive as the justification — local evol^n, local hole, SB sel^c effects and/or photometric effects — are liable to be both type and luminosity dependent.

3. Popular Generic Faint Galaxy Models

Recent popular models to explain the Sd/Irr counts are:

DWARF MODELS, e.g. Driver *et al.* (1994) - These exploit the uncertainty in the space density of Sd/Irrs and increase the normalization and slope of the Sd/Irr LF until an optimal fit is found. Such models give good fits to the counts in all bands and can explain the trend to bluer colours at fainter magnitudes. The models typically fail to match the z-distribution.

EVOLVING FLAT MODELS, e.g. Broadhurst, Ellis & Shanks (1988) - Assumes the LF for Sd/Irrs is an extrapolation of the luminous galaxies, then luminosity evolution is invoked to match the counts. The models match the z-distributions but the luminosity evolution required to match the counts is extreme, $\sim$ 2.5 mags in every Sd/Irr galaxy by z=0.5.

MERGER MODELS, e.g. Broadhurst, Ellis & Glazebrook (1992) - As above except merging is invoked. The merger rate required is extreme and not seen in w(θ). The morphological data also fail to show the density of close companions required. That the E/S0 and Sabc galaxies fit closely to the no-evolution models argues against merging, since if these were the merger products, they should be significantly rarer in the past.

AN EVOLVING DWARF MODEL, e.g. Phillipps & Driver (1995) - Essentially taking the best of the first two models its clear to see that a good half-way solution should be found. The data for Sd/Irrs show $\sim$ 60 % inert, $\sim$ 20 % tidally disrupted and $\sim$ 20 % knotted cores (spontaneous starbursts ?), supporting all the above models to some extent. This model, although contrived, can fit all the counts and the z-distributions.

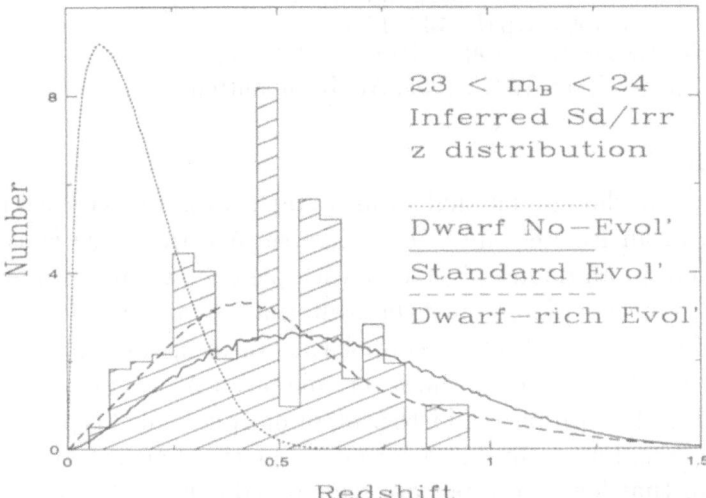

Figure 2. A comparison of the above models to the inferred Sd/Irr redshift distribution. This distribution is derived by subtracting off the predicted distribution for E/S0 and Sabc, assuming no evolution as implied by their counts, from the total redshift distribution of Glazebrook *et al.* (1995).

4. Discussions, Conclusions and Speculations

Perhaps the most far reaching gain, is that now the culprits responsible for the faint blue excess have been isolated, we can return to the original aim of number counts, namely constraining cosmological models. With the mounting evidence that the E/S0 and Sabc galaxies are largely in place by $z \sim 1$, this aim becomes even more attainable. Already the counts presented here rule out any posibility of a large comsological constant (c.f. Driver *et al* 1995) and thereby block the possiblity of using Λ as a way out of the age of the universe problem. Nevertheless such analaysis is still fraught with perils and most surprisingly, it is our knowledge of the local environment which provides the largest remaining obstacles.

References

Broadhurst, T. J., Ellis, R. S., Shanks, T., 1988, MNARS, **235**, 827
Broadhurst, T. J., Ellis, R. S., Glazebrook, K., 1992, NATURE, **355**, 55
Casertano, S., Ratnatunga, K. U., Griffiths, R. E., Im, M., Neuschaefer, L. W., Ostrander, E. J., & Windhorst, R. A. 1995, ApJ, **453**, 000
Driver, S. P., Phillipps, S., Davies, J. I., Morgan, I., Disney, M.J., 1994, MNRAS, **266**, 155
Driver, S. P., Windhorst, R. A., & Griffiths R. E. 1995, ApJ, **453**, 000
Driver, S.P., Windhorst, R. A., Ostrander, E.J., Keel W.C., Griffiths, R. E., Ratnatunga, K.U., 1995, ApJL, **449**, L000
Driver, S.P., Windhorst, R.A., Phillipps, S., Bristow, P.D., 1995, ApJ, submitted
Glazebrook, K., Ellis, R. S., Colless, M. M., Broadhurst, T. J., Allington-Smith, J., Tanvir, N., 1995, MNRAS, **273**, 157
Glazebrook, K., Ellis, R. S., Santiago, B., Griffiths, R.E., 1995, MNRAS, in press
Griffiths, R. E., *et al.* 1994, ApJL, **435**, L019
Mutz, S.B., *et al.* 1994, ApJL, **434**, L055
Phillipps, S., Driver, S.P., 1995, MNARS, **274**, 832
Windhorst, R.A., Keel, W.C., 1995, ApJL, submitted

Questions

WHITE: You showed us that your E/S0 counts are consistent with *no* evolution in an Einstein-de Sitter universe. Are they also consistent with no evolution in an open universe or with passive evolution in either ?
DRIVER: We've explored all of these alternatives and conclude that enough margin of error exists that we cannot rule for or against any of the models you mentioned, yet. We can however rule out a Λ-dominated universe.
WORTHEY: Do you have a histogram of morphological type versus colour.
DRIVER: Yes, but we find little or no correlation between type and colour, other than that few late-types are seen at red colours. This may be due to the poor range of the colour baseline (V_{606} and I_{814}).

FAINT GALAXY NUMBER-COUNTS

N. METCALFE, T. SHANKS, R. FONG AND J. GARDNER
Physics Dept, Durham University
South Parks Road, Durham DH1 3LE, UK

AND

N. ROCHE
Johns Hopkins University
Baltimore, USA

1. Introduction

Observers studying the cosmology and evolutionary history of our Universe through the statistical properties of 'normal' galaxies have four main tools at their disposal. (1) The number-redshift relation. Although a very powerful diagnostic, spectroscopic surveys are currently limited to $B < 24^m$ and significantly incomplete in the range, $23^m < B < 24^m$. (2) Galaxy number-magnitude counts. Although by themselves, they cannot constrain models as tightly as spectroscopy, they can be measured $\sim 4^m$ fainter, where cosmological effects are expected to be significant. (3) Galaxy colours over a wide wavelength range, which provide additional constraints. (4) The dependence of galaxy clustering with magnitude. $\omega(\theta)$ can be measured to the limit of the counts.

Here we report on the latest Durham count and clustering work.

2. Durham Galaxy Counts

At Durham we have been investigating the galaxy number-count and clustering relations for many years. Our original work was done with photographic plates from the 48 in UK Schmidt telescope (limited at $B < 21^m$) (e.g. Stevenson et al. 1986) and from plates taken with the 3.9 m Anglo-Australian telescope ($B < 23.5^m$) (Jones et al. 1991). Since the advent of highly efficient CCD detectors we have concentrated on the deepest possible B-band counts (Metcalfe et al. 1991,1995).

R. Bender and R. L. Davies (eds.), New Light on Galaxy Evolution, 225–228.
© 1996 IAU. Printed in the Netherlands.

2.1. THE DURHAM DEEP FIELD

Our deepest data all lie on one field, the Durham Deep Field, centred approximately at $00^h20^m + 00°00'$ (1950). We have four major exposures:

(1) a 24 hr B-band exposure with an RCA CCD at the prime focus of the 2.5 m Isaac Newton Telescope (INT) (Metcalfe et al. 1995), covering an area of $3.3' \times 5.7'$ (with 0.74" pixels), with $\sim 1.5''$ full-width half maximum (FWHM) seeing. A 3σ detection inside twice the seeing FWHM corresponds to a *total* magnitude of $B = 27.0$ for an unresolved object.

(2) a 10 hr B-band exposure with a Tektronix CCD at the auxiliary port of the 4.2 m William Herschel telescope (WHT) (Metcalfe et al. 1995). This is only a 0.8' radius circular area (0.22" pixels), centred on the INT field, with a FWHM $\sim 0.9''$. With the much improved seeing, the 3σ detection limit becomes $B_{total} = 27.6^m$.

(3) Our latest work, reported here for the first time, consisting of a 30 hr B-band exposure with a Tektronix CCD at the WHT prime focus. This gives a much bigger field of $7' \times 7'$ (0.42" pixel size) which encompasses fields (1) and (2). With a FWHM of $\sim 1.2''$, a 3σ detection is $B_{total} = 27.9^m$

(4) Again reported here for the first time, a 30 hr K-band exposure, covering the 0.8' radius area of (2), taken with IRCAM3 on the 3.8 m UK Infra-Red telescope. The FWHM is $\sim 1.2''$ and the 3σ limit is $K_{total} \sim 22.5^m$.

Combining (1), (2) and (3) creates an ultra-deep field (see fig. 1) equivalent to 50 hours exposure with a 4 m telescope. The 3σ limit on this image, which we believe is the deepest yet, is $B_{total} \sim 28.3^m$. Analysis of this image, and of the K-band data, are still in progress. Ultra-deep U and R exposures on field (3) are scheduled for later in 1995.

2.2. GALAXY CLUSTERING

Roche et al. (1993, 1994) studied the clustering on the $B \sim 25^m$ frames of Metcalfe et al. (1991) and at $B \sim 27^m$ on (1) above. In new work, Roche et al. have measured the clustering to $B \sim 26^m$ and $R \sim 25^m$ over a 400 sq.arcmin. area using 4 hr INT exposures. Work is also underway to determine the clustering at $B \sim 28^m$ on our 30 hr WHT exposure.

3. What do we find?

The main conclusions which can be drawn from this work are summarised as follows (and see fig. 2):

* The $B < 17^m$ counts have too steep a slope for the models. As a result, the normalisation for the models is unclear, leading to an uncertainty in the evolution required to fit the fainter counts.

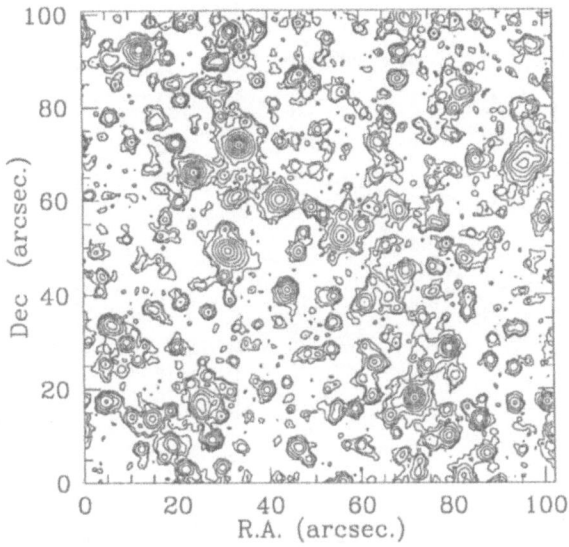

Figure 1. A lightly-smoothed isophotal contour map of our stacked WHT and INT data. The contours are spaced in 1 mag/sq. arcsec apart, with a faintest level of 30 mag/sq. arcsec. Note that there is almost no noise visible in this plot - all the images are real.

* The counts are still rising at $B \sim 28^m$, and are uncomfortably high for a closed universe if galaxy numbers are conserved.

* The counts show signs of flattening off at $B > 25^m$. Such behaviour is indicative of a redshift cut-off or that high redshifts are being reached, where the cosmological volume element starts to increase much less rapidly with redshift. In either case the count slope should then be equal to the faint-end slope of the galaxy luminosity function.

* The B,R,I and K-band counts all seem to have a very similar slope at the limit of observations (~ 0.3 in d(logN)/dm).

* The clustering amplitude is much lower than predicted by non-evolving models. This could be due to a more extended $n(z)$ distribution, lower intrinsic clustering or evolution in the overall clustering strength. There is marginal evidence for a flattening at $B > 25^m$, consistent with (and for the same reason) as the flattening in the counts.

* Redder galaxies are clustered much more strongly that the blue galaxies at $B > 24^m$, as expected if they are unevolved early-type galaxies.

References

Jones, L.R., Fong, R., Shanks, T., Ellis, R.S., Peterson, B.A., 1991, MNRAS, 249, 481
Metcalfe, N., Shanks, T., Fong, R., Jones, L.R., 1991, MNRAS, 249, 498
Metcalfe, N., Shanks, T., Fong, R., Roche, N., 1995, MNRAS, 273, 257
Roche, N., Shanks, T., Metcalfe, N., Fong, R., 1993, MNRAS, 263, 360

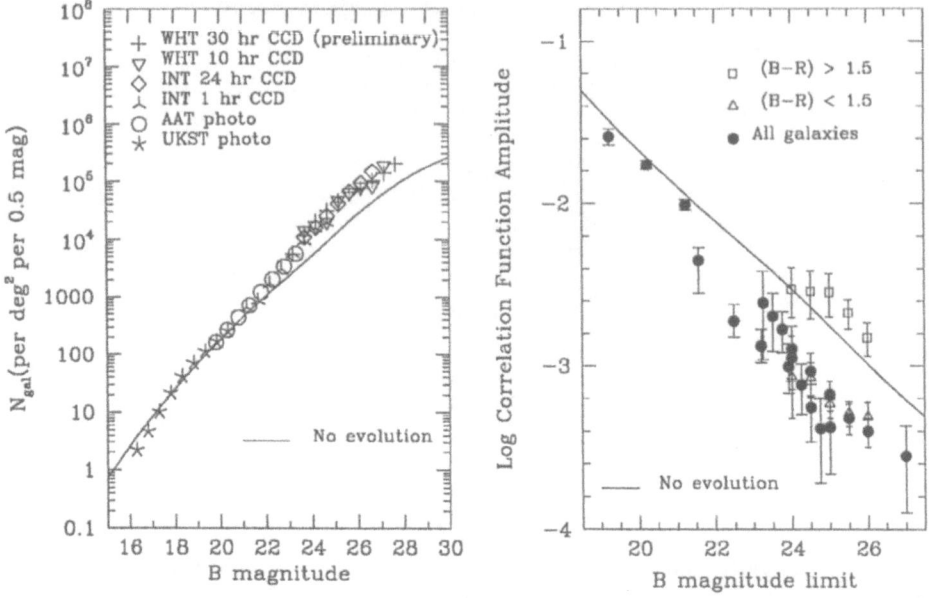

Figure 2. Durham galaxy data: *left,* B-band galaxy number-counts, corrected for confusion and incompleteness. A low q_0 non-evolving model prediction is shown. *right,* B-band clustering amplitude as a function of magnitude, mostly from the same data as used for the counts, again compared with a non-evolving prediction. Our 4 hour INT data is also shown subdivided by colour.

Roche, N., Shanks, T., Metcalfe, N., Fong, R., 1994 in MacGillivray et al., eds, IAU Symp. 161, Astronomy from Wide-Field Imaging. Kluwer, Dordrecht, p. 635
Stevenson, P.R.F., Shanks, T., Fong, R., 1986, in Chiosi, C., Renzini, A., eds, Spectral Evolution of Galaxies. Reidel, Dordrecht, p. 439

Discussion

WINDHORST: The B-band galaxy counts cannot continue for $B_J > 28^m$ with slope $\alpha > 0.4$, or their integrated synchrotron disk-emission at GHz frequencies would exceed the temperature errors on the cosmic background radiation. So what slope do you infer for $B_J > 28^m$?

METCALFE: The count slope has flattened to around 0.3 by the limit of our (confusion corrected) data. In terms of the Schechter luminosity function this would correspond to a faint-end slope of ~ -1.8, which if we really are probing high redshifts might imply that the luminosity function was steeper in the past.

CHARACTERIZING DISTANT BLUE GALAXIES WITH HST IMAGES AND KECK SPECTRA.

GARTH D. ILLINGWORTH

UCO/Lick Observatory
University of California, Santa Cruz, CA 95064 USA

Abstract. Keck spectra and HST images have been used to derive characteristic velocity and length scales for an enigmatic component of the faint, blue, field galaxy population: the compact, narrow emission-line galaxies (CNELGs). These galaxies are very luminous, but have been found to be quite low mass systems (with typical masses $\sim 10^9$ M$_\odot$). Their blue colors and strong emission lines indicate that they are undergoing a major burst of star formation. Following the completion of their current burst they will fade, becoming, in the absence of further major bursts, objects very similar to contemporary spheroidal galaxies. With mean sizes $R_e \sim 1.4$ kpc and Gaussian velocity profiles with mean $\sigma = 45$ km s^{-1}, the length scales and velocity widths of CNELGs are also quite consistent with the measured length scales and velocity widths of current spheroidals.

1. Introduction

Studies of distant galaxies have undergone a revolution in the last few years. Examples are the striking results of the deep, distant galaxy survey by Lilly and Le Fèvre and their collaborators (see Lilly, this conference), and the superb images being obtained from HST and the WFPC2 (see Dickinson, this conference). New large telescopes like Keck also allow us to measure velocity widths in distant galaxies.

The blue galaxies that have been studied at Keck to derive velocity widths, and imaged with HST to provide structural length scales, are members of a set of objects known as compact, narrow emission-line galaxies (CNELGs) first identified by Koo and Kron (1988). Those studied to date typically have redshifts $z \sim 0.1 - 0.7$, and were highlighted because of their non-stellar colors, and their unusually small size for galaxies (they are typically unresolved on 4-m survey plates taken in 1-2" seeing).

R. Bender and R. L. Davies (eds.), New Light on Galaxy Evolution, 229–232.
© *1996 IAU. Printed in the Netherlands.*

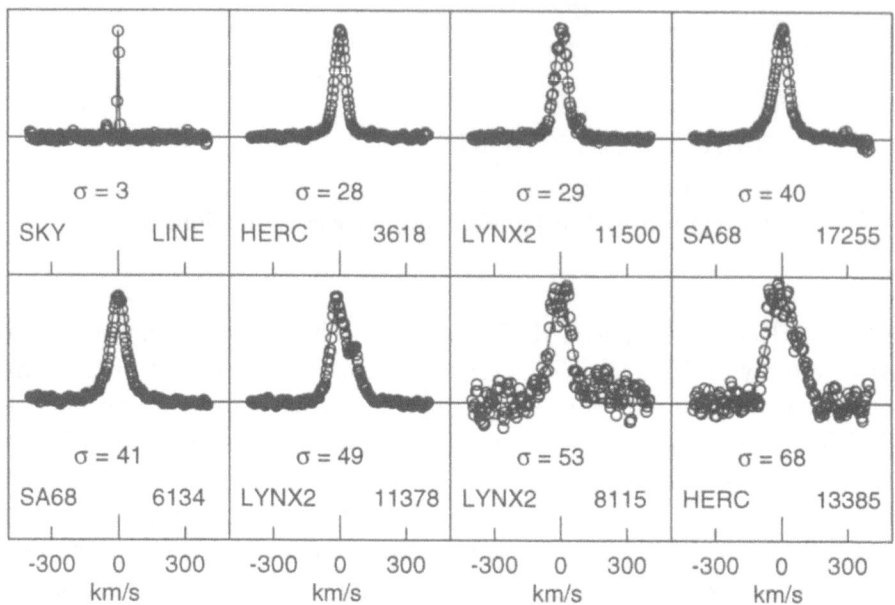

Figure 1. Panel of emission line profiles from a set of CNELGs and compact blue galaxies. The velocity profiles are very well fitted by Gaussians. The instrumental profile is in the upper left. This is a subset of the complete sample.

2. Observations

The HST and Keck Observations of the CNELGs are described in Koo *et al.* (1994), Koo *et al.* (1995) and Guzmán *et al.* (1996). The HST observations were used to derive effective radii, to identify the form of the surface brightness profile and to characterize their morphology. CNELGs can be characterized as centrally concentrated objects, with little substructure, and exponential light profiles. Typical half-light diameters for the seven objects imaged by HST WFPC are 0.65″, corresponding to $R_e \sim 1.4$ kpc ($H_0 = 50$; $q_0 = 0.1$, as used in the above papers).

The Keck spectra were taken with the HIRES spectrograph with a spectral resolution of 8 km s^{-1} FWHM. Velocity widths were measured from the emission lines [OII] 3727 Å, $H\beta$, and/or the [OIII] 4959/5007 Å pair, depending on the line strengths. Since the lines were typically (and interestingly) Gaussian in shape, the velocity width σ was determined by fitting Gaussians (where $\sigma = $ FWHM/2.35). The sample of seven CNELGs that are the focus of the Guzmán *et al.* (1996) paper have a mean $z = 0.22$, with mean rest frame values of $M_B = -20$, $[U - B]_0 = -0.24$, $[B - V]_0 = 0.39$, and a mean $\sigma = 45$ km s^{-1}.

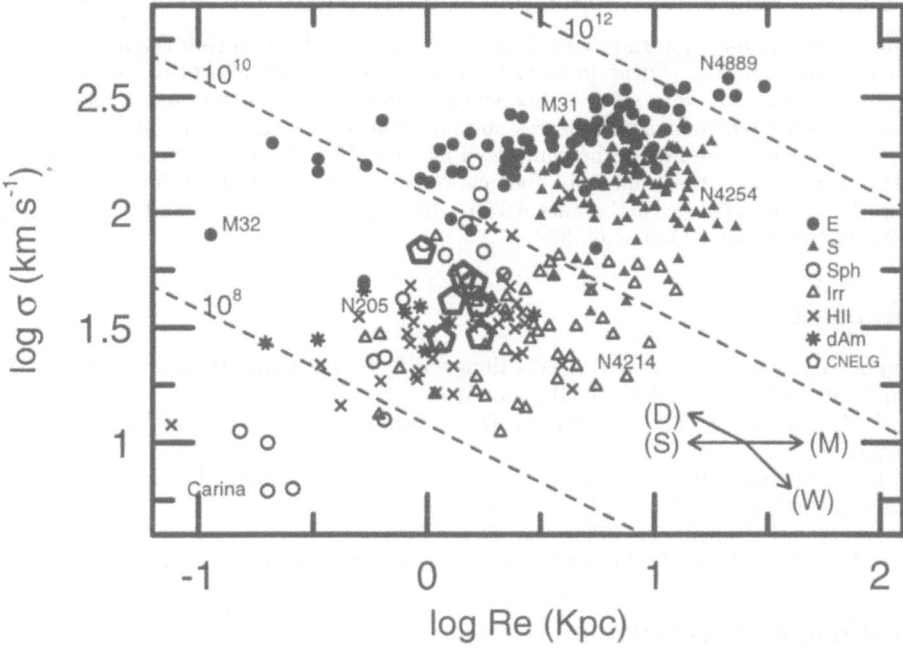

Figure 2. Length scale – velocity width relation (R_e *vs.* σ) comparing the CNELGs with a sample of nearby galaxies from ellipticals to spheroidals. The dashed lines indicate masses of 10^8, 10^{10} and 10^{12} $M_\odot$; surface densities are highest to the upper left. The CNELGs have typical masses around 10^9 $M_\odot$, and clearly fall amongst the local dwarf population in their R_e – σ properties. The labelled arrows correspond to a variety of physical processes that could change the location of galaxies in this figure (namely, mergers – M, tidal stripping – S, galactic winds – W, and dissipation – D). As discussed in Guzmán *et al.* (1996), none of these processes are likely to displace the observed values far enough to move them out of the contemporary dwarf galaxy regime.

3. Results and Discussion

Comparison of the properties of the CNELGs, namely, their high luminosities, low velocity widths, small sizes, strong emission lines, and their
very blue colors, with those of nearby galaxies suggests a close link to the
HII galaxies, a set of low mass, extreme star-forming galaxies (Salzer *et al.*
1989). Koo *et al.* (1995) argued from simple burst models that the velocity
widths, luminosities and surface brightnesses suggested a close link between
CNELGs and current-day spheroidals. This link has been strengthened with
the combined dataset of length scales and velocity widths now in hand for
seven CNELGs. A comparison in the R_e – σ plane shows that the CNELGs
are unambiguously related to contemporary dwarf galaxies in size and velocity scales, and hence in mass. The details of the arguments can be found
in Guzmán *et al.* (1996). These data provide an excellent example of the
power of HST, and telescopes such as Keck, for clarifying the nature of the

generic faint blue galaxy population at intermediate and high redshifts.

I would especially like to acknowledge David Koo for his realization that the Keck HIRES could measure velocity widths in such faint galaxies, and his efforts to assemble the observing program on very short notice when LRIS failed just before our Keck run. In addition, Rafael Guzmán also deserves special mention for his considerable efforts on this project, one of the first to be completed as part of the DEEP program. DEEP is a project sponsored by the Center for Particle Astrophysics at UC Berkeley. Funding for this work was provided by NSF grants AST91-20005 and AST-8858203 and NASA grants GO-2684.04-87A and GO-2684.05-87A.

References

Guzmán, R., Koo, D. C., Faber, S. M., Illingworth, G. D., Kron, R. G., Takamiya, M., and Bershady, M. A. 1996, ApJL, submitted.
Koo, D. C., Bershady, M. A., Wirth, G. D., Stanford, S. A., and Majewski, S. R. 1994, ApJ, 427, L9.
Koo, D. C., Guzmán, R., Faber, S. M., Illingworth, G. D., Bershady, M. A., Kron, R. G., and Takamiya, M. 1995, ApJ, 440, L49.
Koo, D. C., and Kron, R. G. 1988, ApJ, 325, 92.
Salzer, J. J., MacAlpine, G. M., and Boroson, T. A. 1989, ApJS, 70, 447.

Questions and Answers

P. Guhathakurta: What fraction of all $20 < B < 23$ galaxies do CNELGs represent?

G. Illingworth: Very small; roughly $\sim$ 1-2% to $B \sim 21$.

E. Khachikian: According to your data (spectra) it seems to me that your objects look like Liners or Superassociations (Giant HII regions)?

G. Illingworth: They are more like the latter, and not Liners, since the spectra are typical of those seen for giant, star-forming complexes.

I. Appenzeller: Can you say something about the relative frequency or space density of the class of blue galaxies described in your talk, and (if these objects are indeed the precursors of the Dwarf Spheroidals) can one perhaps estimate from such data the poorly known space density of the Dwarf Spheroidals outside the local group.

G. Illingworth: These objects are a few percent of the blue galaxy population. At this stage I think it would be very difficult to use CNELGs to provide any useful constraints on the spheroidal space density because it is likely that they formed over a very wide range of redshifts, and we are only sampling a modest range in redshift. The formation rate probably varied considerably and we have not made any estimate for the formation rate function, nor would any estimate be very reliable at this time.

A. Oemler: It gives me great pleasure to hear that Dave Koo now believes in evolution, but it gives me even greater pleasure to say that I don't believe it. You have shown that the high redshift blue compact emission line galaxies are identical to nearby objects, but it is only an astrophysical inference from gas loss mechanisms to expect that they might evolve.

G Illingworth: It is my mistake if I did not make the argument very clear during my talk. What I wanted to emphasize was that there is a very good set of arguments that indicate that such low-mass bursting galaxies could well be the progenitors of spheroidal galaxies, but that *there is no reason to suggest that the intermediate redshift epoch was unique as the time of formation of such objects*. Since local counterparts to CNELGs exist, it is likely that spheroidals have formed over a large range of epochs. These data tell us little about evolution; surveys such as those of Lilly and Le Fèvre will provide more definitive answers on this.

MEASURING THE EVOLUTION OF THE M/L RATIO FROM THE FUNDAMENTAL PLANE IN CL 0024+16 AT $Z=0.39$

MARIJN FRANX and PIETER G. VAN DOKKUM

Kapteyn Astronomical Institute
P.O. Box 800, 9700 AV Groningen
The Netherlands

Abstract. The existence of the Fundamental Plane of early-type galaxies implies that the M/L ratios of early-types are well behaved. It provides therefore an important tool to measure the evolution of the M/L ratio with redshift. These measurements, in combination with measurements of the evolution of the luminosity function, can be used to constrain the mass evolution of galaxies.

We present the Fundamental Plane relation measured for galaxies in the rich cluster CL 0024+16 at $z=0.391$. The galaxies satisfy a tight Fundamental Plane, with relatively low scatter (15 %). The M/L is 31 ± 12 % lower than the M/L measured in Coma, which is consistent with simple evolutionary models. Hence, galaxies with very similar dynamical properties existed at a $z=0.4$.

More, and deeper data are needed to measure the evolution of the slope and the scatter of the Fundamental Plane to higher accuracy. Furthermore, data on the richest nearby clusters would be valuable to test the hypothesis that the Fundamental Plane is independent of cluster environment.

1. Introduction

Galaxy evolution may be a complex process, with possibly a large role for mergers, interactions, infall, and starbursts triggered by these events. Such processes complicate the interpretation of observations of high redshift galaxies, as galaxies can change rapidly in luminosity (due to starbursts), and can change morphology due to mergers, and infall of gas. The progen-

233

R. Bender and R. L. Davies (eds.), New Light on Galaxy Evolution, 233–240.
© 1996 *IAU. Printed in the Netherlands.*

itors of certain types of galaxies at some redshift may be of different type at some other redshift, and their luminosities may be quite different.

In order to quantify these effects, more information is needed than the evolution of luminosity and color of galaxies. Detailed information on the morphological evolution, and the evolution of the mass function is essential. The evolution of the mass function is possibly the most important, as it gives direct insight into the mass evolution of individual galaxies, and can directly determine when typical galaxies were assembled.

Unfortunately, the total masses of galaxies are notoriously difficult to measure. However, there exist good relations between circular velocity, and velocity dispersion, and photometric parameters: the Tully-Fisher relation for spirals (Tully & Fisher 1977), the Faber-Jackson relation (Faber & Jackson 1976), and the Fundamental Plane for early-types (Djorgovski & Davis 1987, Dressler et al. 1987). These relations are very suitable for evolutionary studies, because their intrinsic scatter is low at $z=0$.

Here we present results on a program to measure the evolution of the Fundamental Plane relation with redshift. The Fundamental Plane is a relation between effective radius r_e, effective surface brightness I_e, and central velocity dispersion σ of the form $r_e \propto \sigma^{1.24} I_e^{-0.82}$ (e.g., Bender et al. 1992, Jørgensen et al 1995, JFK). Its scatter is low, at 17% in r_e (Lucey et al. 1991, JFK). The implication of the Fundamental Plane is that the M/L ratio of galaxies is well behaved (e.g., Faber et al. 1987). Under the assumption that galaxies are a homologous family, the implied M/L scaling is $M/L \propto r_e^{0.22} \sigma^{0.49} \propto M^{0.24}$. Such scaling is sufficient for the existence of the Fundamental Plane, and vice versa. The cause of the variation in M/L with mass parameters is not well understood, but it can be due to variations in metallicity, IMF, dark matter fraction, and age (e.g., Renzini & Ciotti 1993).

2. What can we hope to learn ?

The low intrinsic scatter of the Fundamental Plane might make it suitable for "classical" cosmological tests. These tests would require that the evolutionary effects can be ignored. The following applications can be considered:
• Surface brightness test ? The Fundamental Plane can be used to perform the classical surface brightness test (Kjærgaard et al. 1993). However, this surface brightness test has done by COBE. If surface brightness were not to decrease with $(1 + z)^{-4}$, the Cosmic Background Radiation would deviate from a Planck spectrum on a very short time-scale. Since it satisfies a Planck spectrum to 0.03 % (Mather et al. 1994), the cosmological dimming holds to very high accuracy.
• q_0 - Apparent angle of a standard rod ? We can define as a standard rod

the observed effective radius of some galaxy with fixed surface brightness, and velocity dispersion. The dependence against redshift might provide a constraint on q_0, except that evolutionary effects in the luminosity of galaxies are larger than this cosmological effect.

These cosmological tests are therefore of limited use. The most important application of the evolution of the Fundamental Plane is the measurement of the evolution of the M/L ratio. This measurement can be used to determine the evolution of the mass function, if the evolution of the luminosity function is also well determined. Furthermore, the scatter and the slope of the Fundamental Plane may change with redshift. The measurement of this evolution will provide additional insight into the cause of the dependence of the M/L ratio on mass (Renzini & Ciotti 1993).

3. Models for the evolution of the M/L ratio

The luminosity of a co-eval stellar population is expected to evolve with time. This is due to the fact that much of the light is produced by stars on the giant branch, which is a very short phase in a stellar lifetime. Tinsley (1980) showed that the luminosity evolves like

$$L \propto 1/(t - t_{form})^{\kappa}$$

where $\kappa = 1.3 - 0.3x$, and x is the slope of the IMF. The Miller–Scalo IMF implies $x = 0.25$, and $\kappa \approx 1.2$. Recent studies indicate that the value of κ depend on passband and metallicity (Buzzoni 1989, Worthey 1994). These authors find $0.6 < \kappa < 0.95$ for the V band.

To first order, this evolution implies that the M/L ratio evolves like

$$\ln M/L(z) = \ln M/L(0) - \kappa(1 + q_0 + 1/z_{form})\, z,$$

where z_{form} is the formation redshift (Franx 1995). Hence the logarithm of the M/L ratio is expected to decrease linearly with redshift, and the coefficient depends on κ(IMF), q_0, and z_{form}. This equation is valid for $q_0 \approx 0$, and high z_{form}, but it is a reasonable approximation even for $q_0 = -1$. The rate at which the M/L ratio decreases is therefore a function of several unknown variables, and the interpretation of the observed decrease of the M/L ratio will not be very straightforward.

3.1. COMPLEX EVOLUTION

There is no good reason to assume that all early-type galaxies formed at the same redshift. Furthermore, a single galaxy may have had a complex formation history, with starformation extending over a long time. The evolution of the M/L ratio will be much more complex if such age differences are taken into account.

If galaxies have different mean ages, then the M/L ratios will be different (if other properties are the same). Hence, there will be scatter in the M/L ratios, and the scatter will increase with look-back time, as the relative age difference will increase. Such effects can therefore be found by measuring the scatter in the Fundamental Plane as a function of redshift.

Galaxy evolution can be substantially more complex than that, however, as demonstrated by the Butcher-Oemler effect (Butcher & Oemler 1978, 1984), and the presence of many post starburst galaxies in clusters (Dressler & Gunn 1983, Couch & Sharples 1987). We have constructed simple models to evaluate the effect of such evolution on the observed M/L ratios. We assume that galaxies are assembled at $z=4$, and form stars in a very regular way. This occurs presumably in a disk. Then, at some time, they undergo a strong starburst, in which 10% of their mass is converted into stars, and they cease forming stars. Their spectra will be characteristic of post starburst galaxies for another 1.5 Gyr, and after that they will classified as normal early-type galaxies. This type of evolution implies that the morphologies of galaxies evolve with time, from spiral, to post star burst galaxy, to early-type. This has important consequences, since the set of early-types at higher redshifts will be a special subset of the set of early-types at $z=0$. If we select early-types at higher and higher redshift, we are selecting a subsample that is more and more biased towards the oldest early-types. In short, we may be selecting the oldest galaxies, and find that they are old.

The problem is illustrated in Fig. 1. The typical evolution of 3 galaxies is shown. The thick line is the phase in which they appear as early-types. Clearly, the oldest early-types appear as early-type for the longest time. Fig. 1b demonstrates the effect on the observed L/M ratios of a large sample. At low redshifts, all galaxies appear as early types, and the evolution of the median L/M ratio remains normal. The scatter around the mean increases rapidly with redshift. Around $z = 0.2$, some of the galaxies appear as post star burst galaxies, and they would be excluded. The median L/M ratio is biased towards low values. This effects increases at higher redshifts. The bias is as strong as 30 % at $z=0.5$. As galaxies disappear from the sample, the scatter in the L/M ratio may decrease at higher redshifts.

Many variations of such models can be made, and they will produce similar effects. These models have one common aspect: they predict that the morphologies of galaxies evolve with redshift. The morphologies of galaxies in distant clusters can now be determined accurately from HST images (e.g., Dressler et al. 1994). In combination with good spectral information, the morphological evolution can be measured directly.

The effect of this complex evolution will be similar on other properties such as colors, and linestrengths. The scatter is expected to increase at lower redshifts, and may actually decrease at higher redshifts. The median

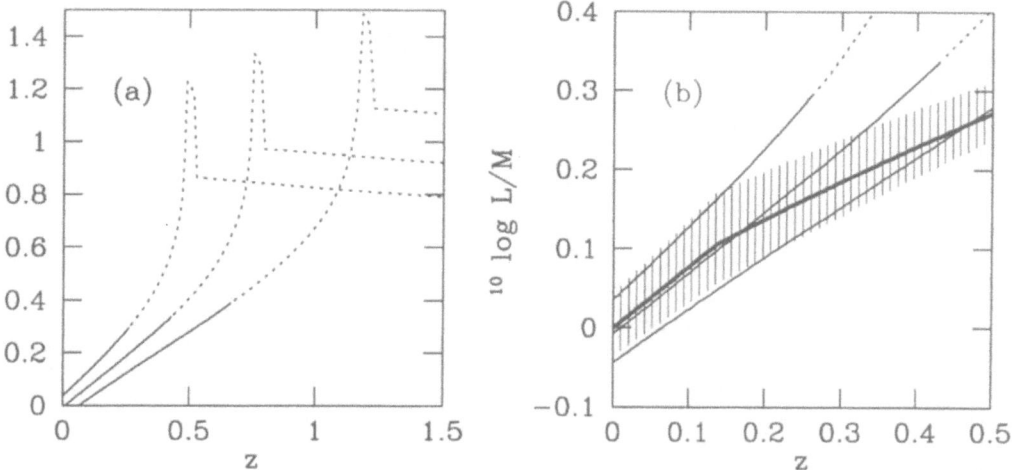

Figure 1. The evolution of galaxies which undergo three distinct phases: I regular star formation in a disk, II starburst, III quiescent evolution. a) shows the luminosity evolution for three such galaxies. The galaxies are classified as regular early-types after 1.5 Gyr after the burst. This is indicated by the thick line. b) the evolution of the mean L/M ratio for a sample of early-types which formed in this complex way. The starburst is assumed to occur at a random time between $z=0.5$ and $z=2$. The line indicates the median L/M, the shaded area is bounded by the upper and lower quartile of the sample. The median L/M ratio bends at $z=0.2$, as more and more galaxies drop out from the sample. The sample becomes more and more biased to the oldest early-types at higher redshifts.

color evolution may show a "break", at the redshift where galaxies start disappearing from the sample.

4. The Fundamental Plane in CL 0024+16 at $z=0.39$

CL 0024 is a rich cluster at $z=0.39$, and has been extensively observed (e.g., Dressler et al. 1985). We have obtained a deep, 19 hour integration at the MMT to measure the internal velocity dispersions of luminous galaxies in the cluster. In total, 13 galaxies were observed, and velocity dispersions were measured for 9. The dispersions were measured in the same way as for low redshift galaxies, with the important difference that the instrumental resolution had to be determined in an absolute way. This is due to the fact that it is impossible to obtain good template star spectra with the same instrumental setup (Franx, 1993a,b). Full details of the observations and the analysis can be found in van Dokkum and Franx (1996).

HST images were used to measure the length scales of the galaxies. The procedure consisted of convolving an $r^{1/4}$ model with the Point Spread Function, and fitting it directly to the data. In this fashion, the parameters r_e and I_e could be determined reliably. Although the errors in r_e and I_e can

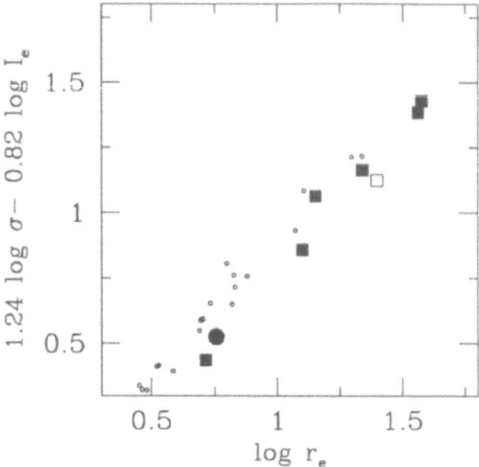

Figure 2.   The Fundamental Plane for galaxies in CL 0024+16 at z=0.391 in the V band. The small symbols are galaxies in Coma. The Fundamental Plane in CL 0024 is very similar to that in Coma, with similar low scatter (15%).

be substantial, the error in $r_e I_e^{0.8}$ is very small (on the order of 5%). The R photometry was reduced to restframe V band photometry, which made the color corrections very small.

Fig. 2 shows the resulting Fundamental Plane. There is a very clear relation, with relatively low scatter (15 %). The slope is very similar to that for nearby cluster galaxies (e.g., JFK). In short, *early-type galaxies exist at z=0.4 which are very similar to galaxies at z=0.* This extends earlier work on the colors, and color-magnitude relation of early-type galaxies (e.g., Bower et al. 1992).

The evolution of the M/L ratio can be determined from the data. Fig. 3a shows the observed M/L ratios for Coma and CL 0024 against the parameter $r_e^{0.22}\sigma^{0.49}$. The Fundamental Plane implies that galaxies lie along a line in the plot. If galaxies undergo only passive evolution, then the parameters r_e and σ remain constant with time. The evolution will cause a vertical shift with redshift. We see a clear offset between the two data sets. The lines indicate fits to both data sets. The mean difference in the M/L ratio is 31 %. The error is dominated by systematic effects, and is estimated at 12 %. It is clear that the sample for CL0024 is biased towards the most massive galaxies, and this selection bias is partly the cause for the systematic uncertainty.

Fig. 3b shows the evolution of the M/L ratio with redshift. The current results are in good agreement with a large formation redshift, and small values of κ. The resulting constraint is $\kappa(1 + q_0 + 1/z_{form}) = 0.84 \pm 0.32$. Obviously, many combinations of q_0 and z_{form} are allowed, given the uncertainty in κ and the observed change in M/L ratio.

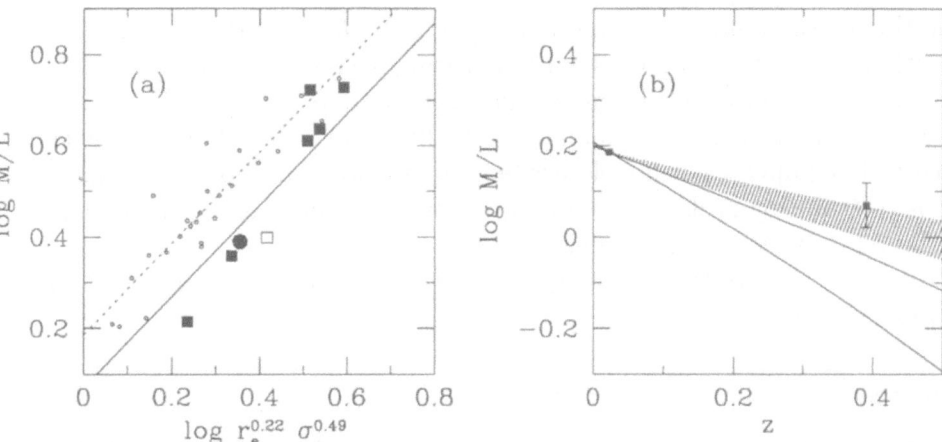

Figure 3.   a) The M/L ratio against $r_e^{0.22}\sigma^{0.49} \propto M^{0.24}$, for CL 0024 and Coma. The lines are fits to the data points. The M/L ratio in CL 0024 is lower by 31 ± 12 %. b) The evolution of the M/L ratio against redshift. The two points are Coma and CL 0024. The hatched area is the area allowed by models with formation redshift $z_{form} = \infty$. The two lines delineate the area allowed by models with $z_{form} = 1$. Such models are marginally consistent with the data.

5. Discussion

We have shown that the equivalent of the Fundamental Plane exists at $z=0.39$, and we have derived the mean evolution of the M/L ratio. The evolution is low at 31%, but is in agreement with models. The uncertainties in the modeling still allow a wide range in formation redshifts z_{form}.

The next step is to observe more and fainter galaxies in distant clusters, and to observe more clusters. Such measurements can address questions about the change in slope of the Fundamental Plane, and the dependence of galaxy evolution on environment.

The results can be compared in a qualitative way to measurements of the evolution of the luminosity function in clusters and the field. The evolution of the luminosity function gives additional information on the luminosity evolution of individual galaxies, with the added uncertainty that dissipationless merging and infall can affect the luminosities greatly. The M/L ratio is much less sensitive to such processes. Aragon-Salamanca et al. (1993) found that the K luminosity function of cluster galaxies does not evolve with redshift. The uncertainties are still large, however, and the measurements are not inconsistent with the evolution of the M/L as measured from the Fundamental Plane. Lilly et al. (1995) and Ellis et al. (1996) found very weak evolution of the bright end of the luminosity function in the field. Lilly et al. took one further step, and derived very weak evolution for red galaxies in general. All these measurements are qualitatively in agreement with the current results. As more data will be gathered on the evolution of the luminosity function and the M/L ratio, the general mass evolution of galaxies may be determined from such data in the future.

Studies of color evolution of galaxies in clusters indicate weak evolution for $0 < z < 0.5$ (Aragon-Salamanca et al. 1993, Rakos & Schombert 1995). Since stellar population models predict rapid color evolution when the luminosity evolution is slow, it is not clear yet how good these measurements can be reproduced by the models.

It is a pleasure to thank the organizers for a stimulating conference, and for financial support.

References

Aragon-Salamanca A., Ellis R. S., Couch W. J., Carter D., 1993, MNRAS, 262, 764
Bender R., Burstein D., Faber S. M., 1992, ApJ, 399, 462
Bower R., Lucey J. R., Ellis R. S. 1992, MNRAS, 254, 589
Butcher H., Oemler A., 1978, ApJ, 219, 18
Butcher H., Oemler A., 1984, ApJ, 285, 426
Buzzoni A., 1989, APJS, 71, 817
Couch W. J., Sharples R. M., 1987, MNRAS, 229, 423
Djorgovski S., Davis M., 1987, ApJ, 313, 59
Dressler A., Gunn J. E., 1983, ApJ, 270, 7
Dressler A., Gunn J. E., Schneider D. P., 1985, ApJ, 294, 70
Dressler A., Lynden-Bell D., Burstein D., Davies R. L., Faber S. M., Terlevich R. J., Wegner G., 1987, ApJ, 313, 42
Dressler A., Oemler A. J., Butcher H. R., Gunn J.E., 1994, ApJ, 424, 79
Ellis R. S., Colless M., Broadhurst T., Heyl J, Glazebrook K., 1996, preprint
Faber S. M., Dressler A., Davies R. L., Burstein D., Lynden-Bell D., Terlevich R., Wegner G., 1987, Faber S. M., ed., Nearly Normal Galaxies. Springer, New York, p. 175
Faber S. M., Jackson R. E., 1976, ApJ, 204, 668
Franx M., 1993a, ApJ, 407, L5
Franx M., 1993b, PASP, 105, 1058
Franx M., 1995, van der Kruit P. C., Gilmore G., eds, Proc. IAU Symp. 164, Stellar Populations. Kluwer, Dordrecht, p. 269
Franx M., van Dokkum P. G., 1996b, in preparation.
Jørgensen I., Franx M., Kjærgaard P., 1993, ApJ, 411, 34
Jørgensen I., Franx M., Kjærgaard P., 1995, MNRAS, accepted [JFK]
Kjærgaard P., Jørgensen I., Moles M., 1993, ApJ, 418, 617
Lilly S. J., Tresse L., Hammer F., Crampton D., Le Fèvre O., 1995, ApJ, in press.
Lucey J. R., Guzmán R., Carter D., Terlevich R. J., 1991, MNRAS, 253, 584
Mather J. C., et al., 1994, ApJ, 420, 439
Renzini A., Ciotti L., 1993, ApJ, 416, L49
Rakos K., Schombert J., 1995, ApJ, 439, 47
Tinsley B. M., 1980, Fundamentals of Cosmic Physics, 5, 287
Tully R. B., Fisher J. R., 1977, AA, 54, 661
van Dokkum P. G., Franx M. 1996, MNRAS, submitted
Worthey G., 1994, APJS, 95, 107

THE LINEWIDTH–LUMINOSITY RELATION
FOR BLUE GALAXIES AT A REDSHIFT OF∼ 0.25

HANS-WALTER RIX
MPIA, Garching and Steward Observatory

MATTHEW COLLESS
MSSSO, Australian National University

AND

PURAGRA GUHATHAKURTA
Lick Observatory, UC Santa Cruz

1. Introduction

This is a progress report on a collaborative effort to probe galaxy evolution by constraining the central mass-to-light ratios of distant field galaxies. Details can be found in a forthcoming paper (Rix, Guhathakurta, Colless & Ing 1996, RGCI). At the moment, the program is focused on the question of how over-luminous the abundant blue field galaxies at $z \sim 0.3$ are (see Koo, K. and Kron, R. 1992), compared to galaxies of the same circular velocity and type at $z \sim 0$. We have measured the integrated [OII] linewidths for a complete sample of distant blue galaxies to establish a linewidth–luminosity–color relation for this population and to compare this relation to the one for local galaxies. In addition, we explore practical strategies to obtain and analyze data so as to best relate the observable quantities (*e.g.* linewidths) to the physical quantities of interest (such as v_{circ} or M/L).

2. Sample and Observations

The target galaxies were a random subset of LDSS-1 targets (Colless *et al.* 1993), with magnitudes of $21 < m_b < 22$ and colors $b - r < 1.2$. Their integrated emission line spectra for the [OII] doublet were obtained with the AUTOFIB spectrograph at the AAT on the night of October 8, 1993. The spectra covered the redshift range of $0.17 < z < 0.37$, at a mean spectral resolution of $\sigma_{inst} = 50$ km/s. The 2″ fibres integrate the flux

R. Bender and R. L. Davies (eds.), New Light on Galaxy Evolution, 241–244.

over an aperture of 9 kpc $\times[\frac{z}{0.25}]$ diameter ($H_0 = 70$ km/s/Mpc). Several arguments suggest (see RGCI) that the 24 detected emission lines include *all* galaxies in the above redshift, magnitude and color range. The emission line widths (LW) were determined by fitting two Gaussians simultaneously to the [OII] doublet, accounting for the instrumental broadening. Most lines were found to be resolved ($\sigma \gtrsim 30$ km/s), but all LW were $\sigma < 100$ km/s. The results of the line fitting for all 24 sample members are displayed in Figure 1.

3. Data Modeling

Given their [OII] line luminosity, the LWs observed here are 2–3 times too large to match the relation between the line luminosity and *turbulent* line width, observed locally for giant HII regions and "HII galaxies". Hence, the LWs most likely arise from orbital motion and can be related to the circular velocity, v_c, characterizing the galaxy.

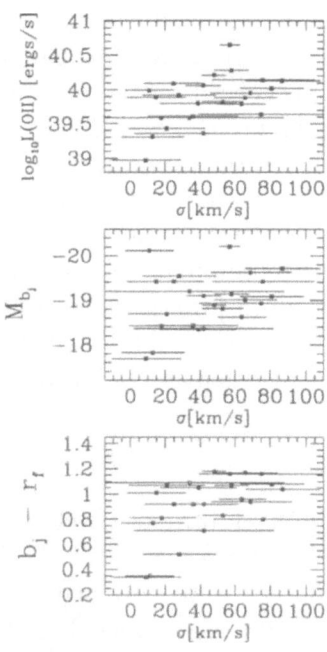

Figure 1.   Global galaxy properties *versus* [OII] linewidth. Integrated [OII] linewidths, σ, were obtained with $2''$ fibres at the AAT for a complete sample of faint ($m_b \sim 22$), blue ($B - V_{rest} < 0.45$) galaxies at $z \sim 0.25$. The three panels compare sigma to the line luminosity, the blue absolute magnitude and the $b - r$ color, respectively. There is a trend for σ to increase with L([OII]), M_b and $b - r$, but the scatter is large. Note that the LW σ has not been corrected for inclination.

Before we can test with these data whether the linewidth–luminosity–color (LWLC) relation in the distant galaxies differs from the local relation, considerable modeling is required:

(1) How is the observed, integrated (Gaussian) width, σ, of the emission lines related to the (asymptotic) circular velocity, v_c, of the galaxy? Aside from their mathematical definition, these two quantities differ because the

ionized gas has a lumpy asymmetric distribution and because not all of it originates from the "flat" part of the rotation curve. Further, from ground-based data we cannot determine the galaxies' inclinations and must instead assume a distribution in $\cos i$. We try to account for all these effects by using Fabry-Perot data cubes of local galaxies — with similar magnitudes and colors — to simulate the observations of the distant galaxies. The velocity fields of these calibrators were fitted and the emission line flux image was deprojected. Subsequently, the velocity fields and flux images were projected at many random angles and for each the line profile was sampled over an aperture corresponding to the $2''$ fibres at $z \sim 0.25$. Gaussians were fit to the line profile and the σ was compared to v_c obtained from the 2D velocity field. This procedure leads to the probability distribution, $p(\sigma|v_c)$, of observing σ with our set-up, given v_c. This distribution peaks at $\sigma \approx 0.55v_c$ and is quite broad (FWHM $\sim 0.45v_c$). In the subsequent analysis we assume that $p(\sigma|v_c) = p(\sigma/v_c)$.

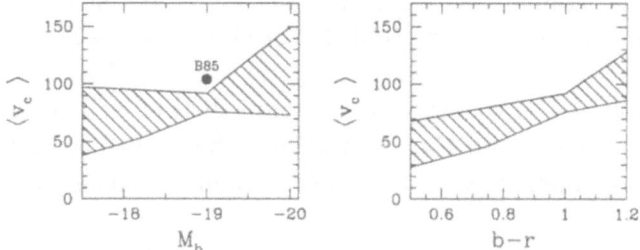

Figure 2. Constraints on the LWLC relation of blue field galaxies at $z \sim 0.25$. The left panel shows the dependence of the mean circular velocity on (blue) luminosity; the two lines enclose the 68% confidence region from a maximum likelihood analysis. The point B85 indicates the LWL relation for a local sample (Bothun *et al.* 1985), after reducing $W_{20}/2/\sin i$ by 15% to match v_c (optical). The right panel shows the $\langle v_c \rangle$ dependence on color; bluer galaxies have smaller linewidths.

(2) We assume that our distant sample galaxies obey a LWLC relation of the form

$$\ln\left[v_c(M_b,\ b-r)\right] = \ln\left[v_c(-19, 1)\right] - \eta\left(M_b + 19\right) + \zeta\left[(b-r) - 1\right]$$

with a fractional velocity scatter of Δ. For each galaxy with $v_c(M_b,\ b-r)$ we should expect to observe σ with a probability $p(\sigma/v_c)$. For each combination of parameters $\left[v_c(-19, 1),\ \eta,\ \zeta\right]$ our set of measurements, σ^i_{obs} and $\Delta\sigma^i_{obs}$ ($i = 1,\ 24$) defines a likelihood, $\mathcal{L}$. The maximum of $\mathcal{L}$ defines the best fit parameter values, and contours offset by various $\Delta\mathcal{L}$ from $\mathcal{L}_{max}$ describe confidence limits for the parameters. Figure 2 shows the results of this analysis: while $v_c(-19, 1) = 85\pm8$ km/s is well constrained, the slopes η and ζ are not. However, the trend that bluer galaxies have smaller linewidths (at fixed b-band luminosity), is significant at the $\geq 95\%$ confidence limit. This analysis also reveals that all the scatter in Figure 1 is consistent with being due to the breadth of $p(\sigma/v_c)$ and due to the observational errors; the intrinsic scatter in the relation, Δ, may be small.

4. Results and Comparison with Local Samples

As Figure 2 shows, these blue $\sim$L$_*$/4 galaxies at $z \sim 0.25$ have linewidths that correspond to circular velocities of ~ 85 km/s. To test for evolution in the LWLC relation with cosmic epoch, we have to define the corresponding local relation. We have attempted to do so by using Tully-Fisher data for nearby galaxies of the same magnitude and color range (*e.g.* Bothun *et al.* 1985). After a 15% downward correction from $W_{20}/(2\sin i)$ to the "optical" v_c, we find that the linewidths for the local galaxies are 15% larger than for the distant sample. For a relation $L_b \propto v_c^3$, this means that blue galaxies of the same linewidth were $\sim 50\%$ more luminous at $z \sim 0.25$ than they are now.

We have presented a novel way to compare distant and local galaxy samples. However, at this point the analysis must be viewed as preliminary and the evolutionary inferences just reached should be taken with several grains of salt:

• Relating the observable σ to v_c requires a large and model dependent correction. This problem may be worst for a fibre measurement, where all spatial information is forsaken, but is likely to plague all ground-based observations with resolutions $\gtrsim 0.5''$. An important source of error could be eliminated by obtaining galaxy inclinations from HST imaging.

• Most local Tully-Fisher studies focus on galaxies with $L >$L$_*$ and have morphological sample selection criteria which cannot be reproduced in the distant samples. Better samples for local comparison are needed.

• Detailed, but telescope time consuming, studies of the spatially resolved velocity fields in a few distant galaxies (ground-based or with STIS) will be extremely helpful in gauging the $p(\sigma/v_c)$ relation.

References

Bothun, G. *et al.*, 1985, ApJS, 57, 423
Colless, M., Ellis, R., Broadhurst, T., Taylor, K., & Peterson, B., 1993, MN, 261, 18.
Koo, D. and Kron, R., 1992, ARAA, 30, 613
Rix, H.-W., Guhathakurta, P., Colless, M. and Ing, K., 1996, MNRAS, *to be submitted*

THEORY OF GALAXY FORMATION

GALAXY FORMATION AND EVOLUTION: WHAT TO EXPECT FROM HIERARCHICAL CLUSTERING MODELS

C.S. FRENK, C.M. BAUGH AND S. COLE
Physics Department
University of Durham, UK

1. Introduction

In hierarchical clustering theories of galaxy formation, galaxies form by gas cooling and condensing into dark matter halos which, in turn, form by a hierarchy of mergers (White & Rees 1978). The context in which this process takes place is specified by a cosmological model that determines the spectrum of primordial density fluctuations and the rate at which they grow by gravitational instability. The best known example of such a model is the cold dark matter (CDM) model (see Frenk 1991 for a review), but a number of alternatives (mostly variants of CDM), have recently become popular in response to new data on large-scale structure and the COBE detection of anisotropies in the microwave background radiation. Regardless of the specific cosmological model that one wishes to consider, there are at least six distinct physical processes that need to be included in any theory of galaxy formation:

- 1. The growth of dark matter halos by accretion and mergers.
- 2. The dynamics of cooling gas.
- 3. Star formation.
- 4. Energy feedback into prestellar gas from the products of stellar evolution.
- 5. Evolution of the stellar populations that form.
- 6. Galaxy mergers.

A number of theoretical tools have been developed over the years to investigate these processes, both individually and collectively. N-body simulations have led to significant progress in understanding process *(1.)*, while the recently developed N-body/hydrodynamic techniques are beginning to

R. Bender and R. L. Davies (eds.), New Light on Galaxy Evolution, 247–254.

address processes *(1-4)* and *(6)*. In addition, semianalytic modelling, a relatively new tool, can treat all six processes together and thus explore the effects of different assumptions on the properties of the galaxy population as a whole. In this review, we will outline some of the areas where progress has been made and highlight some as yet unresolved issues.

2. Physical processes

2.1. EVOLUTION OF DARK MATTER HALOS

The main features of the formation of dark matter halos by hierarchical clustering were already established in N-body simulations carried out a decade ago (eg. Frenk *et al.* 1985, 1988; Efstathiou *et al.* 1988). A protohalo perturbation, initially expanding at a reduced rate, collapses, often into filamentary or sheet-like structures, which subsequently break up into roughly spherical lumps. These merge together producing a centrally concentrated and essentially smooth dark halo. This process is illustrated in Figure 1 which shows the development of a galactic halo in a flat 'low'-Ω CDM model.

One of the main early results from N-body simulations was the realisation that the rotation curves of dark galactic halos in the the standard CDM model are approximately flat, suggesting an explanation for the inferred structure of the halos of spiral galaxies (Frenk *et al.* 1985, Quinn *et al.* 1986). These simulations, however, were limited in particle number and did not resolve the inner regions of the halos where the visible galaxy actually forms. This issue has recently been addressed in a series of high-resolution simulations by Navarro, Frenk & White (1995). The density profiles of galactic halos in the CDM model show noticeable departures from an r^{-2} law, gently sloping from r^{-1} near the centre to r^{-3} near the virial radius. When the gravitational effect of a disk is included, the resulting rotation curves agree well with observations of galaxies, from dwarfs to bright galaxies, provided the disks fulfill two conditions: (i) their stellar mass-to-light ratio increases roughly as $L^{1/2}$ and (ii) the baryon fraction increases with luminosity such that for galaxies with observed circular velocity, $V_c \gtrsim 200 \mathrm{kms}^{-1}$, there is only a weak of correlation between this velocity and total halo mass. It is unclear whether the observed disks of spirals satisfy these conditions.

2.2. THE DYNAMICS OF COOLING GAS

The main ideas here were put forward nearly twenty years ago by Rees & Ostriker (1977), Silk (1977) and Binney (1977). When a dark matter halo collapses, any gas admixed with it will also collapse, but whereas

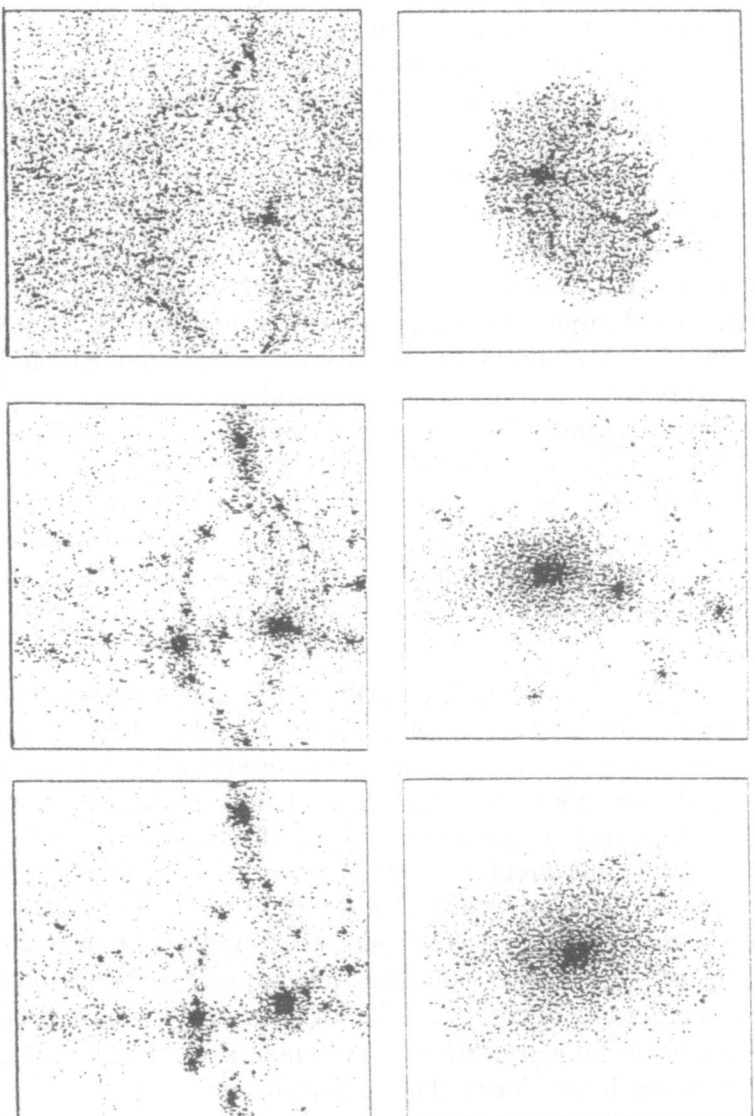

Figure 1. The formation of a galactic dark matter halo in an N-body simulation. The left-hand column shows the projected particle distribution, in comoving coordinates, of a cubical region of present comoving length 27 Mpc. The right hand column shows, now in physical coordinates, the growth of the large clump seen in the bottom left of the region. Each panel on the right hand row has length 3.8 Mpc. From top to bottom the epochs shown correspond to $z = 5, 0.5$ and 0. The parameters of the simulation are: mean density, $\Omega = 0.3$; cosmological constant, $\Lambda = 0.7$; Hubble constant, $H_0 = 100h\,\mathrm{km\ s^{-1}\ Mpc^{-1}}$, with $h = 0.7$; and spectrum normalisation, $\sigma_8 = 1.14$ (as inferred from the COBE data). The simulation followed 262144 particles and was performed the T3D parallel supercomputer at Edinburgh.

the dark matter free streams, the gas is shock heated. These early papers assumed that shocks would heat up the gas to the virial temperature of the halo, an assumption verified – in the non-radiative limit – in the N-body/hydrodynamic simulations of Evrard (1990). These and subsequent simulations (eg. Katz & White 1993, Navarro, Frenk & White 1995) also showed that, in this limit, the gas acquires a density profile that closely parallels that of the dark matter. Rees & Ostriker argued that if the cooling time of virialised gas was shorter than its dynamical time, the gas would collapse to make a galaxy. White & Rees (1978) recognized, however, that in a hierarchical model this simple scheme would lead to a cooling catastrophe since at early times the density is so high that all the gas would cool into subgalactic lumps where it would pressumably turn into stars. This patently did not happen in the universe - there is still plenty of gas around today. White & Rees solved this problem by introducing the idea of feedback, whereby the energy released by supernovae associated with an early generation of stars reheats some of the gas before it has had a chance to condense into halos at high redshift. (Efstathiou 1992 has argued that photoionisation by a UV background at high redshift would have a similar effect.)

Testing these simple physical arguments in numerical simulations is difficult because the propensity of the gas to cool at high density implies that the behaviour of the gas is always determined by the resolution limit of the simulation. This numerical artefact, however, can be turned to advantage if it is loosely interpreted as an effective source of feedback. N-body/hydrodynamic simulations of representative cosmological volumes in which the gas is allowed to cool are still at an early stage (eg Katz *et al.* 1992, Cen & Ostriker 1992, Frenk *et al.* 1995) and show that the behaviour of the gas is more complex than expected in the simple analytic picture. Simulations of the formation of individual galaxies produce disks, often with beautiful spiral arms (e.g. Steinmetz & Muller 1995), but these disks rotate much too slowly. This is because merger events transfer angular momentum from gas fragments to the outer dark matter halo in much the same way as the mergers of collionsless particles do (Navarro, Frenk & White 1995). This angular momentum problem for disks remains a major unresolved issue in studies of galaxy formation. One possible solution may be, again, to invoke some form of feedback which might keep the gas hot and allow it to cool slowly rather to be collected in subclumps.

2.3. STAR FORMATION AND FEEDBACK

Current understanding of star formation and the attendant feedback, in the context of galaxy formation, is laughably poor. All that can be done at present is to try and model these processes in a heuristic fashion. For example, in an N-body/hydrodynamic simulation one can stipulate a number of conditions for gas to turn into stars, eg, that it be cool and dense (ie above the Jeans mass) and that it be inflowing into a halo. Systematic tests of such algorithms are just beginning (e.g. Navarro & White 1993).

2.4. GALAXY MERGERS

Simulations of the merging of individual galaxy pairs or small groups have a long and distinguished history (see Barnes's contribution elsewhere in this volume). Such simulations address issues such as the structure and rotation properties of merger remnants, or the gas flows triggered by mergers. From the point of view of galaxy formation in general, a key issue is the relative timescale for the merging of dark matter halos and the galaxies they harbour. As a consequence of their higher binding energy, galaxies take longer to merge than their halos. Furthermore, the simulations show that galaxies (or at any rate the clumps of cool gas identified with galaxies in the models) merge on a dynamical friction timescale, provided that the mass that is input into Chandrasekhar's classic formula is the total, gas plus dark matter, mass of the merging satellite (Navarro, Frenk & White 1995).

3. Semianalytic models

Our understanding of the full range of complex phenomena listed in the Introduction can be approximated by a set of simple rules. These rules can then form the basis of a semianalytic model for galaxy formation, that follows the collapse and mergers of dark matter halos and the star formation histories of the galaxies. Such models (Kauffman *et al.* 1993, Lacey *et al.* 1993, Cole *et al.* 1994) have been successful in accounting for the general properties of the observed galaxy distribution, such as the shape of the luminosity function, faint number counts and colours. However, a number of fundamental problems remain that appear to suggest that the modelling of the processes (1 - 6) needs to be improved, rather than altering the choice of cosmology (Heyl *et al.* 1994).

In the original model of Cole et al. (1994), the model galaxies were not as red as many observed ellipticals. A study of several stellar population codes (Charlot *et al.* 1995) has led to a revision of the Bruzual and Charlot (1993) models. This has resulted in the model galaxies being typically 0.2 mag redder in $B - K$. An updated comparison of the colour distribution of the model galaxies from Cole et al. with observed colours is given in Figure 2.

We also present an updated versions of the redshift distributions predicted by the model of Cole *et al.* Figure 3 compares the model predictions

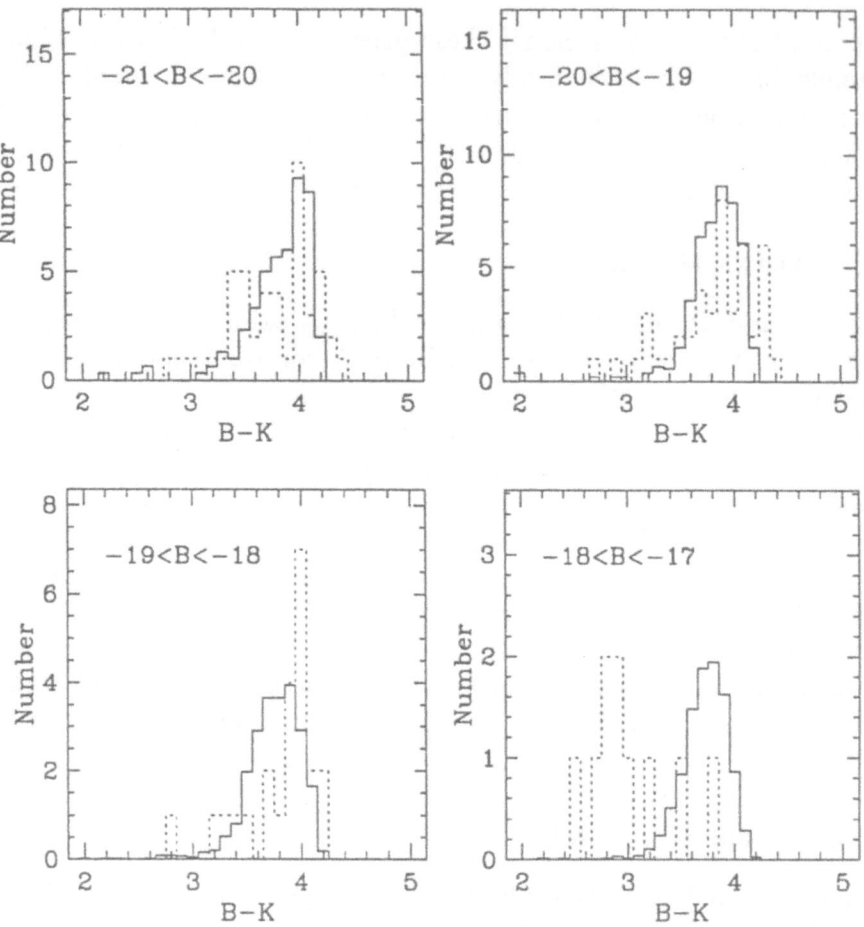

Figure 2.   Histograms of B-K colour distributions for various ranges of B absolute magnitude. The broken lines are data from Mobasher *et al.* (1986) and show the observed number of galaxies in the data set. The model output is shown by the solid lines which have been normalised to enclose the same area as the data.

with the recent redshift survey data from the surveys of Glazebrook *et al.* (1995a,b). The model and observed distributions are in very good agreement.

The luminosity function of Cole et al. was flatter than that achieved by other semianalytic models, because of the strong feedback adopted, which severely restricts star formation in halos of low circular velocity. However, this model still predicts more faint galaxies than are observed (Loveday et al 1992), though recent results indicate that the faint end of the luminosity function is still uncertain (McGaugh 1994, Marzke *et al.* 1994).

In addition, the models reproduce the observed slope and small scatter

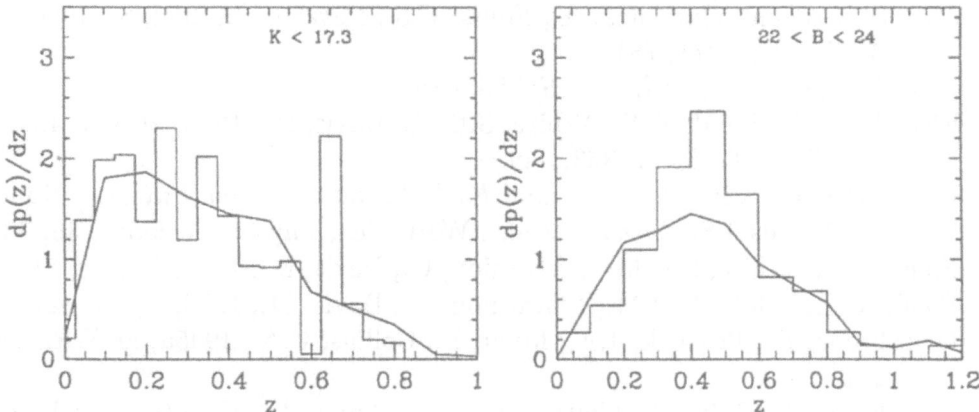

Figure 3. The redshift distribution of model galaxies compared with new data for K and B-band selected samples of Glazebrook *et al.* 1995a,b. The histograms show the observed distributions and the curves show the model galaxy redshifts. The K band data consists of 124 redshifts for galaxies with $K < 17.3$ and is weighted for incompleteness. The B band sample is 70% complete and contains 70 galaxies with $22 < B < 24$.

of the observed Tully-Fisher relation between luminosity and circular velocity, but give galaxies that are either too faint or that have too high a circular velocity for their luminosity. This can be traced back to an overproduction in CDM-like cosmologies of halos typical of those that contain luminous galaxies.

Both the Tully-Fisher and luminosity function problems could be related to surface brightness effects. We plan to incorporate a scale length into the models, allowing us to select only those galaxies above some surface brightness threshold.

We have extended the model to split the light of each galaxy up into a bulge and a disk component, with the bulges formed in violent merger events. This allows us to make a broad morphological classification of our galaxies and make predictions of galaxy properties as a function of bulge to disk ratio. For example we find good agreement with the faint HST counts of Glazebrook *et al.* (1995c) and recover the type of evolution in cluster membership reported by Butcher & Oemler (1978) (Baugh *et al.* 1995).

References

Baugh, C.M., Cole, S., Frenk, C.S., 1995, *in preparation*
Binney, J.J., 1977, *Ap.J*, 215, 483
Bruzual, G., Charlot, S., 1993, *Ap.J*, 405, 538
Butcher, H., Oemler, A., 1978, *A.J.*, 219, 18
Cen, R., Ostriker, J.P., 1992, *A.J.*, 339, 331
Charlot, S., Worthey, G., Bressan, A., 1995 *A.J.*, in press

Cole, S., Aragon-Salamanca, A., Frenk, C.S., Navarro, J., Zepf, S., 1994,
    M.N.R.A.S, 271, 781
Efstathiou, G. 1992, *M.N.R.A.S*, 256, 43p
Efstathiou, G., Frenk, C.S., White, S.D.M., Davis, M., 1988, *M.N.R.A.S*, 235
Evrard, A.E. 1990, *Ap.J.*, 363, 349.
Frenk, C.S. 1991, in *The Birth and Early Evolution of our Universe*, Nobel S
    No 79, eds J.S. Nilsson, *et al.* , World Sci., *Physica Scripta*, T36, 70.
Frenk, C.S., White, S.D.M., Efstathiou, G., Davis, M., 1985, *Nature*, 317, 595
Frenk, C.S., White, S.D.M., Efstathiou, G., Davis, M., 1988, *Ap.J.*, 327, 507.
Glazebrook, K., Peacock, J.A., Miler, L., Collins, C.A., 1995a, *M.N.R.A.S*,
    275, 169
Glazebrook, K., Ellis, R., Colless, M., Broadhurst, T., Allington-Smith, J.,
    Tanvir, N., 1995b, *M.N.R.A.S*, 273, 157
Glazebrook, K., Ellis, R., Santiago, B., Griffiths, R., 1995c, *M.N.R.A.S*, 275,
Heyl, J.S., Cole, S., Frenk, C.S., Navarro, J., 1995, *M.N.R.A.S*, 274, 755
Katz, N., Hernquist, L., Weinberg, D.H., 1992, *Ap.J*, 399, L109
Katz, N., White, S.D.M., 1993, *Ap.J*, 412, 455
Kauffmann, G., White, S.D.M., Guiderdoni, B., 1993, *M.N.R.A.S*, 264, 201
Lacey, C.G., Guiderdoni, B., Rocca-Volmerange, B., Silk, J., 1993, *Ap.J*, 402,
Loveday, J., Peterson, B.A., Efstathiou, G., Maddox, S.J., 1992, *Ap.J*, 390, 3
Marzke, R.O., Geller, M.J., Huchra, J.P., Corwin, H.G., 1994, *A.J.*, 108, 437
McGaugh, S., 1994, *Nature*, 367, 538
Mobasher, B., Sharples, R.M., Ellis, 1986, *M.N.R.A.S*, 223, 11
Navarro, J.F., White, S.D.M., 1993, *M.N.R.A.S*, 265, 271
Navarro, J.F., Frenk, C.S., White, S.D.M., 1995, *M.N.R.A.S*, in press
Rees, M.J., Ostriker, J.P., 1977, *M.N.R.A.S*, 179, 541
Silk, J., 1977 *Ap.J*, 211, 638
Steinmetz, M., Muller, E., 1995, *M.N.R.A.S*, 276, 549
Quinn, P.J., Salmon, J.K. and Zurek, W. 1986, *Nature*, **322**, 329.
White, S.D.M., Rees, M.J. 1978, *M.N.R.A.S*, **183**, 341.

Discussion

McGaugh: There is an extra factor of 2 in the Low Surface Brightness galaxy number density, but probably only $\sim$ 30% in luminosity density. What happens to the models if you turn off feedback?

Frenk: Feedback (and mergers) control the faint end slope of the luminosity function. If feedback is turned off this slope becomes unacceptably steep, with $\alpha \approx -2.5$.

THE STRUCTURE OF COLD DARK MATTER HALOS

JULIO F. NAVARRO
Steward Observatory
University of Arizona
Tucson, AZ 85721
U.S.A

Abstract. High resolution N-body simulations show that the density profiles of dark matter halos formed in the standard CDM cosmogony can be fit accurately by scaling a simple "universal" profile. Regardless of their mass, halos are nearly isothermal over a large range in radius, but significantly shallower than r^{-2} near the center and steeper than r^{-2} in the outer regions. The characteristic overdensity of a halo correlates strongly with halo mass in a manner consistent with the mass dependence of the epoch of halo formation. Matching the shape of the rotation curves of disk galaxies with this halo structure requires (i) disk mass-to-light ratios to increase systematically with luminosity, (ii) halo circular velocities to be systematically lower than the disk rotation speed, and (iii) that the masses of halos surrounding bright galaxies depend only weakly on galaxy luminosity. This offers an attractive explanation for the puzzling lack of correlation between luminosity and dynamics in observed samples of binary galaxies and of satellite companions of bright spiral galaxies, suggesting that the structure of dark matter halos surrounding bright spirals is similar to that of cold dark matter halos.

1. Introduction

Dark matter halos are often modelled as isothermal potential wells whose depth is usually identified with the velocity dispersion of stars in spheroidal galaxies or with the disk rotation speed in spirals. Since velocity dispersion increases with galaxy luminosity, this implies that more massive halos should surround brighter galaxies, a hypothesis that, however, does not

R. Bender and R. L. Davies (eds.), New Light on Galaxy Evolution, 255–258.
© 1996 IAU. Printed in the Netherlands.

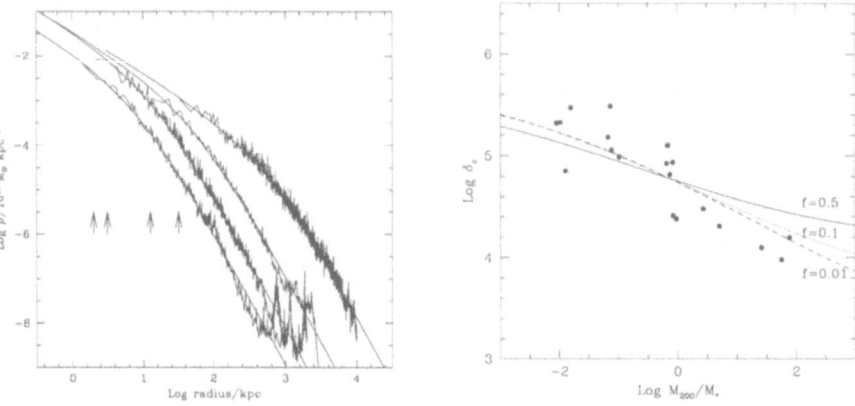

Figure 1. (a–left) The typical density profiles of CDM halos. The leftmost (rightmost) system has a mass $M_{200} = 3 \times 10^{11} M_\odot$ ($M_{200} = 3 \times 10^{15} M_\odot$). Arrows indicate the value of the gravitational softening in each simulation. (b–right) The characteristic overdensity of halos as a function of mass. Masses are expressed in units of the current non linear mass corresponding to the standard biased CDM spectrum, $M_* = 3.3 \times 10^{13} M_\odot$.

seem to be supported by observations. Studies of the dynamics of binary galaxies and of spiral/satellite samples have revealed a distinct lack of correlation between the dynamics and the luminosity of the system (White *et al.* 1983, Zaritsky & White 1994). Motivated by this disagreement, we have decided to investigate the structure of dark matter halos in the standard biased ($b = 1.6$) CDM cosmogony using high-resolution N-body simulations. A total of 19 halos with masses ranging from that of a rich galaxy cluster ($\sim 10^{15} M_\odot$) to that of a dwarf galaxies ($10^{11} M_\odot$) were selected from large cosmological simulations and resimulated individually with higher resolution. (I assume $H_0 = 50$ km s^{-1} Mpc^{-1} for all physical quantities quoted in this paper.) The numerical parameters were carefully chosen so that the numerical resolution of each simulated halo was the same, enabling meaningful comparisons between systems of widely different mass.

2. Results

Figure 1a shows the density profiles of four halos, spanning four orders of magnitude in mass. The solid curves are fits using a density profile of the form suggested by Navarro, Frenk & White (1995a),

$$\frac{\rho(r)}{\rho_{crit}} = \frac{\delta_c}{(r/r_s)(1 + r/r_s)^2}. \tag{1}$$

Remarkably, all the profiles are very well fit by this simple functional form. (Here ρ_{crit} is the critical density and r_s is a "scale" radius.) If radii are

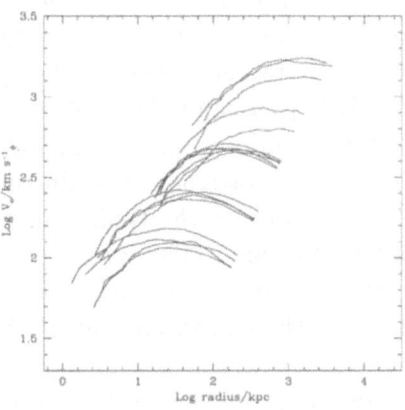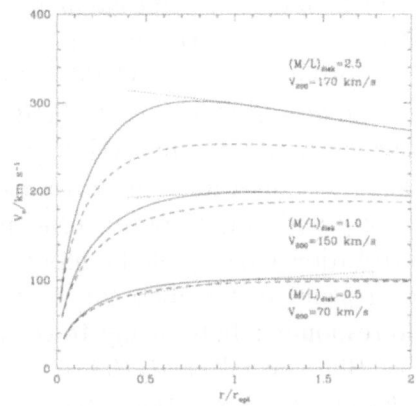

Figure 2. (a–left) The circular velocity as a function of radius for the 19 halos simulated in our series. The curves are truncated at the virial radius, r_{200}, of each system. (b–right) The result of matching the shape of observed disk rotation curves (dotted lines, from Persic & Salucci 1995). The dashed lines are the halo contribution to the total circular velocity (solid lines) in each case. Radii are expressed in units of the optical radius of each galaxy, defined as 3.2 times the exponential radial scalelength. $(M/L)_{disk}$ values are in the B-band.

expressed in units of the virial radius, r_{200}, (which determines the total mass of the system, M_{200}) there is a single free parameter in eq.(1); the characteristic overdensity δ_c. Figure 1b shows that this overdensity correlates strongly with the mass of the halo; less massive systems tend to be denser than their larger counterparts. Such a trend is expected in hierarchical clustering scenarios such as CDM, since lower mass scales collapse at higher redshift and should therefore have typically higher characteristic densities. If we define the formation redshift of a halo of mass M as the first time when half of its final mass was in progenitors with individual masses exceeding some fraction f of M, we can compute analytically the expected dependence of δ_c on M by assuming that characteristic overdensity of a halo merely reflects the density of the universe at the time of formation, i.e. $\delta_c(M) \propto (1 + z_{form}(M))^3$. The curves in Figure 1b show that such identification provides a good description of the results of the numerical experiments for various values of the parameter f, lending support to the conclusion that the characteristic density of a halo is determined primarily by its formation redshift. (See Navarro, Frenk & White 1995b for details.)

The circular velocity, $V_c(r) = (GM(r)/r)^{1/2}$, of all 19 halos is shown in Figure 2a as a function of radius. Consistent with eq.(1), V_c rises near the center and declines near the virial radius. Over a wide range in radius V_c is almost constant, in agreement with the results of previous, lower resolution studies (Frenk *et al.* 1985, 1988, Quinn *et al.* 1986). Is the radial behaviour of V_c consistent with the observed rotation curves of galaxy disks? To compute the rotation curve of disk galaxies with a given rotation speed forming within a CDM halo, only two parameters need to be specified; the disk stellar mass-to-light ratio, $(M/L)_{disk}$, and the halo mass or, equivalently, its circular velocity at the virial radius, $V_{200}^2 = GM_{200}/r_{200}$. (This is because, for a given rotation speed, observations constrain the luminosity and optical radius of a spiral disk, as well as the typical shape of the rotation curve.) Slight modifications to the halo structure caused by the presence of the disk can be taken into account by assuming that the halo responds adiabatically to the assembly of the galaxy (Barnes & White 1984, Blumenthal et al. 1986).

Figure 2b shows the values of these two parameters needed to match the typical rotation curves of spiral disks, as given by Persic & Salucci (1995). Some trends are clear; faster rotating (brighter) disks require larger disk mass-to-light ratios, $(M/L)_{disk} \propto L^{1/2}$, and the asymptotic halo circular velocity is systematically lower than that of the disk. Furthermore, bright galaxies, i.e. those with $V_{rot} > 200$ km/s, are surrounded by halos of very similar mass. Indeed, disks with rotation speeds 200 or 300 km/s should be surrounded by halos whose mean circular velocities differ by only $\sim 10\%$ and which can be up to a factor of two lower than the disk rotation speed. This result agrees well with the estimates of Zaritsky & White (1994), who found from their study of the dynamics of satellite galaxies that the average circular velocity of the halos of bright spirals is between 180 and 200 km/s. The weak correlation between halo and disk circular velocity shown in Fig. 2b also provides a simple explanation for the lack of correlation found between luminosity and dynamics of binary galaxies and satellite/primary pairs. This result is especially encouraging, since the halo properties we infer were chosen to match the shape of the inner rotation curves, and do not use any information about dynamics at the much larger radii probed by binaries or satellite companions. We conclude that the structure of dark halos surrounding spiral galaxies is quite similar to that of Cold Dark Matter halos.

References

Barnes, J. & White, S.D.M., 1984, MNRAS, 211, 753.
Blumenthal, G.R., Faber, S.M., Flores, R., & Primack, J.P. 1986, ApJ, 301, 27.
Frenk, C.S., White, S.D.M., Efstathiou, G.P., and Davis, M. 1985, Nature, 317, 595.
Frenk, C.S., White, S.D.M., Davis, M., and Efstathiou, G.P. 1988, ApJ, 327, 507.
Navarro, J.F., Frenk, C.S., & White, S.D.M. 1995a, MNRAS, 275, 720.
Navarro, J.F., Frenk, C.S., & White, S.D.M. 1995b, ApJ, in press.
Persic, M., & Salucci, P. 1995, ApJSS, in press.
Quinn, P.J., Salmon, J.K., & Zurek, W.H. 1986, Nature, 322, 329.
White, S.D.M., Davis, M., Huchra, J., Latham, D. 1983, ApJ, 203, 701.
Zaritsky, D., & White, S.D.M. 1994, ApJ, 435, 599 .

MERGERS AND THE FORMATION OF DISK GALAXIES IN HIERARCHICALLY CLUSTERING UNIVERSES

M. STEINMETZ
Max-Planck-Institut für Astrophysik
Postfach 1523, D-85740 Garching, Germany

1. Introduction

It is commonly accepted, that galaxies acquire their angular momentum from tidal torques due to surrounding matter (Hoyle 1949). Most of the angular momentum is gained during the linear phase of the collapse (White 1984). After the turnaround point is reached, any further gain of angular momentum is strongly suppressed because it is increasingly difficult for the tidal field to act on the shrinking radius of the forming dark halo.

Due to its high cooling capabilities, the gas within the halo, which at turnaround possesses an angular momentum distribution similar to the dark halo, is virtually isothermal. In the standard picture, as it was essentially drawn by Fall & Efstathiou (1980), this gas cannot be supported by pressure and collapses into a thin, rotationally supported disk. The scale length of the disk is determined by the spin parameter λ of the dark matter halo and the dark matter to baryon ratio f_b within the halo. For reasonable values for the spin parameter ($\lambda \approx 0.07$, Barnes & Efstathiou 1987, Steinmetz & Bartelmann 1995) and $f_b \gtrsim 10$, the expected scale length ($\approx 3\,\mathrm{kpc}$) of a Milky–Way sized galaxy is in good agreement with observations.

This picture, however, only holds if no angular momentum is transported, i.e. the galaxy collapses axisymmetrically. If, however, galaxies form hierarchically, they are successively built up by merging of smaller structures (see, e.g., Lacey & Cole 1993). But mergers are typically identified with the sites where ellipticals are formed, but not spirals. Even more, numerical simulations of merging spirals (see e.g. Barnes, this volume) show that a huge amount of the angular momentum is transported to the dark halo and a very slowly rotating object is left. In a recent study, Navarro, Frenk & White (1995) found, that in simulations which start from cosmological initial conditions the specific angular momentum of the gas in the

R. Bender and R. L. Davies (eds.), New Light on Galaxy Evolution, 259–263.
© 1996 IAU. Printed in the Netherlands.

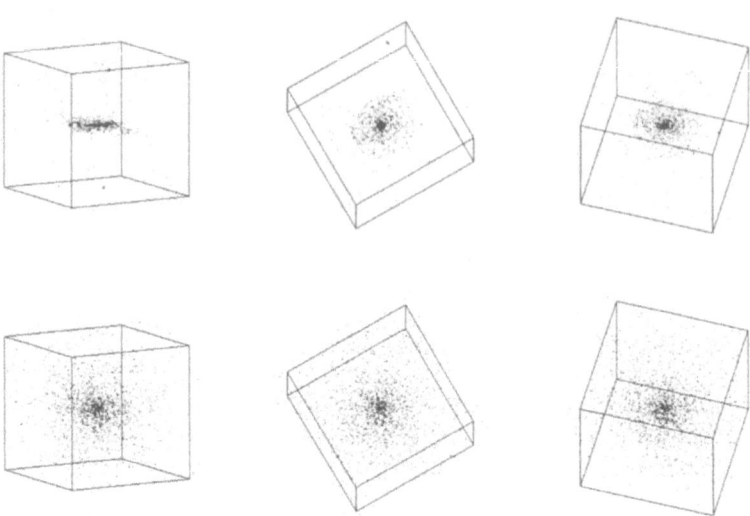

Figure 1. Three dimensional view on the distribution of gas (top) and dark matter (bottom) in a Milky Way sized halo ($v_c = 200\,\text{km/sec}$) for three different projections. The box has a side length of 50 kpc.

central gas knot is only 20% of that of the dark halo. This implies that these disks should possess a much smaller scale length than observed. However, the numerical resolution of these simulations was not sufficient to investigate the inner structure of the central gas knots. Especially it is an open question whether these dense knot of gas represents disks at all.

2. The structure of disk galaxies in gasdynamical simulations

In order to analyze the distribution of gas near the centers of galaxies more detailed, we performed numerical simulations which possesses a roughly 5 times higher resolution than others have done so far. Initial conditions and boundaries were treated similarly to Navarro *et al.* (1995): Based on a coarsely grained, large scale N–body simulation (CDM, $\Omega = 1$, $\sigma_8 = 0.67$, $H_0 = 50\,\text{km/sec/Mpc}$), individual haloes are picked out. Every halo is re-calculated in a high–resolution run. Gas dynamical effects are included using smoothed particle hydrodynamics (SPH). A very nice impression how structures built up and how mergers can cause a huge amount of angular momentum transport can be visualized by a movie (Steinmetz 1996).

Figure 1 shows the distribution of gas and dark matter for a typical halo formed in one of the simulations. At first glance, it seems to be quite similar to observed galaxies: the model galaxy consists of an extended spheroidal halo with a radius of about 300 kpc. The gas is much more centrally condensed and has settled in a thin disk with a diameter of about 50 kpc. A more careful analysis, however, shows that more than 70% of the gas can be found in the central gas clump within a radius of 2 kpc, the gravitational softening length. This gas has almost no angular momentum. The specific angular momentum of the disk, however, is similar to that of the dark halo. Consequently, the specific angular momentum of all the gas, within the disk and the dense gas knot, is a factor of about four smaller than that of the halo, in agreement with the simulations of Navarro *et al.* (1995). Compared to observations, the numerical models predict the formation of disks, which have the right size, but far too much mass is acquired by the central "bulge".

A deeper look at the formation history of the galaxy shows that the disk is mainly formed at low redshift ($z < 1$) by accretion of diffusely distributed gas. The dense gas knot, however, forms preferentially at higher redshifts ($z > 2$) by several merging events. In order to solve the angular momentum problem, one has to take care, that more gas is diffusely accreted.

3. Summary and discussion

Up to now, there is no self–consistent numerical simulation which avoids this angular momentum problem. However a variety of different processes are currently under discussion, which might be able to solve this problem. In my personal view, the first three can already be ruled out.

1. *Numerical imperfections.* The numerical viscosity which is usually used in SPH to describe shock waves (Gingold & Monaghan 1983) do not vanish in pure shear flows and, therefore, may cause an artificial angular momentum transport. We used a modified formulation which was shown not to exhibit a significant viscous angular momentum transport over several Hubble times (for details see Steinmetz 1995a and references therein).

2. *Changing the cosmological parameters* Ω_0, Λ_0. This probably has rather little influence: Although the merging rate in the near past can be changed a lot, the merging history, expressed in terms of number of mergers and mass distribution of progenitors, is quite similar (Lacey & Cole, 1993).

3. *Photoionisation by a UV background field.* A UV background with a strength as inferred from quasar observations at high redshifts ($z = 1 \ldots 4$) may suppress the formation of small structures ($v_c \lesssim 50$ km/sec, Efstathiou 1992) and the gas might fall in more diffusely. Therefore, the angular momentum transport to the halo might be reduced. We have performed numerical simulations including a UV background field.

It turns out that the formation of the dense central clump is only marginally affected: Most of the gas in the clump collapses at high redshift, i.e. at high densities, where the influence of a UV background is small (Steinmetz 1995b). However the background field is able to prevent cooling of the late and diffusely infalling gas, which possesses a low density. Therefore, the formation of the disk found in the simulations without a UV background is almost completely suppressed.

4. *Feedback processes*. The most likely solution to the angular momentum problem is related to star formation and supernovae feedback. As a first approach we performed a simulation, where regions of rapidly cooling and convergent gas flow are assumed to form stars. In this model supernovae increase the thermal energy of the surrounding gas. However, due to the very high cooling capability of this gas, most of the energy is radiated away and so the efficiency of the supernovae feedback is rather low. As a result, the formation of small lumps of gas is not prevented. The main influence of star formation is to transform a dense knot of gas into a slightly more diffuse lump of stars, but the extensive transport of angular momentum to the dark halo is not overcome. It is conceivable that momentum input due to supernovae might have a much stronger effect (Navarro & White 1993): In contrast to thermal energy, kinetic energy cannot immediately be radiated away resulting in a much higher efficiency of the supernova feedback.

References

Barnes, J., Efstathiou, G. 1987, ApJ, 319, 575.
Efstathiou, G. 1992, MNRAS 262, 43p.
Fall, S.M., Efstathiou, G., 1980, MNRAS, 193, 189.
Hoyle, F. 1949, in: *Problems of Cosmical Aerodynamics*, Dayton, Ohio: Central Air Documents Office, p.195
Lacey, C.G., Cole, S., 1993, MNRAS, 262, 627.
Monaghan, J.J., Gingold, R.A., 1983, JCP, 52, 374.
Navarro J.F., White, S.D.M., 1993, MNRAS 265, 271.
Navarro, J.F., Frenk, C.S., White S.D.M. 1995, MNRAS, 275, 56.
Steinmetz, M. 1995a, MNRAS, in press.
Steinmetz, M. 1995b, Proc. Texas Symposium on Relativistic Astrophysics 1994, Munich, in press.
Steinmetz, M., Bartelmann, M. 1995, MNRAS, 272, 570.
Steinmetz, M., 1996, ApJ (video) to be submitted.
White, S.D.M. 1984, ApJ 286, 38

PUSTIL'NIK: What mass fraction is left in you models in gas clumps which are not in dark matter haloes:

STEINMETZ: Almost all gas clumps are hosted by a dark halo. Only those objects which are going to be accreted by a more massive galaxy have sometimes tidally stripped their dark halo. It may, however, well be a numerical artifact. These objects are very small in mass and are represented by a few dozens of particles. In the case of a higher numerical resolution, the maximum possible phase space increases and the dark halo may (partially) survive the encounter.

VÖLK: If it is just a question of angular momentum *redistribution*, would not the easiest way to achieve this be an outflow (with magnetic field)?

STEINMETZ: As I already mentioned I agree that the solution to the angular momentum problem has most likely to do with feedback processes (which also may produce outflow of gas or galactic winds). Note, however, that the main redistribution is transport of angular momentum from the gas to the dark matter.

RENZINI: Can the code make both, spirals *and* ellipticals.

STEINMETZ: In the simulations I showed in the movie, star formation was not included. Therefore, galaxies always end up as spiral like objects. However, you have seen a halo which was involved in a very violent merging event which was even able to flip the spin of the galaxy. In a simulation including star formation such a merger would most probable form an elliptical galaxy.

Carlos Frenk and Renzo Sancisi

THE AGES OF ELLIPTICAL GALAXIES IN A MERGER MODEL

G. KAUFFMANN
Max–Planck–Institut für Extraterrestrische Physik
D–85740 Garching, Germany

1. Introduction

There have long been two competing views on the formation history of the ellipticals galaxies we see today. One is that most of the stars in present-day galactic bulges and ellipticals were produced during a relatively short, early phase of intense star formation at high redshift. The second view is that elliptical galaxies are relative latecomers, having been produced as the result of the merging of disk galaxies drawn together by gravity as their surrounding dark matter halos coalesced.

In recent years, the second view has come to enjoy increasing popularity. Detailed numerical simulations have shown that mergers between two spiral galaxies of comparable mass lead to the production of spheroidal merger remnants with physical characteristics such as density profiles, gravitational radii, mean velocity dispersions and surface brightnesses that are quite comparable to to observed ellipticals (see Barnes & Hernquist 1992 for a review). In addition, observational evidence has begun to accumulate that mergers and interactions are rather common in the universe, particularly at higher redshift. Hubble Space Telescope images indicate that at redshifts $\simeq 0.4$, 25-30 % of all galaxies show disturbed morphologies (Griffiths *et al* 1995). At lower redshifts, almost all the most luminous IRAS galaxies have turned out to be interacting systems, leading to the conclusion that the merging process may fuel very intense bursts of star formation.

The fact remains, however, that the stellar populations of present-day elliptical galaxies appear to be rather homogeneous. The colours, mass-to-light ratios and Mg_2 absorption strength of ellipticals are correlated only with their central velocity dispersions, and the scatter about these relations appears to be surprisingly small. Bower, Lucey & Ellis (1992) show that the intrinsic scatter about the U-V and V-K colour-magnitude relations for early-type galaxies in the Coma and Virgo clusters is less than 0.05

R. Bender and R. L. Davies (eds.), New Light on Galaxy Evolution, 265–273.

mag. This leads to the conclusion that elliptical galaxies must have formed the bulk of their stars at redshifts greater than 2, or else the observed homogeneity in the colours would require a very precise synchronization of formation epochs and star formation histories for these galaxies. Likewise, Bender, Burstein & Faber (1993) find a very tight relationship between the strength of the Mg_2 index at the centre of ellipticals and their central velocity dispersions, σ.

The tightness of these limits calls into question the tenability of the merger model, as merging might be expected to mix together stellar populations of quite disparate ages (Renzini 1994). In addition, a starburst accompanying the merging event would increase the amount of light contributed by young stars. However, detailed predictions have not yet been made for any realistic theory of galaxy formation via merging. Here, I present results calculated for a $b = 1.5$ cold dark matter (CDM) universe using semi-analytic techniques. In this model, elliptical galaxies are produced when two progenitor galaxies of roughly equal mass merge with each other. Spiral galaxies are formed in a two-step process. Two galaxies of equal mass must first merge with each other to form the bulge component. Gas from the surrounding dark matter halo then cools and settles to form a new disk. I first consider the stellar populations of elliptical galaxies in clusters. I also extend the analysis to the stellar populations of field elliptical galaxies. Finally, I make some predictions for the scatter in the colours of cluster elliptical galaxies at high redshift.

2. Semianalytic Models of Elliptical Galaxy Formation

The semi-analytic models we employ are described in detail in Kauffmann, White & Guiderdoni (1993, hereafter KWG) and Kauffmann & White (1993). Application of the model to the evolution of the galaxy population in clusters at high redshift is discussed in Kauffmann (1995).

To summarize:
I use an algorithm based on an extension of the Press-Schechter theory due to Bower (1991) and Bond et al (1991) to generate Monte Carlo realizations of the merging paths of dark matter halos from high redshift until the present. Dark matter halos are modelled as truncated isothermal spheres and it is assumed that as the halo forms, the gas relaxes to a distribution that exactly parallels that of the dark matter. Gas then cools and condenses onto a *central galaxy* at the core of each halo. Star formation and feedback processes take place as described in KWG. In practice, star formation in central galaxies takes place at a roughly constant rate of a few solar masses per year, in agreement with the rates derived by Kennicutt (1983) for normal spiral galaxies.

At a subsequent redshift, a halo will have merged with a number of others, forming a new halo of larger mass. All gas which has not already cooled is assumed to be shock heated to the virial temperature of this new halo. This hot gas then cools onto the central galaxy of the new halo, which is identified with the central galaxy of its *largest progenitor*. The central galaxies of the other progenitors become *satellite galaxies*, which are able to merge with the central galaxy on a dynamical friction timescale. If a merger takes place between two galaxies of roughly comparable mass, the merger remnant is labelled as an "elliptical" and all cold gas is transformed instantaneously into stars in a "starburst".

Note that the infall of new gas onto satellite galaxies is not allowed, and star formation will continue in such objects only until their existing cold gas reservoirs are exhausted. Thus the epoch at which a galaxy is accreted by a larger halo delineates the transition between active star formation in the galaxy and passive evolution of its stellar population. The stellar populations of elliptical merger remnants in clusters hence redden as their stellar populations age Central galaxy merger remnants in the "field" are able to accrete new gas in the form of a disk to form a "spiral" galaxy consisting of both a spheroidal bulge and a disk component. As demonstrated in KWG, this picture is able to account for the observed numbers and luminosity distributions of galaxies of different morphologies, both in clusters and in lower-density environments.

The spectrophotometric models of Bruzual & Charlot (1993) are used to translate the predictions of the models into observed quantities such as magnitudes and colours, which may be compared directly with the observational data. Results are presented for a $b = 1.5$ CDM universe with $\Omega = 1$ and $H_0 = 50$ km s^{-1} Mpc^{-1}.

3. The Ages and Colours of Elliptical Galaxies

Figure 1 shows the V luminosity-weighted mean stellar age of the galaxies in a cluster of $10^{15} M_\odot$. The classification into morphological type is based on the B-band disk-to-bulge ratio of the galaxy, as given in Table 3B of Simien & de Vaucouleurs (1986). As can be seen, the stellar populations of spiral galaxies span a very wide range in age, but those of early-type galaxies are confined to ages between 8 and 12.5 Gyr. For our adopted cosmological parameters, this means that the bulk of stars in elliptical and SO galaxies were formed at redshifts exceeding 1.9.

The age distribution of early-type galaxies found in halos of mass $10^{13} M_\odot$ is plotted in figure 2. These galaxies either occur in small groups, or have no companion of comparable luminosity, and thus can be regarded as "field ellipticals". In order to get a a sizeable sample of these objects, it was nec-

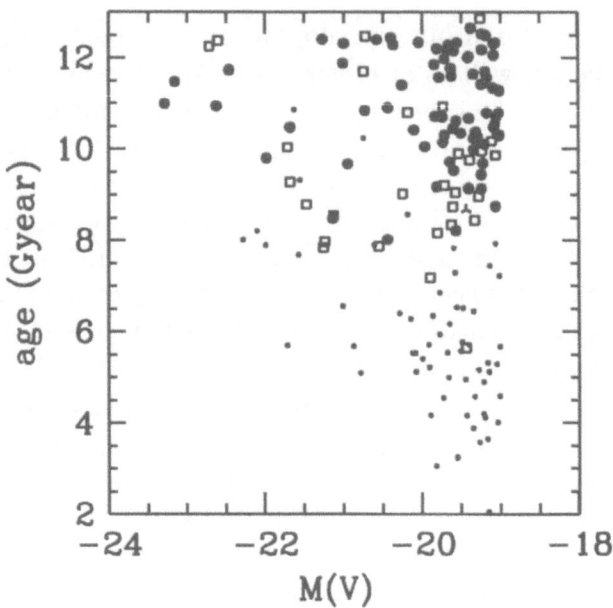

Figure 1. The V-luminosity weighted mean stellar ages of galaxies in a cluster of mass $10^{15} M_\odot$. Large filled circles represent ellipticals, open squares are S0s, small filled circles are spirals and 3-pronged pinwheels are elliptical galaxies that were formed by a major merger in the past 1.5 Gigayears.

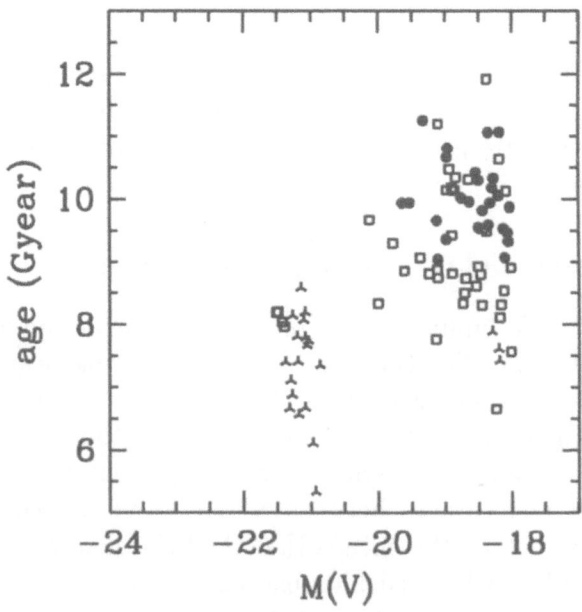

Figure 2. The mean ages of the early-type galaxies in halos of mass $10^{13} M_\odot$. Symbols are as in figure 1.

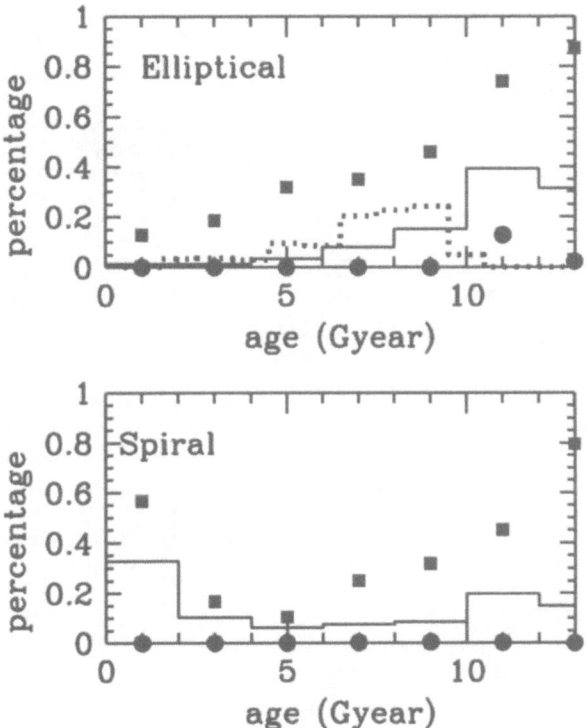

Figure 3. The fraction of the total V-band light of cluster ellipticals and spirals contributed by stars which form at a given epoch. The solid line is the average star formation history. The squares and circles represent the maximum and minimum contributions to the total light for any galaxy at that epoch. The dotted line is a histogram of the percentage of elliptical galaxies that experienced their last major merger at a given epoch.

essary to combine the populations of 100 such low-mass halos. As can be seen, the field elliptical population splits up into two distinct classes. The faint field ellipticals have stellar ages comparable to early-type galaxies in clusters. There is also a population of bright field ellipticals that have much younger stellar populations (mean ages between 5 and 8 Gigayears). These bright galaxies are all *central* galaxies and thus are still in the process of accreting cold gas. They are also mostly recent mergers. The fact that we see these galaxies as ellipticals is simply because they have not yet had the time to accrete a new disk following their last big merging event.

Figure 3 is a histogram of the fraction of the total V-band light of a cluster elliptical or spiral galaxy contributed by stars which form at a given redshift. A large fraction of the V-band light of spirals comes from stars that formed in the past 5 Gyr, whereas *on average*, this contribution is negligible in ellipticals. The dotted line in the top panel of figure 3 illustrates the epoch at which elliptical galaxies typical underwent their last major

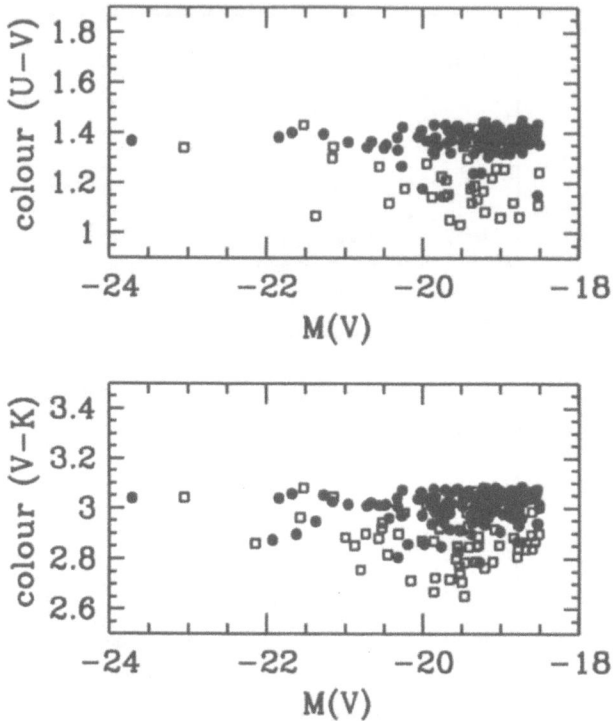

Figure 4. The distribution in U-V and V-K colour of early-type galaxies in a cluster of $10^{15} M_\odot$. Filled circles are ellipticals and open squares are SOs.

merging event. As can be seen, most ellipticals are formed by a merger that occurred between 5 and 10 Gyr ago. It is also apparent from this diagram that a substantial fraction of the stars in an elliptical galaxy may be formed *before* the merging event that actually causes the transformation in morphology. Star formation may take place at a constant rate in the two spiral progenitors of an elliptical for several gigayears before they are able to merge. It is thus important to distinguish clearly what is meant by "the formation epoch of ellipticals". It can be defined either as the epoch when the bulk of the stars were formed or as the epoch when the merging event took place.

Finally, in figure 4, I have used the evolutionary synthesis models of Bruzual & Charlot (1993) to generate a scatterplot of the U-V and V-K colours of early-type galaxies in clusters. In U-V, the average colour of an elliptical is 1.37 with a rms standard deviation, σ, of 0.034. In V-K, the mean is 3.01 and σ is 0.045. If SO galaxies are included, the rms increases to $\simeq 0.07$. Note that the colour- magnitude relation is *flat*. As shown in figure 1, cluster elliptical galaxies of all luminosities have roughly the same mean stellar age, so any trend in colour would have to result from metallicity differences between galaxies of different masses. The rms scatter in the colour of the model ellipticals is well within the upper limit for the intrinsic scatter in the colour of Coma and Virgo ellipticals quoted by Bower, Lucy & Ellis (1992).

4. Ellipticals in Clusters at High Redshift

Present-day elliptical galaxies contain the oldest stars in the universe. The repair of HST permits the direct identification of elliptical and S0 galaxies at significant cosmic depth. The hope is, therefore, that studies of the stellar populations of ellipticals at early epochs will place strong constraints on cosmology.

Ellis (1995) has measured the rest-frame U-V colours of elliptical galaxies in the cluster 0016+16 at z=0.54 and finds a rms scatter of 0.07. In figure 5, model predictions for the U-V colours of cluster ellipticals are shown at a series of redshifts. There are two trends that should be noted. Firstly, the "red envelope" of the distribution shifts bluewards with increasing redshift. The magnitude of this shift is consistent with what one would expect from the passive evolution of a 12 Gyr old burst, as given by the models of Bruzual & Charlot (1993). Secondly, the scatter in the colour distribution increases with redshift and most of this increase arises from a growing population of galaxies which have merged in the 1.5 Gyr prior to the epoch of observation. Note, however, that the scatter is not dramatically large until redshifts in excess of 1. This may be understood as follows. As pointed out by Charlot & Silk (1994), the colours of stars formed in a short burst evolve very rapidly for the first 3-4 Gyr. After that, the evolution slows down and the colours redden only gradually over time. So long as most of the stars in the elliptical population of a cluster are formed 3-4 Gyr prior to the time the cluster is observed, no great conspiracy is required to produce a small scatter in the colours. At redshifts greater than 1, this is no longer possible (assuming $\Omega = 1$ and $H_0 = 50$ km s^{-1} Mpc^{-1}) and the scatter becomes very large. Spectroscopic age indicators may turn out to be more sensitive tests of cosmology.

5. Discussion & Conclusions

Two conclusions follow from this analysis. One is that star formation in cluster ellipticals formed by mergers occurs at high enough redshifts so that the predicted scatter in the colour- magnitude relation of these systems falls within observational bounds. That ellipticals in clusters form at high redshift should come as no surprise. For a random Gaussian initial density field, it is well-known that the redshift of collapse of density peaks on

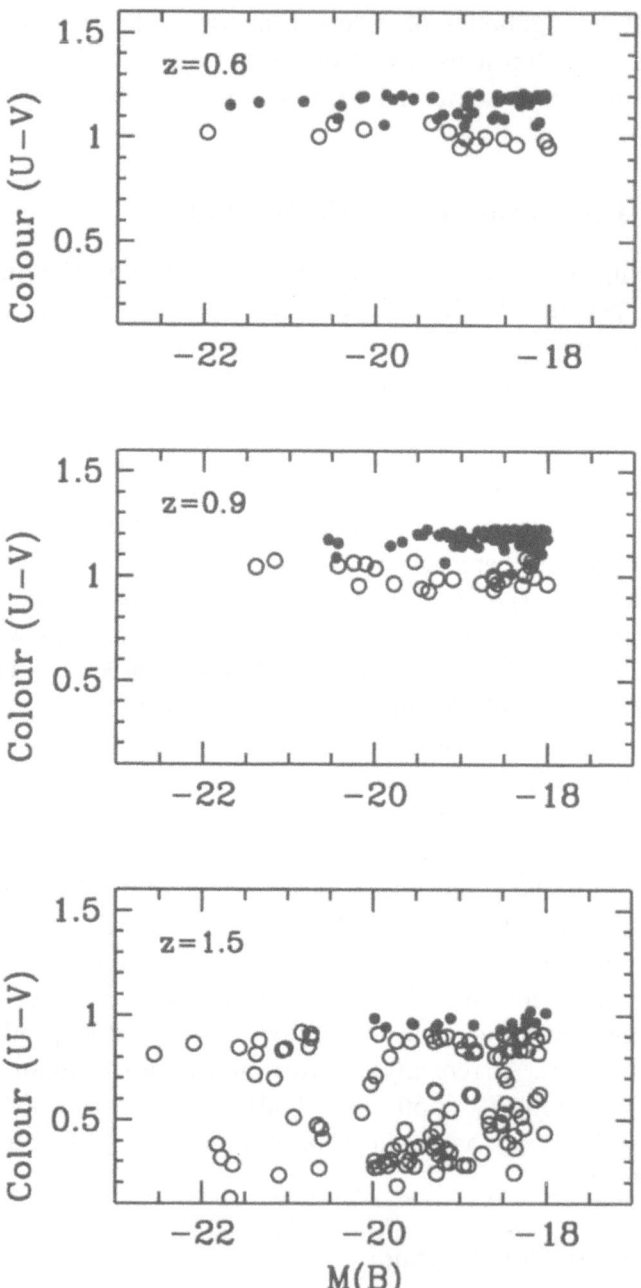

Figure 5. Rest-frame U-V colour-magnitude relations for elliptical galaxies in clusters of $10^{15} M_\odot$ observed at z=0.6, 0.9 and 1.5. Large open circles denote ellipticals that have been formed by a major merging event in the past 1.5 Gyr.

galaxy scales is boosted by the presence of the surrounding, larger-scale overdensity destined to collapse to form the cluster at z=0 (Bardeen et al 1986). Galaxies forming in peaks collapsing at $z > 2$ quickly merge together within groups and their star formation is truncated when their gas is used up. Thus in the hierarchical clustering picture, galaxies that form at high redshifts are in some sense *preselected* to end up as cluster elliptical galaxies. What is shown here, is that this process can work in detail.

The second conclusion is that bright elliptical galaxies in the field are *intrinsically* different to cluster ellipticals. These galaxies have undergone recent merging and star formation. It is only because we "catch them at the right time" that we observe them as elliptical systems. In a few gigayears, they will have grown new disks and have turned back into spirals. Finally, I derive predictions for the colour distributions of elliptical galaxies in clusters at high redshift. The scatter is predicted to increase very substantially at redshifts greater than 1.

References

Bardeen, J.M., Bond, J.R., Kaiser, N., Szalay, A.S., 1986, APJ, 304, 15
Barnes, J.E., Hernquist,L. 1992, Ann.Rev.Astr.Ap., 30, 705
Bender, R., Burstein, D., Faber, S.M. 1993, ApJ, 411, 153 (BBF)
Bond, J.R., Cole, S., Efstathiou, G., Kaiser, N. 1991, ApJ, 379, 440
Bower, R. 1991, MNRAS, 248, 332
Bower, R., Lucey, J.R., Ellis, R.S. 1992, MNRAS, 254, 601
Bruzual, G., Charlot, S. 1993, ApJ, 405, 538
Charlot, S., Silk, J., 1994, APJ, 405, 538
Ellis, R.S., 1995, in Unsolved Problems in Astrophysics, eds Ostriker, J.P. & Bahcall, J.N., Princeton, in press
Griffiths, R.E. *et al* 1995, ApJ, in press
Kauffmann, G., White, S.D.M. 1993, MNRAS, 261, 921
Kauffmann, G., White, S.D.M.,Guiderdoni, B. 1993, MNRAS, 264, 201 (KWG)
Kauffmann, G. 1995, MNRAS, 274,161
Kennicutt, R.C. 1983, ApJ, 272, 54
Renzini, A. 1995, in Stellar Populations, ed Gilmore, G. & van der Kruit, P. (Dordrecht:Kluwer), in press
Simien, F., De Vaucouleurs, G. 1986, ApJ, 302, 564

GERHARD: Both you and Carlos Frenk had a baryon density $\Omega_b = 0.1$ in your models. How much of this is tranformed into stars and how is this constrained?

KAUFFMANN: The total mass of stars that is formed is constrained by requiring that the mass and lumininosity of the *central* galaxy in a halo of circular velocity 220 km/s be equal to that of our own Milky Way. The total Ω in stars is about 0.01-0.02.

Martin Rees and Colin Norman

GALAXIES IN FORMATION

THE QUEST FOR PROTOGALAXIES

S.G. DJORGOVSKI
Palomar Observatory, Caltech
Pasadena, CA 91125, USA

Abstract.
The current state and the future prospects of searches for protogalaxies (PGs) are reviewed. Many high-redshift objects are now known, mostly associated in some way with AGN, and at least some of them may be young galaxies. Quasars at $z > 4$ and high-z quasar pairs may mark galaxy formation sites. Deep field surveys for Lyα luminous galaxies powered by star formation alone have failed so far to reveal a large population of such objects, and the observed limits are in conflict with simple model predictions by three orders of magnitude. Some extinction by dust can account for this. However, strong limits from COBE place severe constraints on models of completely obscured PGs. New searches in the near IR are now beginning to probe the relevant line flux and number density regime, and first interesting PG candidates are being discovered. Searches from mid-IR to mm wavelengths would complement these efforts.

1. Introduction

Understanding of galaxy formation is one of the central goals of modern cosmology. Thanks to the great advances in observational technologies in the last few years, the quest for protogalaxies (PGs) is now becoming one of the most exciting and lively fields of extragalactic astronomy.

We have a reasonably good, if mostly qualitative understanding of the processes leading to galaxy formation (cf. White & Frenk 1991, Silk & Wyse 1993, and references therein). Most models and simulations deal with the assembly of dark matter into galaxy-sized chunks, a process which is not directly observable. The weakest link in the models now seems to be the physics of star formation and other dissipative processes in PGs.

R. Bender and R. L. Davies (eds.), New Light on Galaxy Evolution, 277–286.

Observations are now reaching the flux levels and the volume coverage at high redshifts where models can be tested. An excellent recent review of the subject is by Pritchet (1994); see also Djorgovski & Thompson (1992) or Djorgovski (1992). This review is intended mostly as an update to them.

The relevant observable parameters of PGs are discussed in these reviews, as well as in a number of the recent survey papers, listed below. Briefly, depending on the cosmology and formation redshifts, we expect to find ~ 1 to 10 PGs/arcmin^2, depending on the net total duration of the formation phases, with typical visible/near-IR continuum magnitudes $\sim 23^m - 26^m$ for proto-L_* objects, depending on the bandpass, and with recombinant line fluxes (including Lyα) of $\sim 10^{-17\pm1}$ erg/cm^2/s, if they were unobscured. Most of the energy released would come from the young, massive stars which generated the metals we see in the spectra of old stellar populations today along with the corresponding amounts of helium. Up to $\sim 10\%$ of the total luminosity of PGs would be contributed by the release of the binding energy of PGs themselves, and the protostars within them. AGN, if present, would make all of the fluxes higher.

Redshifts of formation ranging from $z \sim 0$ to $z \sim 100$ have been proposed, but probably the most likely range is from $z \sim 2$ to $z \sim 5$ or 10 (cf. Peebles 1989 for a discussion). Deep field redshift surveys and absorption line studies indicate that most (at least non-dwarf) galaxies do not evolve very much out to $z \sim 1$ or 1.5. The comoving number density of quasars peaks near $z \sim 2.5$, and falls of steeply at $z > 3$ or 4; this may be indicative of a general merging history in the universe. Yet, quasars at $z \sim 4$ or 5 seem to reside in already chemically evolved galaxy cores, and some galaxy formation must have started at $z > 5$.

The key problem is that of the recognition of PGs. Unwise and unsubstantiated claims have been made in the literature about ostensible discoveries of PGs or even protoclusters, simply on the basis of the objects being faint, or being blue, or being red, or being numerous, or being rare, etc. There is no substitute for real redshifts and spectra which can be used to constrain the physical nature of the candidates, e.g., if an AGN is present.

The observable aspect of galaxy formation is the collapse of the baryonic component and star formation – dark matter is dynamically important and may be driving the process, but it is hard to observe! An observer's definition of PGs may be: massive galaxies undergoing their first, major bursts of star formation at high redshifts. This probably means most ellipticals and massive bulges; disks seem to form their stars at a nearly uniform pace over the Hubble time scales. Some dwarf galaxies, e.g., objects like I Zw 18 may be only forming now. The early evolution of disk systems can be probed indirectly through absorption line studies (cf. Wolfe 1993, Steidel & Dickinson 1995, and references therein).

Even within the restricted class of young ellipticals and bulges there was likely some spread of star formation histories. The AGN phenomena may be also ubiquitous among PGs (Djorgovski 1994). The real phenomenology of galaxy formation may be very diverse, with a complicated interplay of astrophysical processes at a range of redshifts. PGs are unlikely to be a simple, uniform class of objects.

2. High Redshift Galaxies and Protogalaxy Candidates

Most likely, PGs have already been discovered. Aside from an obvious possibility that some or all high-z quasars reside in PGs, there are many high-z galaxies now known, almost all of them associated in some way with AGN. Proto-disks may have been discovered indirectly, as damped Lyα absorbers (DLAs) and/or metallic line and Ly limit absorption systems.

Powerful radio galaxies remain the most distant non-QSO objects known, and some of them may well be objects in early stages of formation. Some examples include: 3C 326.1 at $z = 1.825$ (McCarthy et al. 1987), B2 0902+34 at $z = 3.395$ (Lilly 1988, Eisenhardt & Dickinson 1992), 4C 41.17 at $z = 3.800$ (Chambers et al. 1990), or 8C 1435+635 at $z = 4.25$ (Lacy et al. 1994). However, it remains impossible to disentangle the effects of AGN from those of star formation. For a good review, see McCarthy (1993).

The first galaxy at $z > 3$, the companion of the quasar PKS 1614+051 at $z = 3.215$ (Djorgovski et al. 1985, 1987), was discovered using the Lyα narrow-band imaging technique, which remained the method of choice for PG searches ever since. Other high-z quasar companions include galaxies near QSO 1548+0917 at $z = 2.749$ (Steidel et al. 1991), Lyα objects near the gravitational lens MG 2016+112 at $z = 3.273$ (Schneider et al. 1987), and several cases of Lyα and possibly also continuum nebulosities near radio-loud quasars at $z \sim 2 - 3$ (Heckman et al. 1991ab, Hu et al. 1991, Moller & Warren 1993). Several cases of Lyα companions of high-z radio galaxies are also now known (cf. Windhorst et al. , this volume).

Narrow-band imaging was also used to look for Lyα galaxies at the redshifts of DLAs near the lines of sight towards background quasars. The best cases so far include active galaxies (high-ionization lines are present) at $z = 2.309$ (Lowenthal et al. 1991), and at $z = 3.428$ (Macchetto et al. 1993, Giavalisco et al. 1994). A possible Lyα nebulosity at $z = 2.466$ was reported by Wolfe et al. (1992), but it remains unconfirmed. Other cases have been found recently (Bechtold et al. , in prep.; Djorgovski et al. , in prep.). A related approach has been made by Francis et al. (1996), who discovered a possible group of galaxies at $z = 2.38$, associated with a proposed cluster of C IV absorbers; their one object with a secure redshift also contains an AGN. Obviously, DLA and other absorber fields present very promising targets for future searches. The success of these studies also suggests that some clustering is present at $z \sim 2 - 3$ (Wolfe 1993).

3. Deep Lyα Searches and the Question of Dust Obscuration

Objects containing or located near AGN may be unrepresentative of the general PG population (but see Sect. 5). A considerable effort has been expended in deep field searches for PGs powered by star formation alone. There, one is looking for a large *population* of PGs, rather than a few interesting but rare objects.

Some emission line signature may be necessary, both as a probe of redshifts and physics (i.e., AGN vs. star formation), and Lyα has always been the line of choice. From case-B recombination and the semiempirical conversion of star formation rate to Hα luminosity (Kennicutt 1983), we get a conversion of $L(\text{Ly}\alpha) \sim 10^{42}$ erg/s or slightly less for $SFR = 1 M_\odot/\text{yr}$, assuming a normal IMF (a top-heavy IMF would be more efficient in producing the Lyα photons). Additional Lyα flux may come from shock ionization, or cooling of the first stars. However, its detection requires that PGs were not obscured by dust. For a discussion, see Charlot & Fall (1993).

Both narrow-band and long-slit searches are reaching typical line fluxes $F_{\text{Ly}\alpha} \sim 10^{-17}$ erg/cm^2/s, or even fainter, which, at the redshifts of interest translates into the rest-frame line luminosities of $L(\text{Ly}\alpha) \sim 10^{42}$ erg/s, and unobscured SFR of a few $M_\odot/\text{yr}$, over typical comoving volumes surveyed of $\sim 10^5 h_{75}^{-3}$ Mpc^3 (De Propris *et al.* 1993, Thompson *et al.* 1995, Thompson & Djorgovski 1995; cf. Pritchet 1994 for a review and further references). Some faint PG candidates *have* been found in these surveys, but none have been confirmed yet, due to their extreme faintness. Assuming that no PGs are found, the observed limits are some 3 orders of magnitude in conflict with simple model predictions, for unobscured PGs. The key question is then, were PGs obscured by dust?

We see prominent Lyα emission from many high-z galaxies, but most or all of them seem to involve AGN. The earliest stages of a PG must have been dust-free, on the account of absence of metals in the ISM; that phase might have lasted up to $\sim 5 \times 10^7$ yr. On the other hand, most starbursts at $z \sim 0$ tend to be dust-obscured, and their counterparts do exist at high-z, e.g., IRAS FPS 10214+4724 (Rowan-Robinson *et al.* 1991), even if AGN are also involved. Thermal emission from dust has been detected from several high-z radio galaxies and quasars (McMahon *et al.* 1994, Chini & Krugel 1994, Ivison 1995, and references therein). Lyα emission has been seen from some low-z starforming dwarfs, but usually at a level lower than what is expected from the simple case-B photoionization by young stars (cf. Terlevich *et al.* 1993, Calzetti & Kinney 1992, and references therein). The data suggest that the extinction from these objects can be modeled effectively as a patchy dust foreground screen, so some Lyα leakage is expected. UV radiation from young starbursts or AGN can also destroy dust (Hartquist *et al.* 1995, Calzetti *et al.* 1995). Lyα emission may be also filling up stellar Lyα absorption lines in PG spectra (Valls-Gabaud 1993).

The strongest limits on dusty PGs come from the COBE FIRAS experiment (Mather *et al.* 1994, Wright *et al.* 1994). The integrated energy density today generated by stars which produced the observed metals and helium in old, metal-rich stellar populations is of the order of a few percent of the energy density of the CMBR, viz., $\sim 10^{-15}$ erg/cm^3 (cf. Djorgovski 1992, or Djorgovski & Thompson 1992 for a discussion). A comparable integrated background is expected from AGN at all redshifts (Chokshi & Turner 1992). Yet the maximum allowed deviations from the pure blackbody CMBR spectrum in the sub-mm regime, where an integrated background from putative dusty PGs would be observable now, are $\sim 0.03\%$ of the CMBR peak intensity. This suggests, in a fairly model-independent way, that at most a few percent of all star formation in PGs was obscured by dust. A more detailed modeling by Blain & Longair (1993b) suggests that a sub-mm background may be just in conflict with the COBE FIRAS limits. However, Puget *et al.* (1995) have claimed a detection of a sub-mm background in the very same COBE data, but with a different analysis approach. This is indicative of the difficulty of the problem. PGs, evolving galaxies, and AGN should all contribute to an integrated FIR/sub-mm background, and even if a positive detection is made, disentangling the PG contribution may be hard.

An approach complementing Lyα based surveys is to select PG candidates (which then must be followed up spectroscopically) on the basis of their stellar continuum Lyman break at 912Å (Guhathakurta *et al.* 1990, Steidel & Hamilton 1992, 1993, DeRobertis & McCall 1995). This promising technique is now yielding results (Steidel, priv.comm.).

4. Searches in the Near Infrared

Emission-line searches in the near-IR are the natural next step: nebular oxygen and Balmer lines are much less affected by the dust. One can imagine a scenario in which PGs contain modest amounts of dust, sufficient to extinguish the Lyα emission, but not dusty enough to generate a substantial sub-mm background, detectable by COBE. The modern NIR imaging technology is now good enough to make such searches viable.

An early attempt, and a discussion of observable parameters of near-IR PGs were presented by Thompson *et al.* (1994). Mannucci & Beckwith (1995) provide a detailed discussion of detectability of PGs in the near-IR. The early results from the Keck were presented by Pahre & Djorgovski

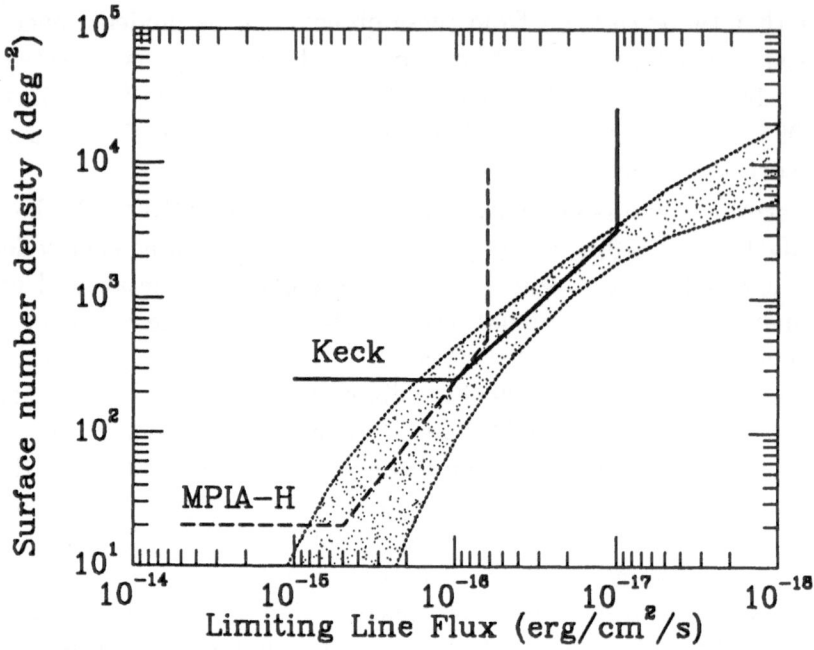

Figure 1. Estimated line flux limits and sky coverage in K band PG surveys. The dotted area indicates the range predicted by the simple models from Mannucci & Beckwith (1995). Estimated limits as of late 1995 are indicated for two of the most extensive surveys to date: dashed line = MPIA-H group (Beckwith & Thompson, priv. comm.); solid line = Keck (Djorgovski & Pahre, in prep.). Both groups are now probing the relevant portion of the parameter space, and both have found some PG candidates.

(1995); as of late 1995, their coverage has been increased by about a factor of 4, and a few good candidates have been found. Another, as yet unconfirmed candidate has been reported by Malkan *et al.* (1995). The object found by Lowenthal *et al.* (1991) was detected in Hα both by Hu *et al.* and by Bunker *et al.* (1995). Most of the work so far has been in the K band, looking for the Hα, Hβ, [O III] 5007, and [O II] 3727 lines. An optimistic, albeit unsuccessful search for the Lyα line at $z \sim 7 - 9$ in the J band was reported by Parkes *et al.* (1994). The MPIA-H group (Beckwith *et al.*) is continuing their search over a wide area, down to moderate fluxes; this is a good strategy if one believes that PG starbursts would be rare, but intense, so that a large volume should be covered. Figure 1 summarizes some of the IR limits to date.

These searches are already starting to produce interesting candidates, and may be the most promising method to look for PGs at the present. It is also possible that some PGs have been detected in the deepest K band surveys to date (cf. Djorgovski *et al.* 1995, and references therein); their expected magnitudes are $K \sim 23^m - 24^m$. An integrated NIR background from PGs is now being constrained by the COBE DIRBE (cf. Hauser

1995 for a review). Observations in the as yet poorly explored cosmological window at $\lambda \sim 3 - 6$ μm are particularly interesting. Deep surveys with HST+NICMOS, ISO, and SIRTF, perhaps in a conjunction with the ground-based spectroscopy, may lead to great advances in this area.

5. AGN and Galaxy Formation

The same kind of astrophysical processes, dissipative merging and infall of gas, are believed to be essential both for massive starbursts and the onset and fueling of AGN. It is thus natural to expect that at least some, and perhaps all PGs would undergo an immediate AGN phase, which could mask the underlying starburst. The idea that high-z quasars *are* PGs was proposed by Meier (1976) and expanded by Terlevich (1992; see also Terlevich & Boyle 1993, and references therein). In a less extreme form, both a traditional AGN and a starburst can coexist in young PG, and primordial starbursts can lead to a formation of the first quasars (cf. Norman & Scoville 1988, Williams & Perry 1994, Djorgovski 1994).

It is perfectly possible that most or all ellipticals and luminous bulges today contain remnants of the former AGN, and that every one of them has undergone an early AGN phase, at least once (Chokshi & Turner 1992, Small & Blandford 1992, or Haehnelt & Rees 1993, and references therein). There is a kinematical evidence for the quiescent central mass concentrations in the nuclei of several nearby galaxies, including M31, M32, and others (cf. Kormendy & Richstone 1995 for a review).

Metallicities in quasars at $z > 4$ suggest that they reside within already considerably chemically evolved stellar populations, presumably cores of giant ellipticals (Hamann & Ferland 1993, Matteucci & Padovani 1993). The very existence of numerous $z > 4$ quasars suggests that galaxy formation must have started at higher redshifts (Turner 1991).

The fate of a protogalaxy may be severely affected by an early AGN, if one is rapidly formed, while the bulk of its host is still gaseous. The feedback from an AGN could have regulated some of the global properties of its host galaxy, e.g., the total mass, density, etc. (Ikeuchi & Norman 1991). Quasar-endowed PGs may even create an inverse bias and suppress star formation in their vicinity, by ionizing the ISM in companion objects.

Another intriguing possibility is that the high-z quasars can be used as probes of the primordial large scale structure. The first object should be forming at the highest peaks of the initial density field, and such peaks should be strongly clustered (Kaiser 1984). For a CDM cosmogony, quasars at $z > 4$ should be clustered as strongly as the bright galaxies today (Efstathiou & Rees 1988). If this is true, then practically all $z > 4$ quasars should be pointing towards protoclusters in the early universe, and the optimal sites for galaxy formation. This would also be consistent with their interpretation as being the nuclei of young giant ellipticals.

Possible examples of such structures include three quasar pairs at $z > 3$, found in a complete survey by Schneider, Schmidt & Gunn (1994). The quasars are separated by a few Mpc, and the Poissonian probability of finding these 3 pairs in their survey is $P \simeq 10^{-13}$! The implied clustering length is $r_0 \sim 50h^{-1}$ Mpc, comparable to the richest Abell clusters today.

6. Conclusions and Future Prospects

There may be some PGs among the various high-z objects already discovered, including quasars (especially those at $z > 4$), their companions, radio galaxies, and other AGN at $z > 3$ or 4. Deep Lyα based surveys have placed severe limits on a population of unobscured, luminous PGs, but some of the faint candidates may yet turn out to be PGs. Searches based on the 912 Å break, or in the vicinity of DLAs are also starting to yield interesting candidates. Modern near-IR searches based on the nebular oxygen and Balmer lines are also starting to probe the relevant flux and number density regime, and to yield the first candidates.

A virtually unexplored territory lies from the mid-IR through sub-mm regime, where dusty PGs may be hiding (cf. Blain & Longair 1993a, Franceschini *et al.* 1994, Braine 1995, and references therein). Analysis of integrated extragalactic backgrounds is already providing interesting constraints on galaxy formation, but eventually one must resolve and study individual sources of this emission. New bolometer arrays and receivers from the ground, and ISO and SIRTF from space may lead to great advances in this area. The [C II] 154 μm line, and possibly molecular lines of H_2 and CO, may be good redshift indicators for dusty PGs.

Gravitational lenses may yet prove to be an important tool in discovering and studying distant galaxies or even PGs to faint to detect otherwise. Examples may be the Lyα nebulosities near MG 2016+112 (Schneider *et al.* 1987), the notorious IRAS FPS 10214+4724 (Graham & Liu 1995), the object discovered by Warren *et al.* (1995), some cluster arcs, etc.

Observational study of galaxy formation is now a reality, at a range of wavelengths and search strategies, even if we are not quite sure what to look for: PGs were probably a very diverse bunch. We have probably already found some of them, and hopefully we shall find many more.

I wish to thank my collaborators, and in particular D. Thompson, M. Pahre, and J. Kennefick, and to the staff of Palomar and W. M. Keck observatories for their expert help. This work was supported in part by the NSF PYI award AST-9157412, and the Bressler Foundation.

References

Blain, A., & Longair, M. 1993a, MNRAS, 264, 509
Blain, A., & Longair, M. 1993b, MNRAS, 265, L21
Braine, J. 1995, A&Ap, 300, 20
Bunker, A., Warren, S., Hewett, P. & Clements, D. 1995, MNRAS, 273, 513
Calzetti, D. & Kinney, A. 1992, ApJ, 399, L39
Calzetti, D., Kinney, A. & Storchi-Bergmann, T. 1995, ApJ, in press
Chambers, K., Miley, G. & van Breugel, W. 1990, ApJ, 363, 21
Charlot, S. & Fall, S.M. 1993, ApJ, 415, 580
Chini, R., & Krügel, E. 1994, A&Ap, 288, L33
Chokshi, A. & Turner, E. 1992, MNRAS, 259, 421
De Propris, R., Pritchet, C., Hartwick, F.D.A., & Hickson, P. 1993, AJ, 105, 1243
DeRobertis, M. & McCall, M. 1995, AJ, 109, 1947
Djorgovski, S., Spinrad, H., McCarthy, P., & Strauss, M. 1985, ApJ, 299, L1
Djorgovski, S., Strauss, M., Perley, R., Spinrad, H., & McCarthy, P. 1987, AJ, 93, 1318
Djorgovski, S. 1992, in: Cosmology and Large-Scale Structure in the Universe, ed. R. de
    Carvalho, ASPCS vol. 24, 73
Djorgovski, S. & Thompson, D. 1992, in: IAU Symp. #149, The Stellar Populations of
    Galaxies, eds. B. Barbuy & A. Renzini, Dordrecht: Kluwer, p. 337
Djorgovski, S., Thompson, D. & Smith, J.D. 1993, in: First Light in the Universe, eds.
    B. Rocca-Volmerange et al. , Gif sur Yvette: Editions Frontières, p. 67
Djorgovski, S. 1994, in: Mass-Transfer Induced Activity in Galaxies, ed. I. Shlosman,
    Cambridge: CUP, p. 452
Djorgovski, S., et al. 1995, ApJ, 438, L13
Efstathiou, G. & Rees, M. 1988, MNRAS, 230, 5P
Eisenhardt, P. & Dickinson, M. 1992, ApJ, 399, L47
Franceschini, A., Mazzei, P., De Zotti, G. & Danese, L. 1994, ApJ, 427, 140
Francis, P., et al. 1996, preprint
Giavalisco, M., Macchetto, F., & Sparks, W. 1994, A&Ap, 288, 103
Graham, J. & Liu, M. 1995, ApJ, 449, L29
Guhathakurta, P., Tyson, J.A. & Majewski, S. 1990, ApJ, 357, L9
Haehnelt, M. & Rees, M. 1993, MNRAS, 263, 168
Hamann, F. & Ferland, G. 1993, ApJ, 418, 11
Hartquist, T., et al. 1995, ApJ, 453, 77
Hauser, M. 1995, in: IAU Symp. #168, Examining the Big Bang and Diffuse Background
    Radiations, ed. M. Kafatos, Dordrecht: Kluwer, in press
Heckman, T., Lehnert, M., van Breugel, W., & Miley, G. 1991a, ApJ, 370, 78
Heckman, T., Lehnert, M., Miley, G., & van Breugel, W. 1991b, ApJ, 381, 373
Hu, E., Songaila, A., Cowie, L., & Stockton, A. 1991, ApJ, 368, 28
Hu, E., Songaila, A., Cowie, L., & Hodapp, K. 1993, ApJ, 419, L13
Ikeuchi, S. & Norman, C. 1991, ApJ, 375, 479
Ivison, R. 1995, MNRAS, 275, L33
Kaiser, N. 1984, ApJ, 284, L9
Kennicutt, R. 1983, ApJ, 272, 54
Kormendy, J. & Richstone, D. 1995, ARAA, 33, 581
Lacy, M., et al. 1994, MNRAS, 271, 504
Lilly, S. 1988, ApJ, 333, 161
Lowenthal, J., et al. 1991, ApJ, 377, L73
Macchetto, F., Lipari, S., Giavalisco, M., Turnshek, D. & Sparks, W. 1993, ApJ, 404, 511
Malkan, M., Teplitz, H. & McLean, I. 1995, ApJ, 448, L5
Mannucci, P. & Beckwith, S. 1995, ApJ, 442, 569
Matteucci, F,. and Padovani, P. 1993, ApJ, 419 485
McCarthy, P., et al. 1987, ApJ, 319, L39
McCarthy, P. 1993, ARAA, 31, 639

McMahon, R., *et al.* 1994, MNRAS, 267, L9
Meier, D. 1976, ApJ, 203, L103
Moller, P., & Warren, S. 1993, A&Ap, 270, 43
Norman, C. & Scoville, N. 1988, ApJ, 332, 124
Pahre, M.A. & Djorgovski, S.G. 1995, ApJ, 449, L1
Parkes, I., Collins, C. & Joseph, R. 1994, MNRAS, 266, 983
Peebles, P.J.E. 1989, in: The Epoch of Galaxy Formation, eds. C. Frenk *et al.*, Dordrecht:
    Kluwer, p. 1
Pritchet, C. 1994, PASP, 106, 1052
Puget, J.-L., Abergel, A., Bernard, J.-P., Boulanger, F., Burton, W., Désert, F.-X. &
    Hartmann, D. 1995, A&Ap, in press
Rowan-Robinson, M., *et al.* 1991, Nature, 351, 719
Schneider, D., *et al.* 1987, AJ, 94, 12
Silk, J. & Wyse, R. 1993, Phys. Rep., 231, 295
Small, T. & Blandford, R. 1992, MNRAS, 259, 725
Steidel, C., Sargent, W., & Dickinson, M. 1991, AJ, 101, 1187
Steidel, C. & Hamilton, D. 1992, AJ, 104, 941
Steidel, C. & Hamilton, D. 1993, AJ, 105, 2017
Steidel, C. & Dickinson, M. 1995, in: Wide Field Spectroscopy and the Distant Universe,
    eds. S. Maddox & A. Aragon-Salamanca, London: World Scientific, p. 349
Terlevich, E., Diaz, A., Terlevich, R., & Vargas, M. 1993, MNRAS, 260, 3
Terlevich, R. 1992, in: The Stellar Populations of Galaxies, IAU Symp. #149, eds. B.
    Barbuy & A. Renzini, Dordrecht: Kluwer, p. 271
Terlevich, R. & Boyle, B.J. 1993, MNRAS, 262, 491
Thompson, D., Djorgovski, S. & Beckwith, S. 1994, AJ, 107, 1
Thompson, D., Djorgovski, S. & Trauger, J. 1995, AJ, 110, 963
Thompson, D. & Djorgovski, S. 1995, AJ, 110, 982
Turner, E. 1991, AJ, 101, 5
Valls-Gabaud, D. 1993, ApJ, 419, L7
Warren, S., *et al.* 1995, MNRAS, in press
White, S., & Frenk, C. 1991, ApJ, 379, 52
Williams, R., & Perry, J. 1994, MNRAS, 269, 538
Wolfe, A., Turnshek, D., Lanzetta, K. & Oke, J.B. 1992, ApJ, 385, 151
Wolfe, A. 1993, ApJ, 402, 411
Wolfe, A. 1993, Ann. N.Y. Acad. Sci., 688, 281

7. Discussion

Windhorst: If the luminous galaxies of today formed by the rapid mergers of $\sim 20 - 50$ smaller subclumps at high z, shouldn't you be looking for different things?

Djorgovski: Such a scenario cannot account for the observed properties of elliptical galaxies and their stellar populations today, viz., their high metallicities and luminosity and phase space densities, the existence of the mass – metallicity relation, metallicity gradients, etc. These properties are easily understood if *most* star formation in ellipticals happened within their own potential wells. Disks and dwarfs are another matter: there you could have a sporadic flicker of star formation and a gradual assembly.

SEARCH FOR PRIMEVAL GALAXIES WITH THE "CALAR ALTO DEEP IMAGING SURVEY"

H. HIPPELEIN, K. MEISENHEIMER, E. THOMMES,
R. FOCKENBROCK AND H.-J. RÖSER
Max-Planck-Institut für Astronomie
Königstuhl 17, D–69117 Heidelberg, Germany

1. Introduction

Understanding the formation of galaxies is certainly one of the major challenges in astrophysics. Observationally, the most important insight will be provided by the detection and study of primeval galaxies. With the large telescopes and the efficient detectors available today, it seems feasible to observe those early stages of galaxy evolution directly. Despite considerable efforts with different techniques (Cowie *et al.* 1990; De Propris *et al.* 1993; Pritchet & Hartwick 1990; Thompson *et al.* 1995; Djorgovski *et al.* 1993) no primeval galaxy is found yet, in contradiction with predictions by model calculations (e.g. Baron & White 1987). What could be the reason for this lack of sucess ?

First, it seems that former searches were done at too low redshifts. Although Cold Dark Matter models tend to place the epoch of galaxy formation at $2 \lesssim z \lesssim 4$, there are reasons to believe that it took place at higher redshift: Quasars and radio galaxies are found out to $z \gtrsim 4$ and show strong metal lines; damped Ly-α systems, detected out to $z \simeq 4$ also contain a considerable amount of heavy elements; the old stellar population of elliptical galaxies at intermediate redshifts must have formed at $z > 3$ (Ziegler & Bender, this conference, poster 53).

Second, the predicted number of observable galaxies is grossly overestimated since the models did not take into account that the duration of the "primeval galaxy" phase could be short-lived. After several 10^8 years the first generation of stars may enrich the primordial material with metals causing significant UV extinction by dust (Thommes & Meisenheimer, poster 124, this conference).

R. Bender and R. L. Davies (eds.), New Light on Galaxy Evolution, 287–290.

2. Search Strategy for Primeval Galaxies

How do we recognize a primeval galaxy? By definition, its main characteristics are an almost pure hydrogen emission line spectrum over a faint continuum ($R \gtrsim 25$ mag) with Lyman break. The brightest line is Ly α with an expected equivalent width of the order of $W_{obs} \gtrsim 50$ nm at $z > 4$. At this redshift all other strong hydrogen lines are shifted beyond $2.4\,\mu m$ where groundbased observations are very difficult. Although at later stages in the galaxy formation process Ly α might be suppressed strongly by dust extinction, this complication is of little relevance for the first burst of star formation which we are looking for.

Any search for emission line objects of unknown redshift requires a trade-off between spectral coverage and field of view. In principle a longslit spectroscopic survey has the same efficiency as a narrow-band imaging survey with the same spectral resolution. In practice, however, narrow-band imaging has the advantages that the optimum sampling aperture can be chosen for each object individually when analysing the exposures, and that background subtraction is much more accurate on imaging data. In our approach only three narrow windows in the atmospheric spectrum, around $\lambda = 700$, 820, and 918 nm will be searched for emission lines.

Where shall we look for? Biased searches around known radio galaxies or quasars are subject to unknown selection effects and thus may even reduce the success rate. In order to get astrophysical insight into the galaxy formation process we need the unbiased statistics of primeval galaxies and therefore have to carry out the survey in the general field.

Along these lines we have developed an optimized strategy to search for primeval galaxies: the Calar Alto Deep Imaging Survey (CADIS). Though this survey project is specifically designed to detect primeval galaxies it will in addition produce a large data base for investigations of faint galaxies and quasars, by combining a very deep emission line survey with deep broad- and medium-band photometry. So the effort will not be lost even if no primeval galaxies can be found.

3. The CADIS Concept and First Results

The search for Ly α emission from primeval galaxies is done with an imaging Fabry-Pérot-Interferometer (FPI) in the three windows mentioned above, corresponding to redshifts of $z \sim 4.7, 5.7$, and 6.5. The spectral resolution is $\delta\lambda \sim 2.0$ nm and each window will be covered by 10 wavelength settings.

In most cases the emission line objects detected in the FPI images will be foreground galaxies seen in H_α, [O III] $\lambda 500.7$, *etc.* In order to recognize these cases we carry out additional medium-band observations at those wavelengths where one expects a second line for the above identifications. These filters are called veto filters because a signal in them excludes Ly-α.

Moreover, since most foreground objects have spectral energy distributions (SED) which significantly differ from that expected for a primeval galaxy we do observations with a set of 9 medium-band filters covering the wavelength range from 450 to 920 nm plus three broad band filters, including a deep K-band observation (limiting magnitude for 10% photometry: $R \simeq 23^{\mathrm{m}}5$, $K \simeq 20^{\mathrm{m}}5$). For more details see Meisenheimer et al., 1995). In order to avoid color selection effects in the data reduction, all filter images have the same priority. Object lists are produced independently for every filter and FP image by using FOCAS and are merged into a master list, where objects found within $1''$ of each other on different frames are assumed to be the same. The SED for every object is then determined from accurate CCD photometry on each individual image.

In the entire survey we expect to discover several tens of primeval galaxies, > 1000 emission line galaxies in several narrow redshift bins between $z = 0.24$ and $z = 1.4$ at least 500 early type galaxies in the range $0.5 < z < 1.2$), with redshifts determined within $\sigma(z) \lesssim 0.02$) with the multicolor method developed by Belloni et al. (1995) , and several hundred faint quasars, most of them at redshifts $z > 3$. Thus, CADIS will not only probe the space density of primeval galaxies to unprecedented depth and redshift but also provide a unique data base to study the evolution of the luminosity function and clustering properties of faint field galaxies.

The survey is carried out with the focal reducers at the Calar Alto 2.2 m and 3.5 m telescopes, with a 2k×2k optical CCD (field of view $\gtrsim 10' \times 10'$) and a 1k×1k NICMOS array for the NIR observations. We will observe 10 fields (total field 0.28 $\square°$) distributed over the northern sky in order to assure that at least one field is observable at any time. The fields have been selected on the IRAS maps and the Palomar Sky Survey plates to minimize galactic extinction and the presence of bright stars.

The attempted limiting flux for Ly-α is 3×10^{-20} W m^{-2} (5σ-detection). equivalent to a star formation rate of 15 M$_\odot$ yr^{-1} at $z = 6.4$ (for $H_0 = 50$ and $q_0 = 0.5$). This limiting flux is reached after an integration of 15 ksec with the focal reducer CAFOS 2.2 at the 2.2 m telescope (or 6 ksec at the 3.5 m-telescope). The required integration per medium-band filter to reach a (5σ) flux limit of 1.5 μJy is typically 15 ksec. The total integration time per field is thus 130 hours (at the 2.2 m telescope, the 3.5 m is 2.5× faster). So we need about 120 clear nights distributed among both telescopes for the project which will be distributed over a span of about 3 years.

Due to delays in getting the 2k×2k CCDs, only a few preparatory observations with a 1k×1k CCD (field of view 8'×8') at the 2.2 m telescope on two survey fields have been done so far. In Fig. 1 we present typical SEDs

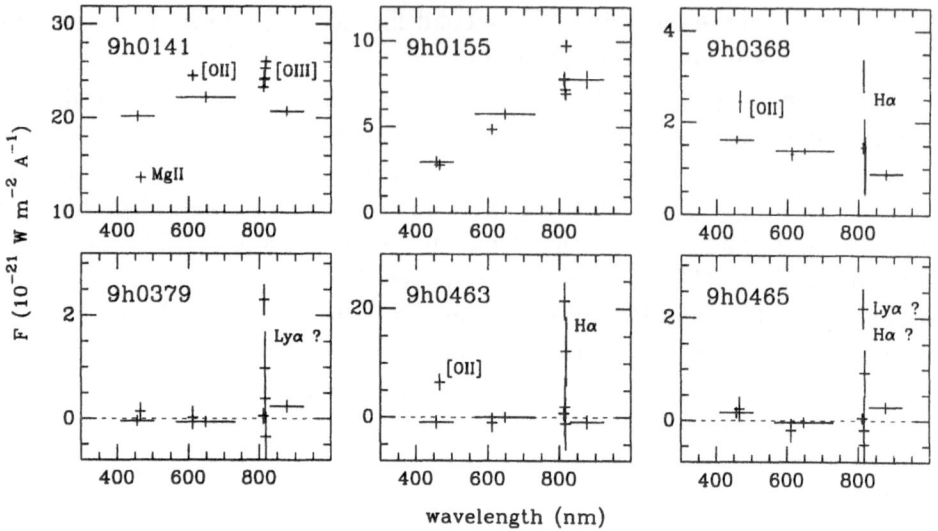

Figure 1.   SEDs of emission line objects in the 9h field. Object 9h0141 is probably a spiral at redshift 0.62 with distinct oxygen lines and magnesium absorption falling right in the 466 nm filter window. 9h0155 could be an M star. Object 9h0368 has a faint blue continuum ($R = 23.1$ mag) and two relatively strong emission lines and therefore is most probably a faint blue galaxy at redshift ~ 0.25. The strong line in the spectra of 9h0379 and 9h0465 may be Ly α which would make these objects candidates for primeval galaxies. In the case of object 9h0463, the presence of a second line in the 466 nm filter indicates an H II galaxy at redshift 0.25. Since only half of the data set is completed, the classification is still somewhat ambiguous.

for some of the ~ 400 emission line objects sofar detected in our 9h field. These spectra do not only demonstrate that we are able to identify the most likely contaminants (faint galaxies at $z < 1$) but also show that we indeed detect a considerable number of promising PG candidates which have been overlooked by previous surveys. Even if many of them will be identified with HII galxies at redshifts $\lesssim 1$ we regard their detection as first proof that the CADIS concept does find the right type of objects.

References

Baron, E. & White, S.D.M. (1987), *Astrophys.J.* **322**, 585

Belloni, P., Bruzual, G., Thimm, G. & Röser, H.-J. (1995), *Astron.Astrophys.* **297**, 61

Cowie, L.L., Gardner, J.P., Lilly, S.J. & Mc Lean, I. (1990), *Astrophys.J.* **360**, L1

De Propris, R., Pritchet, C.J., Hartwick, F.D.A. & Hickson, P. (1993), *Astron.J.* **105**, 1243

Djorgovski, S., Thompson, D. & Smith, J.D. (1992), in *First Light in the Universe*, B. Rocca-Volmerange *et al.* (eds.), p. 67

Meisenheimer, K., *et al.* (1995), in *Galaxies in the Young Universe*, H. Hippelein *et al.* (eds.), p.273

Pritchet, C.J. & Hartwick, F.D.A. (1990), *Astrophys.J.* **355**, L11

Thompson, D.J., Djorgovski & S., Trauger, J. (1995), *Astron.J.* **110**, 963

RESULTS FROM SEARCHES FOR HIGH-REDSHIFT GALAXIES AROUND DAMPED LYα SYSTEMS

S. J. WARREN
Astrophysics, Blackett Laboratory
ICSTM, Prince Consort Rd, London, SW7 2BZ, UK

AND

P. MØLLER
Space Telescope Science Institute,
3700 San Martin Drive, Baltimore, MD 21218, USA
on assignment from the Space Science Department of ESA

1. Introduction

There was not much said at this conference about the damped Lyα absorbers, yet locally and at high redshift nearly all the neutral hydrogen in the Universe lies in the damped systems, so they may have much to tell us about when and how galaxies formed. This talk is concerned with the results of narrow-band imaging searches for Lyα emission from, and around, damped Lyα absorbers. There have been many such searches, but very few detections. Here we concentrate on two groups of galaxies at high redshift, $z > 2$, found by this technique. One of these is a recent discovery (Francis et al. 1995), and for the other we discuss some new observations (Warren and Møller 1995). A study of these two groups provides some clues to the nature of the damped absorbers at high redshift, and the rôle they play in the formation of galaxies.

2. Damped Lyα absorbers

The damped systems form the high-column end of the distribution of Lyα absorbers, $N(\text{H I}) > 10^{20} \text{cm}^{-2}$. The shape of the column-density distribution means that virtually all the neutral gas in the Universe is contained in the damped systems, so surveys for damped absorbers in the spectra of quasars can be used to measure the cosmic density of neutral gas Ω_{DLA}.

291

R. Bender and R. L. Davies (eds.), New Light on Galaxy Evolution, 291–294.

Lanzetta et al. (1995) provide a plot of Ω_{DLA} in the redshift range $0 < z < 3$. There is a rough correspondence between the mass in neutral gas at $z = 3$, and the mass in stars in spiral galaxy disks today, and the standard picture has been that the damped systems at $z = 3$ are the fully formed gaseous disks of spiral galaxies, that turn into the stellar disks of today. The disks would have been a factor > 2 larger in the past, in order to explain the large number of absorbers found. However, the mass in stars today is not known to better than a factor of 2, and given the low metallicity of the damped systems at $z > 2$, we could instead be seeing the gas from which all stars formed, i.e. in both disks and spheroids. Nevertheless, accounting for the effects of dust in the damped systems may be critical in such considerations (Pei and Fall 1995). This needs to be quantified better before drawing firm conclusions about the history of star formation in the damped absorbers.

3. Lyα searches

The theme of this talk is the comparison between what we observe at high redshift, and the predictions of hierarchical clustering theories (Frenk, this volume). In the simulations the constituents of a galaxy today reside at $z = 3$ in a small number of pieces, which can lie far apart. A nice illustration of this is given by Navarro et al. (1995). In their Fig. 1 we see at $z = 3$ two large clumps of cold gas separated by a proper distance of $100h^{-1}$ kpc, that merge at $z \sim 1$ to form a disk galaxy. Therefore in searching for emission around damped systems at $2 < z < 3$ we should be thinking in terms of sub–units of galaxies at angular separations from the line of sight to the quasar of, say, 30 arcsec (1 arcsec $\equiv \sim 4h^{-1}$ kpc, $2 < z < 3, q_o = 0.5$).

3.1. A GROUP OF GALAXIES AT $Z = 2.38$ TOWARDS THE QUASAR 2139–4434

This group was discovered at the end of 1994. Francis and Hewett (1993) describe the spectra of two high-z quasars separated by 8 arcmin, in which in each spectrum common high-column-density absorption is seen at each of two redshifts, $z = 2.38$, $z = 2.85$. This common absorption may be due to a galaxy supercluster, and follow-up narrow-band imaging in Lyα has begun in an effort to map out the structure. Francis et al. (1995) describe the results from imaging towards the quasar 2139–4434, targeted at the absorber at $z = 2.38$. This absorber has a column density 7.0×10^{18} cm^{-2}, so in fact it is a Ly-limit system rather than a damped system. The narrow-band image revealed two candidate high-redshift galaxies, and subsequent spectroscopy has confirmed Lyα emission from each. The brighter source, $f = 8 \times 10^{-16}$ erg s^{-1} cm^{-2}, lies at 22 arcsec from the line of sight to the quasar. It is extended over some 5 arcsec, and probably shows C IV emission. In many respects it resembles the high-redshift radio galaxies, although it is radio quiet. The second source, $f = 3 \times 10^{-16}$ erg s^{-1} cm^{-2}, lies at 63 arcsec separation. The two galaxies are certainly bright enough to have been discovered in the wide-field blank-sky surveys by Thompson and Djorgovski, had they lain in one of their fields.

There are three objects at $z = 2.38$ in this field: the Lyα absorber, and the two emission-line galaxies. The main feature that we draw attention to is the fact that the three sources are closely aligned. We return to this point below.

3.2. EMISSION-LINE GALAXIES AT $Z = 2.81$ TOWARDS PKS0528-250

The second absorber we discuss is the unusual damped system at z=2.81 seen in the spectrum of the quasar PKS0528-250. The absorber lies sufficiently close that the quasar's Lyα emission line is completely absorbed. Møller and Warren (1993) imaged the system at Lyα, in an attempt to measure the size of the absorber. The idea was that the quasar would light up the backside of the absorber, and the periphery of the cloud would be visible as a ring of Lyα emission. However the narrow-band image did not reveal a ring. Rather three blobs of emission were found at angular separations of 1.2, 11, and 21 arcsec from the quasar.

Subsequent spectroscopy and deep imaging of this field are presented in Warren and Møller (1995). The interpretation of what we are seeing here is not straightforward. Briefly, based on the morphology of the three blobs, the line velocity widths, and the line equivalent widths, we are led to the conclusion that we are seeing star formation rather than photoionisation by the quasar. As with the field studied by Francis et al. the three sources are closely aligned. One possibility is that they are three regions of star formation in an edge on disk. They span a distance of $80h^{-1}$ kpc, and this would accord with the picture outlined above of spiral galaxy disks being larger in the past. However the velocities are inconsistent with systematic rotation: the middle blob is redshifted with respect to the outer two blobs. Therefore we seek an alternative explanation.

We put forward the idea that the alignments of galaxies seen in these two fields correspond to the filamentary structures found in simulations of hierarchical theories of galaxy formation: See e.g. Evrard, Summers, and Davis (1994) for good plots showing this phenomenon. At this conference Windhorst et al. presented a third group of high-redshift galaxies. There are four objects which apparently show emission lines in a narrow-band image. Three of these (objects A, B, R) have redshifts confirmed by spectroscopy, and once again they are closely aligned. We predict that such alignments will prove common, as more high-redshift galaxies are discovered.

The mass of the gas in the 0528 absorber is estimated to be $1 \times 10^9 h^{-2} M_\odot$, and the total mass $4 \times 10^{10} h^{-1} M_\odot$. The dynamical mass of the group is $\sim 3 \times 10^{11} h^{-1} M_\odot$, and the dynamical time a few times 10^8 yrs. The group resembles groups seen in the computer simulations, that coallesce at lower redshift. The overall picture that emerges is of star formation in a sub-$L_\star$ galaxy, at an early stage in its evolution, during the process of assembly.

4. Remarks on the nature of galaxies at high redshift $z > 2$

1. The sensitivity of emission-line searches for high-z galaxies may be compromised by the fact that today's galaxies are in pieces at high redshift.

2. There are three confirmed groups of galaxies at $2 < z < 3$. All show alignment, which may correspond to the filamentary structures seen in simulations of galaxy formation.

3. At $z \sim 3$ at least some of the damped systems are probably sub-components of galaxies in the process of assembly, rather than the fully-formed progenitors of spiral-galaxy disks.

References

Evrard, A. E., Summers, F. J. and Davis, M. (1994) ApJ 422, 11
Francis, P. J. and Hewett, P. C. (1993) AJ 105, 1633
Francis, P. J., Woodgate, B. E., Warren, S. J., et al. (1995) ApJ in press
Lanzetta, K. M., Wolfe, A. M. and Turnshek, D. A. (1995) ApJ 440, 435
Møller, P. and Warren, S. J. (1993) A&A 270, 43
Navarro, J. F., Frenk, C. S. and White, S. D. M. (1995) MNRAS 275, 56
Pei, Y. C. and Fall, S. M. (1995) ApJ in press
Warren, S. J. and Møller, P. (1995) A&A in press

DISCUSSION:

Fritze-v. Alvensleben: Is the COBE upper limit on dust in agreement with estimates of dust in the damped systems, made from measurements of the reddening of QSOs, and from the Cr to Zn ratio?

Warren: Yes. The dust-to-gas ratio in the damped systems at $2 < z < 3$ is only about 1/10th of that in the local group, but this would be sufficient to extinguish Lyα effectively, if resonant scattering is important.

Rees: The aligned emission regions are slightly reminiscent of what is seen in high-z radio sources. Do you think there is any relation?

Warren: I suppose it is just possible, since two of the fields contain radio sources. None of the three aligned objects in the Francis field is a radio source, however.

Windhorst: In the case of our group at z=2.40, the other Lyα emitters (objects A and B) occur in a direction $\sim 40°$ different from the radio axis of 53W002 (object R).

GALAXY EVOLUTION FROM QSO ABSORPTION–LINE SELECTED SAMPLES

MARK DICKINSON

STScI, 3700 San Martin Dr., Baltimore MD 21218, USA

AND

CHARLES C. STEIDEL

Caltech Astronomy, MS 105-24, Pasadena CA 91125, USA

1. QSO Absorption Lines and Galaxies

Although a connection between certain classes of QSO absorption line systems and gas associated with galaxies was hypothesized long ago, the first systematic evidence supporting this was provided by Bergeron & Boissé (1991). Observing QSOs with known MgII absorption lines at $z_{abs} < z_{QSO}$, they identified galaxies near the QSO sightline and spectroscopically confirmed that their redshifts matched those of the MgII absorption doublet.

Since that time, we have been carrying out surveys aimed at (1) characterizing the nature of the galaxies selected by this method, and (2) using these galaxies to study the evolution of the field galaxy population at high redshift. We refer to our sample as one "selected by gas cross–section," since a MgII rest–frame equivalent width $W_0 > 0.3$Å is essentially equivalent to a neutral hydrogen column density $N(HI) > 10^{17}$ cm^{-2}. For the purposes of studying field galaxy evolution, this is useful primarily because the selection depends only on a robust and easily measured *rest frame* property (W_0(MgII)), and not on any *observed frame* characteristic such as apparent magnitude, color, surface brightness, etc. The method is thus free of many potential biases which affect deep magnitude–limited redshift surveys (although it may be subject to its own!). Moreover, once the gas halo cross–sections are understood (see below), the resulting sample is volume–limited, which is greatly advantageous when studying luminosity functions and the like. If the nature of absorption–selected galaxies can be firmly established, then comparison between our samples and those from the deep

R. Bender and R. L. Davies (eds.), New Light on Galaxy Evolution, 295–298.
© 1996 IAU. Printed in the Netherlands.

redshift surveys (such as the CFRS sample discussed by Lilly in the present volume) may lead to new insights about galaxy evolution.

Our first survey studied MgII absorption systems at $0.2 < z < 1.0$, and is essentially finished, with nearly complete imaging identification of candidate absorbers and spectroscopic redshift confirmation for $\sim 80\%$ of these candidates. We are presently pushing for 100% redshift completeness using the W.M. Keck 10m telescope. The second survey covers the range $1 < z < 2$ and is now underway using infrared and optical imaging from KPNO, Palomar and Keck, as well as selected spectroscopic follow–up from Keck. Space limitations prevent more than a cursory summary of results from these surveys, and no elaboration of the methods, procedures, data, or analyses is offered. For these, the reader is directed to previously published descriptions, including Steidel & Dickinson 1992, Steidel, Dickinson & Persson 1994, Steidel & Dickinson 1995, and Steidel 1995.

For the range $0.2 < z < 1.0$, we find that *all* galaxies with luminosities $L_K > 0.05 L_K^*$ are potentially MgII absorbers, provided that they fall within a particular impact parameter of the QSO sightline. This impact parameter (effectively, the gaseous halo radius) scales weakly with luminosity as $R_{\rm halo} = 38 h^{-1} (L_K/L_K^*)^{0.15}$ kpc, (where $h = H_0/100{\rm km/s/Mpc}$). Conversely, we find no cases of bright intervening galaxies within this impact parameter limit which do *not* produce absorption. The only interlopers have turned out to be dwarfs, mostly very blue. Apparently, the MgII systems are dominated by "big" galaxies (within a few magnitudes of L^*). The "faint blue galaxies" which dominate deep number counts do not contribute appreciably to the gas cross–section of the Universe at $N({\rm HI}) > 10^{17}$ cm^{-2}. The K–band luminosity function of the absorbing galaxies follows a Schechter-like distribution with L^* (at $\langle z \rangle \approx 0.65$) indistinguishable from the present-day value, but with a high normalization consistent with values measured from the deep field galaxy surveys (e.g. the CFRS) at similar redshifts.

Interestingly, the halo radius vs. luminosity scaling is better behaved when computed with K–band magnitudes (instead of optical photometry). Moreover, we see no relation between a galaxy's color and its nature as an absorber. The absorbers span the range of normal galaxy colors from flat-spectrum "Magellenic irregulars" to red ellipticals. There is no evidence for color evolution in the absorber population out to $z = 1$, although Lilly et al. (1995) have shown that our results are not formally inconsistent with the evolution of the field galaxy population seen in the CFRS. Evidently, to first order, the presence of an extended gaseous halo does not depend on the current star formation rate in a galaxy, but rather on the luminosity of its older stellar population. Because L_K, for evolved galaxies, roughly traces the total stellar mass, we might therefore suppose that the halo diameter primarily reflects the *mass* of the galaxy.

Our extension of this survey to the redshift regime $1 < z < 2$ has only begun, but the preliminary indications are that nothing is dramatically different from the situation at $z < 1$ – the galaxy luminosities, colors, impact parameters, and space densities are roughly the same. Overall, this suggests that the "big" (massive?) galaxy population has been, for the most part, remarkably stable over a very long span of cosmic time. The apparent *absence* of luminosity evolution might seem to contradict expectations for simple passive evolution of stellar populations. However, this may be interpreted as implying a roughly constant star formation rate with redshift when averaged across the absorber population.

2. MgII Absorber Morphologies

Recently, we have begun obtaining *HST* WFPC2 images of MgII absorbers in order to evaluate their morphologies and to measure their orientations and inclination angles relative to the quasar line of sight. In Cycle 4 we imaged 3C 336, a $z = 0.92$ quasar with the largest number of foreground absorbers (5!) along any single line of sight in our survey. In Cycle 5, we are imaging 12 additional fields. All images are taken through the F702W filter ($\lambda_{eff} \approx 7000$Å).

Figure 1 shows a montage of the absorbers imaged to date. Most are fairly ordinary galaxies spanning the range of normal Hubble Types, from bulge–dominated systems (e.g. the $z = 0.318$ and 0.660 absorbers) to late-type disks (e.g. the $z = 0.442$ and 0.723 galaxies). A few exotic objects are also found – the $z = 0.525$ absorber looks like the highly elongated, peculiar galaxies seen in many deep *HST* images and emphasized by Cowie et al. 1995. This galaxy, however, is extremely red, suggesting that it is a highly reddened, edge–on disk – extinction may account for its odd morphology. Several absorbers are highly inclined (e.g. the $z = 0.891$ galaxy), reinforcing other evidence that MgII absorption arises from halo material rather than from gas in the disk. The few $z > 1$ galaxies we have imaged look somewhat peculiar, but this may only reflect the fact that the R–band WFPC2 images sample their emitted–frame ultraviolet continuum.

M.D. would like to thank the organizers for their hospitality and travel support, as well as STScI for additional funds. This work was partially supported by NASA/STScI grants GO–5304 and GO–5984.

References

Bergeron, J., and Boissé, P. 1991, A&A, 243, 344.
Cowie, L.L., Hu, E.M., and Songaila, A. 1995, Nature, 377, 603.
Lilly, S.J., Tresse, L., Hammer, F., Crampton, D., and Le Fèvre, O. 1995, ApJ. (in press).
Steidel, C.C., and Dickinson, M. 1992, ApJ., 394, 81.

Steidel, C.C., Dickinson, M., and Persson, S.E. 1994, ApJ., 437, L75.

Steidel, C.C., and Dickinson, M. 1995, in *Wide Field Spectroscopy & the Distant Universe*, eds. S. Maddox and A. Aragón–Salamanca, World Scientific, p. 349.

Steidel, C.C., in *QSO Absorption Lines*, ed. G. Meylan, Springer, p. 139.

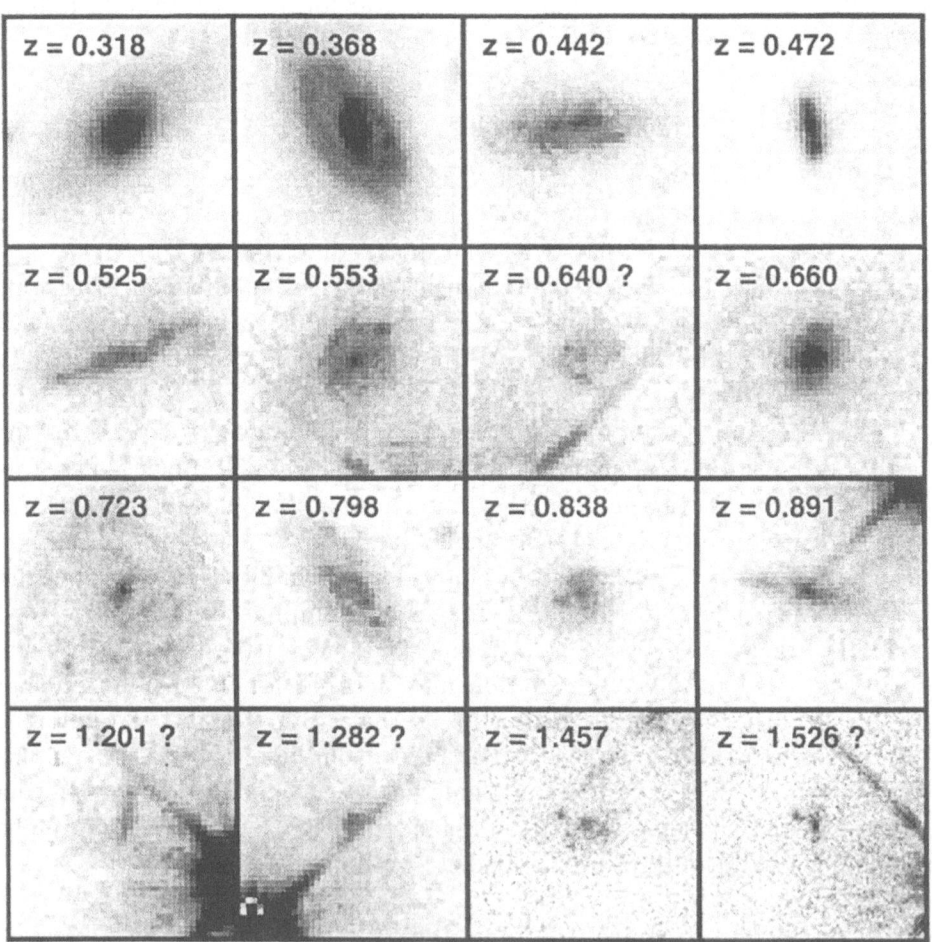

Figure 1:   Montage of *HST* WFPC2 images of MgII absorbing galaxies, arranged by redshift. Each panel is 4 arcsec on a side. Cases where the galaxy redshifts have not been confirmed spectroscopically are indicated with a question mark.

THE ECOLOGY OF QUASAR ABSORPTION SYSTEMS

DAVID TYTLER
University of California San Diego
CASS 0111, La Jolla, CA 92093-0111, tytler@ucsd.edu

1. Introduction

Gas along the lines of sight to QSOs produces narrow absorption lines from ions of the most abundant elements: H, He, C, N, O, Si, S, Mg and Fe. Most lines have rest wavelengths of a few hundred Åbut some have wavelengths $\lambda \geq 900$ Å, sufficient that they can be redshifted into the optical. The absorbing gas is normally transparent, except for Lyman continuum absorption, and dust extinction at the highest columns, so we can see hundreds of gas clouds along a line of sight. We see more absorption from higher redshifts, because the universe was denser then and the absorbing "clouds" have roughly constant proper size.

Understanding of absorption systems is advancing rapidly now, with observation of low redshift absorption by the *Hubble Space Telescope*, with high signal-to-noise spectra from the W.M. Keck observatory, and with better understanding of the Lyα forest from computer simulations.

2. Types of Absorbers

Absorption is seen from a wide variety of environments, from tenuous low density gas in the intergalactic medium and the very outer regions of galaxies, to the inner regions of galaxies and the immediate surroundings of QSOs. In Table 1 I give rough values for the neutral Hydrogen column densities N(H I), the ionization, and the number of absorbers seen per unit redshift $N(z)$.

We are most likely to see absorption from gas which is very spatially extended. This gas usually also has a low density, so it is very sensitive to its environment. The sensitivity makes it extremely hard to predict the properties of the absorbing gas prior to observation, but instead we are learning to use observations to deduce the properties of the environment.

R. Bender and R. L. Davies (eds.), New Light on Galaxy Evolution, 299–302.
© *1996 IAU. Printed in the Netherlands.*

TABLE 1. Types of Absorption System

Origin	Name	N(H I) (cm^{-2})	Ionization (H II/H I)	$N(z)$
IGM	Gunn-Peterson absorption	$< 10^{12}$	$> 10^6$	$> 10^3$
IGM	Lyα forest	$10^{12} - 10^{14.5}$	10^5	10^3
Outer galaxy halos	Lyα forest	$10^{14.5} - 10^{17}$	10^4	10^2
Inner galaxy halos	Lyman Limit Systems^a	$10^{17} - 10^{20}$	10^2	1
Galaxy Disks	Damped Lyα	$10^{20} - 10^{22}$	$1 - 10$	0.1
Near QSOs	BAL & Associated	$< 10^{17}$	high	1^b

^{a} Including most metal lines systems with C IV and Mg II absorption.
^{b} Rate per QSO, not per unit z; about 10% of QSO spectra show broad absorption lines, and most have associated absorption which arises within Mpc of the QSO.

Consider the gravitational concentration of baryons into increasingly dense objects. In order of increasing gas density we have: the smooth intergalactic medium (the IGM is revealed in Gunn-Peterson absorption), filaments and foam in the IGM (seen as Lyα forest lines), the extreme outer parts of galaxies (Lyα absorption at $< 160\ h^{-1}$kpc), the inner parts of galaxies (Lyman Limit Systems – LLS – and metal line absorption from $< 30h^{-1}$kpc), inner parts of galaxies (damped Lyα absorption), and absorption near to QSOs (BAL – Broad Absorption Lines from gas ejected by QSOs, and associated absorption from QSO host galaxies or other near by galaxies). Since most gas is photoionized, as the density rises, the recombination time and level of ionization fall. Frequency of observation also falls with increasing N(H I), because denser gas clouds are smaller and rarer.

The BAL and associated absorption is different because it comes from a special environment. The intense radiation near to QSOs produces unusually high ionization, and the frequency of occurrence is a property of the QSOs, rather than the cosmological distribution of baryons.

3. Connection to Galaxies: Baryon Flow

As baryons flow from the smooth IGM into denser and denser objects there are three important feedbacks: radiation, heat and metals.

Ion ratios show that the absorbing gas is mostly photoionized by UV radiation. Much of this comes from QSOs and AGN and is "hard" with abundant far UV photons. Some near UV light also leaks out from young stars in galaxies, but the amount of this is unknown. Gas far from QSOs receives radiation which has a softer spectrum and reduced intensity because of absorption by intervening gas. The radiation determines the ionization of the gas, and the gas opacity shapes the spectrum and intensity of the radiation.

Star formation in galaxies heats and ejects gas, along with metals, into galaxy halos and presumably the IGM, where it is seen in absorption. We do not know whether most baryons are in galaxies and their halos, or in the IGM, since this depends in large part on the feedback of heat from star formation in galaxies. There will be more gas and metals in the IGM if the initial mass function favors high mass stars and many of these stars form together and over a short period of time in large bursts.

The radiation energy released by the stars which made the carbon seen in Lyα forest clouds with N(H I) $> 10^{14.5}$ is also energetically significant (Madau & Shull 1996), and might have helped re-ionize the IGM.

There is presumably dark matter associated with absorption systems, but we have not measured the ratio of baryonic to dark matter.

How sensitive are the properties of the absorbers to the parameters of galaxy formation: Ω_b, the shape of the perturbation spectrum, the epoch and mechanism of IGM re-ionization, and the parameters of the background radiation field (origin, intensity, shape, spatial and temporal variation)? The answer should be obtained from detailed and accurate computer simulations of galaxy formation. In the long run we expect that observations of the absorbers (eg. rise in abundances and clustering with epoch, change in Ω in hot and cold gas) will constrain the free parameters of galaxy formation.

4. Cosmological Distribution of Baryons and Metals

In Table 2 I give a rough indication of the cosmological distribution of baryons and metals. The uncertainty in the level of ionization affects both the total baryon contents and the metal abundance, which is expressed in log solar units: [M/H] $\equiv log(M/H) - log(M/H)_{solar}$. The ionization depends on the gas density, and on the intensity and shape of the ionizing spectrum, which are not well known.

Entries in the table are from observations at different epochs. If the abundances in the gas seen in damped Lyα absorbers rises ten times from $z = 2$ to today, and if galaxies eject gas so that the abundances in their halos and the IGM also rise ten times, then today the metals in the Lyα forest and IGM would be about 0.3 of those in galaxies, and perhaps most metals are outside the inner regions of galaxies. This is reasonable because 90% of the gas and metals in clusters of galaxies are in the inter-cluster medium. If galaxies eject metals on their own, e.g. in supernova driven winds, then most galaxies, both in and out of clusters, could eject most of their metals.

TABLE 2. Cosmological Distribution of Metals

Origin	$\log\Omega_b$	[M/H]	$\log\Omega_{metals}$ [a]	z
IGM & Lyα	≤ -1.5	-2.5	-6.0	2.8
LLS	-3.0	-2.0	-7.0[b]	3
Damped Lyα	-2.5[c]	-1.0	-5.5	2
galaxies	-2.5	0	-4.5	0

[a] $\log\Omega_{metals} \equiv \log\Omega_b + [M/H] - 2$, all approximate values for $h = 1$.
[b] For LLS both Ω_b and [M/H] depend on the ionization. Ω_{metals} is better known than Ω_b, or [M/H], or the ionization. Data from Steidel (1990).
[c] The value given is twice that observed (Storrie-Lombardi & Wolfe 1995), to correct for ionized hydrogen and systems missing because they contain dust which hides the background QSOs (Pei & Fall 1995). Abundances from Pettini et al (1994).

5. Missing Information

The following are four areas where we expect important advances.
1. Measurement of the three dimensional shapes and sizes of absorbers; from observations of double and triple QSOs, UV spectra of spatially resolved galaxies, and the detection and mapping of Hα emission from LLS.
2. Determination of the ionizing spectrum; from the calibration of the proximity effect using faint QSOs and metal lines, and the measurement of the shape of the background spectrum using metal line ions.
3. Low z identifications have great potential. Do all absorption systems with N(H I) > $10^{14.5}$ arise in galaxies? Are metals always seen in absorption from the outer regions of galaxies? Are all damped systems disks? What types of galaxies produce each type of absorber? How is the gas moving relative to the galaxies?
4. Several "old" problems must be cleaned up: the effects of dust and lensing on damped Lyα absorbers at low z, the fraction of lines (e.g. narrow Lyα and metals) from ejected gas, measurement of temperatures from the comparison of the relative width of ions of different atomic mass, leading to assessment of collisional ionization, and the modeling of inhomogeneous absorbing regions.

References

Madau, P. & Shull, J.M. 1996, ApJ, Feb 1
Pei, Y.C. & Fall, S.M. 1995 ApJ
Pettini, M., King, D., Smith, L.J. & Hunstead, R.W. 1994 in *QSO Absorption Lines* ed. G. Meylan, p.71
Steidel, C.C. 1990 ApJS, **74**, 37-91
Verner, D.A., Barthel, P.D. & Tytler, D. 1994, AApS, **108**, 287

WHAT QUASARS TELL US ABOUT GALACTIC EVOLUTION

MARTIN J REES

Institute of Astronomy

Madingley Road, Cambridge, CB3 OHA

Abstract. Quasars offer three types of clue to galactic evolution. (i) The formation of massive black holes in the centres of young galaxies; (ii) Clues to subgalactic structure from quasar absorption spectra; (iii) Implications for redshifts higher than 5. These are briefly summarised in this text.

1. AGNs and the Stellar Cores of Galaxies

The most remote quasar so far detected has z=4.89 (Schneider, Schmidt and Gunn 1991). The population genuinely seems to thin out at redshifts above 2.5-3: the comoving density of quasars falls by at least 3 for each unit increase in z (see Shaver, 1995 for a review). Recent progress in the study of active galactic nuclei (AGNs) bring into sharper focus the question of how and when supermassive black holes formed, and how this process relates to galaxy formation.

When the stellar core of a galaxy forms, part of the gas may collapse into a black hole. The formation and growth of the hole then manifests itself as a quasar; the peak in the quasar population (i.e. redshifts in the range 2 - 3) signifies the era when large galactic bulges were forming. This process involves complex gas dynamics and feedback from stars; we are still a long way from being able to make realistic calculations. At the moment, the most compelling argument that a massive black hole forms comes from the implausibility of the alternatives. To stop such an outcome, either:

(i) The formation of stars during the infall must be nearly 100 percent efficient; moreover, the stars would all need to be of low mass (so that no material is expelled out again).

or (ii) Gas must remain centrifugally supported in a self-gravitating disc for hundreds of orbital periods, without the onset of any instability that

303

R. Bender and R. L. Davies (eds.), New Light on Galaxy Evolution, 303–311.

redistributes angular momentum and allows the inner fraction to collapse.

Neither of these options seems at all likely – the first would require an IMF quite different from what is actually observed in the inner parts of galaxies; the second is contrary to well-established arguments that self-gravitating discs are dynamically unstable.

We first need to know at what stage in its progressive concentration towards the centre the gas stops being able to form stars (because of radiation pressure, magnetic fields, or whatever) and evolves instead into a supermassive object. A crude argument is the following.

When a self-gravitating object releases its binding energy in a dynamical (or free fall) timescale, the luminosity is

$$L_{diss} = \left(v_{virial}^5/G\right) \simeq 10^{59} \left(v_{virial}/c\right)^5 \text{ erg s}^{-1}.$$

When v_{virial} gets high enough, this becomes comparable with the Eddington limit. Fragmentation is then impossible: instead, the gas is 'puffed up' by radiation pressure into a supermassive object. This would happen whenever $L_{diss} \gtrsim L_{Ed}$, i.e. when

$$v_{virial} > 300 \left(M/10^6 M_\odot\right)^{\frac{1}{5}} \text{ km s}^{-1}.$$

Note that the above condition, while sufficient, is by no means necessary: fragmentation may stop at a substantially earlier stage. For instance, photoionization cross-sections can greatly exceed the Thomson cross-section , allowing pressure support for a lower L_{diss} (Haehnelt 1995). Dust opacity could also be important.

Magnetic stresses could also prevent fragmentation. Full equipartition requires a field of

$$\sim 0.1 \left(M/10^8 M_\odot\right)^{-1} \left(v_{virial}/300 \,\text{km s}^{-1}\right)^4 G;$$

but even a field 1000 times weaker than this would exert enough pressure to inhibit the condensation of stellar-mass subunits; the free-electron fraction maintained by the intense UV radiation would be too high to permit ambipolar diffusion (which permits stars still to form, despite dynamically-important magnetic fields, within dense interstellar clouds in present-day galaxies). Quite apart from these constraints, fragmentation would be inhibited by tidal effects when the gas developed a sufficient central concentration.

Once a large mass of gas became too condensed to fragment into stars, it would continue to contract and deflate like a spinning supermassive star; viscosity would provide internal dissipation, and redistribute angular momentum. Some material would be shed during the contraction to carry away the angular momentum, but a substantial fraction could continue contracting until it underwent complete gravitational collapse – for example, if 10 percent of the mass had to be shed in order to allow contraction by a factor of 2, about 20 per cent would remain to form a black hole.

2. Quasar Remnants

Considerations of AGN 'demography' , by now well known, suggest that a typical galaxy could have passed through a quasar phase, and that by $z = 2$ (2-3 billion years) most had developed central holes of $10^6 - 10^9 M_\odot$. Van der Marel has reviewed the evidence for central dark masses – dead quasars – based on studying the spatial distribution and velocities of stars in the central core.

Much the most compelling evidence for a central black hole, however, recently came from a quite different technique: probing gas motions by measuring the 1.3 cm line of H_2O in the nearby peculiar spiral galaxy NGC 4258 (Watson and Wallin 1994; Miyoshi et al. 1995). The spectral resolution in this microwave line is high enough to pin down the velocities with accuracy of 1 km/sec. Such observations, combined with VLBA mapping with a resolution of 0.5 milliarc seconds (100 times sharper angular resolution than the HST) have revealed, right in the galaxy's core, a disc with rotational speeds following an exact Keplerian law around a compact dark mass. The circumstantial evidence for black holes has been gradually growing for 30 years, but this remarkable discovery clinches the case completely.

Should we be surprised that the putative holes in nearby galaxies are generally quiescent? Their environment could be almost free of gas, so that very little gets accreted; moreover, when the accretion rate is low, he cooling is so slow (because of the low densities) that only a small fraction of the binding energy gets radiated. The radio sources in the centres of some ellipticals, and perhaps also their X-ray emission, are consistent with accretion in a low-efficiency mode (Fabian and Rees, 1995 and references cited therein).

The case for a black hole in our own Galactic Centre was until recently ambiguous, but is now strengthened by the discovery of stars within the central 0.1 pc (Krabbe et al. 1995). The case would be clinched if some of these stars turned out to be moving as fast as (say) 1000 km/sec. If there is indeed a hole of a few million solar masses in our Galactic Center, the most recent tidal disruption of a star (even if it happened tens of thousands of years ago) may have left traces that could still be be detectable. Up to 10^{53} ergs of ionizing radiation could be released by accretion of the captured debris – more photons than would be emitted by steadier UV sources in the entire $\sim 10^5$ years interval between successive disruptions. Moreover the half of the star that is ejected may have left traces in some of the strange patterns of the gas within the central 2 parsecs.

3. Effects of Binary Black Hole Coalescence

Are the putative black holes actually described by a Kerr metric, as relativity predicts? Some important progress on this question has come recently from the ASCA X-ray satellite, which offers sufficient spectral resolution to reveal lines. Whereas optical emission lines come, for straightforward thermodynamic reasons, from a volume much larger than the hole itself, thermal X-ray emission can come from the innermost parts of an accretion flow. The lines should therefore display substantial gravitational redshifts, as well as large doppler shifts. There is already one convincing case (Tanaka *et al.* 1995) of a broad asymmetric emission line indicative of a relativistic disc viewed almost face-on, and others should soon follow

The most impressive test of general relativity would be detecting gravitational radiation from mergers of supermassive binaries: this involves no physics other than that of spacetime itself. Some events of this kind are expected. Quasar statistics suggest that most galaxies should harbour black holes that formed at $z > 2$. Moreover, many galaxies have experienced a merger since that time. When that happens, the holes in the two merging galaxies would spiral together, emitting, in their final coalescence, up to 10 per cent of their rest mass as a burst of gravitational radiation.

This burst would be in a frequency range around a millihertz – too low to be accessible to ground-based detectors. Space-based detectors are needed. One such being proposed, by the European Space Agency, is the Laser Interferometric Spacecraft (LISA) – six spacecraft on solar orbit, configured as two triangles, with a baseline of 5 million km whose length is monitored by laser interferometry.

LISA could detect the mergers of supermassive holes, even whose occurring at high redshifts, with high signal-to-noise. The bad news is that the event rate is low. Even out to $z = 5$, there could be less than one event per decade involving holes above $10^6 M_\odot$, even if there are enough black holes altogether to account for all the quasars (Haehnelt 1994). This is of course a lower limit. There could be lower-mass holes in small galaxies that are more common and underwent more mergers. The sensitivity of LISA is such that it could even detect waves from a stellar-mass object orbiting a supermassive hole.

LISA is at the moment just a proposal – even if it is funded, it is unlikely to fly before 2015. Is there any way of learning, before that date, something about gravitational radiation? The dynamics (and gravitational radiation) when two holes merge has so far been computed only for cases of special symmetry. The more general problem – coalescence of two Kerr holes with general orientations of their spin axes relative to the orbital angular momentum – is one of the US 'grand challenge' computational projects. When this challenge has been met (and it will almost certainly not take all the time until 2015) it will tell us not only the characteristic wave form of the radiation, but also the recoil that arises because there is a net emission of linear momentum.

This recoil could displace the hole from the centre of the merged galaxy – it might therefore be relevant to the low-z quasars that seem to be asymmetrically located in their hosts (and which may have been activated by a recent merger). The recoil might even be so violent that the merged hole breaks loose from its galaxy and goes hurtling through intergalactic space.

Density profiles in the centres of ellipticals would be altered by such an event. The inward spiralling of a binary black hole via dynamical friction would substantially change the orbits of stars whose total mass is at least of the order of the binary's own mass. This would 'puff up' the stellar distribution (as well as expelling roughly its own mass of stars completely from the galaxy when it becomes a 'hard' binary). The binding energy of the stellar core would be further reduced if the binary were completely expelled from the system by the gravitational-wave recoil during final coalescence.

4. Quasars and the end of the 'Dark Age'

The Universe took about half a million years to cool down to 3000 K (corresponding to $z = 1000$). Thereafter, further expansion shifted the primordial radiation into the infrared; a 'dark age' began, which persisted until the first nonlinear perturbations developed into bound systems and released enough nuclear or gravitational energy to light up the universe again. The high-z quasars tell us that, after about one billion years, the dark age had certainly ended. The lack of complete 'Gunn Peterson' absorption in the spectra implies that there had by then been enough energy input to re-ionize the primordial material; the existence of the quasars implies that black holes of as much as $10^8 M_\odot$ had accumulated.

What happened during the timespan from a million to around a billion years? The answer depends on the relation between quasars and galaxies, and on when galaxy formation began. One need only recall three much-studied options for structure formation (none of which can yet be ruled out) to indicate the current level of uncertainty:

(i) According to the simplest CDM model, the first structures – loosely bound systems of subgalactic scale – start to form at redshifts as high as 20 or 30. The galaxies themselves only form more recently, but early enough to have provided enough 'hosts' for the high-z quasars. More detailed studies (e.g. Haehnelt and Rees 1993) suggest that the z-distribution of quasars can be fitted if the mass of the hole that forms within each dark halo depends in a plausible way on the depth and profile of its potential well. If the CDM model were correct, the quasar density should fall off steeply beyond $z = 5$; the intergalactic medium would have been originally ionized, perhaps as early as $z = 20$, by stars in shallow potential wells of subgalactic scale.

(ii) If at least one species of neutrino has a mass of a few e.v, then the CDM model is modified by the presence of 'hot' dark matter. In the hybrid 'mixed dark matter (MDM) model, fluctuations on small scales have lower amplitude, relative to fluctuations on large scales, than for 'pure' CDM; there is therefore less structure at early times. The existence of quasars at $z = 5$ is a severe constraint on the fraction of 'hot' dark matter. Indeed the inferred ionization of the IGM at high redshifts is then itself a problem (even if it isn't due to quasars), since the fluctuation spectrum in MDM has less power on small scales, so not even subgalactic structures form early.

(iii) The so-called primordial isocurvature baryon (PIB) model could lead to non-linear structures soon after recombination. Bound systems that condense out early, and lose their angular momentum by Compton drag, could evolve directly into black holes (Umemura, Loeb and Turner, 1993), even before the virialisation of galactic-scale potential wells which seem a prerequisite for black hole formation in CDM and MDM.

Discovery of quasars at much higher redshifts than 5 would push back our estimates of when galaxies formed, unless the fluctuations are non-gaussian, or we adopt the radical view (which could be maintained if the PIB model were right) that these quasars are not closely connected with galaxies.

5. Subgalactic Structure

In the CDM model, weakly bound clouds of very low mass would be the first objects to condense out, and virialise. If their virial temperature were below about 500K, they would be unable to cool. The first non-linear structures that could inject energy into the Universe, at $z \gtrsim 10$, would have baryonic masses $\sim 10^5 M_{\odot}$. These virialise at temperatures exceeding 500K – hot enough for H_2 cooling to allow continuing collapse and fragmentation. We don't know the IMF of these first stars. If they were predominantly of low mass, they would be inefficient at generating energy, but could create a pre-galactic 'macho' component (which would thereafter behave like CDM, and therefore constitute a fraction of the dark matter in galactic halos) – this is an interesting possibility. But any O or B stars (or very massive objects) would produce UV, and perhaps soft X-rays, that could photoionize the medium, raising the Jeans mass to $\gtrsim 10^8 M_{\odot}$. (Complete photoionisation of the IGM would, as a byproduct, produce by $z = 5$ a universal heavy element abundance of 1 per cent of solar.)

Photoionisation must have been almost completed before $z = 5$. Just how quickly it would happen, even in a specific CDM model, is not clear. The amount of matter that needs to have condensed to provide the UV depends on the IMF in these subgalactic systems. However it has recently been realised that the first UV could have interesting feedback effects on H_2 – depending on its spectrum it could either enhance the H_2 formation, and therefore the cooling, or else destroy the molecules (Heiman, Rees and Loeb 1995) before photoionization has been achieved. These effects determine what adiabat the gas is on at $z = 5$, something which is important for interpreting quasar absorption lines.

6. The Lyman Forest

Quasars offer a wealth of information on galaxy formation and pregalactic structures via their rich absorption spectra. Two decades of effort by many theorists to understand the Lyman forest have bequeathed us a great deal of historical baggage, which confuses the subject even more than necessary. These absorption lines now seem a natural consequence of current ideas on hierarchical structure formation – if they had only just been discovered, this is surely the interpretation that would suggest itself to most of us. But when they were discovered, far less was known about the high-redshift universe, and there were fewer constraints on, for instance, a hot intergalactic medium.

The photoionized medium would be influenced by any inhomogeneities in the dark matter distribution ('minihalos') whose virial temperature exceeded its own temperature of few times $10^4 K$. The medium would then no longer produce a smooth Gunn-Peterson trough: nonuniformities in its density (and in its velocity field as well) would imprint weak Lyman forest lines on any UV continuum passing through it.

Most of the lines described by Tytler at this meeting correspond to HI column densities of only $\sim 10^{12}\,\mathrm{cm\,s^{-1}}$ and optical depths no more than unity even in the line centre. If a homogeneous IGM has a Gunn-Peterson optical depth of (say) 0.1, then (since the neutral fraction increases linearly with density) material need only be compressed one-dimensionally by a factor 10 to produce such a line.

The regions responsible for the weak lines are therefore definitely in a dynamical state: the relevant gas is being pulled by gravity, but has not yet evolved into a quasi-static virialised cloud (and maybe never will). Any cloud that has achieved virial equilibrium must have at least several hundred times the mean density, and a line of sight through it would intercept an HI column density of more than 10^{15} cm s^{-1}. (Miralda-Escudé *et al.* 1995 and references cited therein).

The observed lines are so close that it is unrealistic to think in terms of widely- separated discrete clouds with a uniform medium between them – just as it would be unrealistic to model an ocean surface as completely flat except for a few isolated high waves. Between the overdense regions, the density falls well below that of a uniform IGM, because matter tends to drain into the potential wells.

We expect deviations from Voigt profiles, because the lines will be Doppler-broadened partly by bulk motions. It is therefore important to set the lowest possible upper limit to the *thermal* broadening. This tells us the temperature (and therefore the form of the UV background). More important, it tells us something about the thermal history of the gas at even higher redshifts than those directly observed. This is because the recombination times at intergalactic densities are so long that adiabatic cooling in the expanding IGM (and subsequent adiabatic heating when a proto-cloud has started to collapse) affects the temperatures; so also, at high redshifts, does Compton cooling on the microwave background (see Miralda-Escudé and Rees, 1994 for some illustrations of such effects). The temperature therefore depends on when and how the gas was first ionized.

7. Realistic Limits of Computations

The purely gravitational clustering of non-dissipative dark matter is being computed with higher and higher resolution. It is now also possible to model the gas dynamics, including shocks and cooling, on all scales from the Lyman forest clouds up to clusters of galaxies. If the primordial gas were heated only by shocks when it fell into a bound system or formed a pancake, then the problem would be well-posed, and calculations would be feasible throughout all the non-linear stages.

In principle, the formation of the first stars and quasars is determined by the initial conditions. But there is of course no realistic chance of modelling the internal processes of star formation (not the central 'active nuclei') within bound systems. The physics and feedback effects from star formation, etc – which are important at the pregalactic era as well as within forming galaxies – involve parameters that cannot be computed a priori. Indeed the importance of the simulations lies in the hope that we will learn about these complex processes by modelling a wide parameter space and testing which assumptions yield the best agreement with the data on galaxies and their evolution – data whose volume and quality are advancing at least as fast as computational techniques.

References

Fabian A.C. and Rees, M.J. 1995 MNRAS 277, L55.
Haehnelt, M 1994 MNRAS 269, 199
Haehnelt, M 1995 MNRAS 273, 249
Haehnelt, M and Rees M.J. 1993 MNRAS 269, 168
Krabbe, K *et al.* 1995 Astr. Astrophys (in press)
Miyoshi et al 1995 Nature 373, 127.
Miralda-Escudé, J, Cen, R., Ostriker, J.P. and Rauch, M, 1995 Ap J.(in press)
Miralda-Escudé, J. & Rees, M.J. 1994 MNRAS 266, 343.
Schneider, D., Schmidt, M and Gunn, J.E. 1991 A.J. 102, 837
Shaver, P., 1995 Ann. N.Y. Acad. Sci (in press)
Tanaka, Y. *et al.* 1995 Nature 375, 659
Umemura, M., Loeb, A. and Turner, E.L. 1993 Ap. J. 419, 459.
Watson, W.D. and Wallin, B.K. 1994 Ap.J. 432. L35

C.S. Frenk: You mentioned that it is unlikely that many low mass galaxies would grow a $10^8 M_\odot$ black hole. A hint that these galaxies may be different from bigger ones is that their inferred dark matter halos have large core radii. Could you comment on this possible connection and on the physical processes that would distinguish small galaxies from big ones?

Yes, that might be part of the story – dissipative effects would be more efficient in a compact high-density core. However, I think the most important difference between high-mass and low-mass galaxies is that the former have deeper potential wells.

R. van der Marel: In studying the stellar dynamical effects of black holes in galactic nuclei it is often assumed (mainly for mathematical convenience) that the black hole grows adiabatically in an already existing stellar system. Do you consider this a plausible and viable assumption?

My guess would be that this is a rather poor assumption (though a profile with slope close to 3/2 may still be a good approximation for other reasons, and is an obvious 'template' to use). Stars might still have been forming at the same time as the hole was growing. The cusp profile could also have been modified if the central black hole had merged with another.

CONFERENCE SUMMARY: GLIMPSE OF A NEW FUTURE

S. M. FABER
UCO/Lick Observatory,
University of California, Santa Cruz, CA 95064

Abstract. Research on galaxy formation is about to be inundated with data from the distant Universe and from highly realistic model simulations. First glimpses from this and other work are already showing the need for new paradigms of galaxy formation.

1. Introduction

Some conferences are remembered as landmark events where consensus on important new paradigms is achieved. This conference in contrast may be remembered as the first glimpse of a new future for galaxy evolution studies – an era dominated by incredibly realistic simulations and data from the distant Universe. The need for traditional, local data will not diminish – to the contrary, the new techniques will stimulate comprehensive surveys of the local Universe as never before. At this conference we saw the first trickle of such new results – the first drops splattering down from the bursting dam above. Shortly a flood of new results will be upon us, carrying us in its wake. Such floods have a tendency to scour the landscape, and this one promises to be no different. Heidelberg may be the last time we stood and viewed the forest as it once seemed to be.

2. Interactions and Mergers

A major topic at this conference was galaxy interactions. Freeman, Sancisi, and Kennicutt all argued for continuous merging of small satellite galaxies up through recent times. We badly need to describe this process quantitatively: what fraction of the baryonic mass of a normal field spiral like the Milky Way arrived how recently, and in what-sized pieces? Although the quantitative impact is still not clear, incontrovertible evidence such as kinematic subsystems and tidally stripped satellites of the Milky Way

313

R. Bender and R. L. Davies (eds.), New Light on Galaxy Evolution, 313–317.
© *1996 IAU. Printed in the Netherlands.*

establish that at least *some* material arrived in this form. An implication is that an object like the Galactic halo, built up via many accretions over time, may not be describable by simple two-integral dynamical models.

Barnes emphasized that merger remnants can retain detailed memory of the kinematics of the individual progenitors. Just a few percent of gas appears to have a profound regularizing effect on the dynamical structure of the remnant by erasing box orbits and reducing triaxiality. Triaxial galaxies such as certain massive ellipticals must evidently have formed in mergers that were nearly gas-free.

Important new light was shed on dynamical interactions of galaxies in clusters. Moore argued that galaxies in clusters are subject to repeated high-velocity bombardment at speeds in excess of the internal escape velocity. A fascinating video by Moore and collaborators illustrated this "harassment," showing outer disks being whittled away and perturbed gas falling to the galaxy centers. Both tendencies would be helpful in explaining cluster S0's. In the conference's best one-two punch, this talk was followed by Sharpless' description of an S0-dominated sub-cloud in Coma. The high Balmer strengths in this cloud may have resulted from harassment-triggered starbursts as the cloud amalgamated with the main body of the cluster.

3. Dynamics of Spheroids and Bulges

"Mass easily finds the center" was an important theme of several talks. Burkert emphasized a puzzle with hierarchical clustering, which tends to produce central cusps of dark matter even without gasesous dissipation. Gas exacerbates the problem, as illustrated in videos of merging, forming, and harassed galaxies by Barnes, Steinmetz, and Moore. Disturbed gas in the outer parts falls unerringly to the center in just a few timescales. Why, then, aren't the centers of galaxies denser than they are? Does the gas supply simply run out, or does feedback from star formation or AGNs somehow truncate further infall?

It was once possible to regard the centers of galaxies as separate entities with a distinct evolutionary history from the main body. That view is less tenable now as we see that dense central structures survive mergers and thus in some sense preserve the formation history of the whole galaxy. The elliptical core parameter relations, described by Kormendy based on new HST images, are thus doubly puzzling – what process impressed a core length-scale separate from the global radius, and why aren't these cores of the classic type (as in the isothermal sphere) but rather shallow power laws? The Universe, surprisingly, does not seem to make classic cores in galaxies.

This conference saw a landmark paper (Jeske et al.), which broke the longstanding degeneracy on elliptical galaxy mass measurements. A wide variety of mass profiles are consistent with given photometric and dispersion profiles provided the radial velocity-dispersion anisotropy is adjusted. Jeske et al. added kurtosis, h_4, and obtained, for the first time, clear evidence of dark matter at about $R_e = 1$. The new technique opens the way to mass profiles of ellipticals with simple orbital structure. Similarly, Scorza showed how h_4 plus h_3 (skewness) can be used to decompose disky ellipticals and S0's into disk and spheroid components. Line profile shape is clearly developing into the key dynamical tool that many hoped it would be.

De Zeeuw discussed new work linking HST core mass profiles with global galaxy dynamics. Merrit and Fridman (preprint) have shown that stars in box orbits (as in triaxial galaxies) will eventually be scattered by a steep central density cusp. This means that small ellipticals, which always show steep central cusps (Kormendy) *cannot* be triaxial, in keeping with their high rotation and disky isophotes. The clear triaxiality of certain massive ellipticals may be slowly dissolving with time as the galaxies evolve toward spherical symmetry. Potentially this could be used to date interactions.

Major new results on bars and bulges were presented. Several speakers suggested that bars may be cyclical, forming from disk instabilities and dissolving slowly to form bulges. There was no clear consensus on whether the Galactic bulge was part of our newly discovered bar, the inner puffed-up disk, or simply the inner part of the metal-poor spheroid. Peletier stressed that bulges are more similar in color to their associated disks than they are to one another, consistent with formation from the same disks. Perhaps there are two kinds of bulges: the classic type related to elliptical spheroids (as in the Sombrero) and smaller bulges in later-type spirals formed from disks.

Lest anyone hope that spheroid dynamics is getting simpler, Bender reminded us of the important elliptical NGC 4365, which shows two kinematically decoupled stellar substreams in its inner spectra yet at most only small departures from regular isophotes. A detailed formation scenario for this Rosetta-Stone object is badly needed.

4. Stellar Populations

Age-dating of old stellar populations was the subject of considerable controversy. Davies and Worthey reviewed Hβ evidence that light-weighted mean ages in some ellipticals, particularly small ones, are as low as a few Gyr. According to this interpretation, the sequence of colors and spectral line strength along the Fundamental Plane owes as much to age differences as to metallicity. An objection is that the Fundamental Plane should be underpopulated at low luminosities in distant clusters, which has so far not been observed. The narrowness of the FP locally is also problematic if there is a large age spread at each location. Chiosi presented UV spectra based on models of the hot horizontal branch and showed that ($1550\text{-}V$) color is helpful in distinguishing young single-burst populations from mixtures of old and young stars. On that basis he suggested that M32's stars are a mix of ages rather than exclusively young.

Worthey reminded us of how serious current model uncertainties are in light of the strongly non-solar element ratios that are found in massive ellipticals. Since model spectra are even more sensitive to Z than to age, it may be critical to model these ratios faithfully to obtain reliable absolute calibrations from models.

Bruzual described recent intercomparisons and tests of population models by Bruzual-Charlot, Worthey, and Bressan et al. Disagreements are present – at typically the 25% level in M/L and age – but the progress made in the last 5 years is significant. All zeropoints are questionable, but relative trends vs. age and Z are now well established. The next challenge is to model the element ratios, and until that is explored, all fitted ages and other absolute quantitities should be regarded as uncertain.

5. Distant Galaxies

Data on distant galaxies were to me the most novel and exciting aspect of the conference. Definitive results of the landmark Canada-France redshift survey (Lilly) are now emerging, several talks and posters described morphological results from HST, and the first results on Keck redshifts were presented.

A major breakthrough seems to be some level of agreement on the long-running faint-blue-galaxy problem. Several groups now concur (Lilly, Driver, Marzke) that the excess of faint blue galaxies is due to two effects: more such galaxies locally than previously known, plus a *modest* increase in either the number density or luminosities of such galaxies in the past. This apparent evolution in the *tail* of the Hubble sequence seems to be counterbalanced by relative stability in the luminosity function (Lilly) and linewidth properties (Rix, Franx) of *early types*. Examples of normal-looking ellipticals and spirals have been identified at redshifts as early as 0.8 (Koo).

Several speakers discussed morphologies of faint galaxies from HST. Oemler echoed the thoughts of many in noting that distant galaxies look distinctly different from local Hubble types – balanced, symmetric spirals are much rarer – but that some vague correspondence with standard Hubble types seems to be retained. A speculation is that we are seeing back to a time when the foundations of the present Hubble sequence were being laid, which opens all sorts of new avenues for exciting research. Interestingly, the morphological peculiarities of field and cluster galaxies look hauntingly similar despite their different environments. A more quantitative characterization of "peculiar" is clearly needed, however.

In one of the rare theoretical papers of the conference, Kauffmann described her program to model galaxy formation schematically using Press-Schechter clustering statistics for the merging hierarchy. These are toy models with many assumptions, but they do allow one to explore quickly the consequences of varying assumptions about star formation, gaseous infall, and other variables. They account quite nicely for the morphology-density relation and predict, among other things, that stars in ellipticals formed over a wide range of epochs during repeated bouts of merging. An interesting prediction worth testing is that small satellites are accreted constantly onto L^* galaxies, reminiscent of the evidence from local galaxies mentioned in Section 1.

I close with a rampant speculation that is not well grounded but might be able to organize much of the observational data on both local and distant galaxies reported at this conference. The idea is that the Hubble sequence is really two distinct phenomena. The first part consists of the classic Hubble types at the head of the sequence ending at about Sc (where Hubble originally truncated the sequence). These bright galaxies all have about the same average luminosity, mass, and surface brightness, show tight structural scaling relations, and are evidently rather stable in properties and number densities out to early epochs. The second half is the irregular tail of the sequence, which shows a wider range of luminosities and surface brightnesses, a distinct Hubble type-luminosity correlation, and is apparently strongly evolving over recent epochs. Talks by Kennicutt, Sancisi, and McGaugh listed other properties that seem consistent with this crude dichotomy. Perhaps the explanation is simply that the first class formed from strong density perturbations early in the Universe, while the second class consists of latecomers whose formation dribbles on to the present. Clearly, the coming flood of realistic simulations and distant survey data will be crucial in answering this and other key questions of galaxy formation.

CONFERENCE PHOTOGRAPHS

Rob Kennicutt, Roelof de Jong and Guy Worthey

Bianca Poggianti and Nobuo Arimoto

Mauro d'Onofrio, Luca Ciotti, Silvia Pellegrini and Nicola Caon

Tom Richtler and Beatriz Barbuy

John Kormendy, Sandy Faber and David Koo

Ed Khachikian and Immo Appenzeller

In the castle

On the boat

In the castle

On the boat

Ralf Bender and Roger Davies

Garth Illingworth and Alvio Renzini

Mark Dickinson and Steve Warren

Ben Moore and Gus Oemler

Simon Driver, Stacy McGaugh and David Sprayberry

Leslie Maxfield and George Djorgovski

Regina Schulte-Ladbeck, Stefan Wagner and Ulrich Hopp

Roeland van der Marel and Ed Khachikian

Paola Belloni and Sandy Faber

Ralf Bender and Simon White

Gaby Ramge, Birgit Hoffmann, Elke Bär from the LOC

Cesare Chiosi and Claus Möllenhoff

Michael Rosa and Joachim Krautter

In the castle

Pascale Jablonka, Javier Gorgas and Ben Dorman

Gaby Ramge who took all the pictures except this one

POSTER PAPERS

MAPPING STELLAR EVOLUTION IN THE WAKE OF DENSITY WAVES IN RING GALAXIES

P. N. APPLETON[1], C. STRUCK-MARCELL[1], M. A. BRANSFORD[1],
V. CHARMANDARIS[1], A. P. MARSTON[2], K. BORNE[3] AND R. LUCAS[3]
[1] *Iowa State University, Ames, Iowa 50014, USA*
[2] *Drake University, Des Moines, Iowa 50311, USA*
[3] *Space Telescope Science Institute, Baltimore, MD 21218, USA*

1. Summary of the Project

We have been conducting multi-color observations of a sample of classical ring galaxies with the aim of using them to study the formation and evolution of massive stars. We compare theoretical predictions for the expected color of the material inside the rings assuming that massive stars are created in the wake of the expanding wave. We present ground based data for VIIZw466 and HST data for IIZw28 and the Cartwheel which show strong color gradients.

Color models of star cluster evolution can be mapped onto the disk as the wave expands. For a flat rotation curve in the target, the ring expansion velocity is roughly constant with time, and features such as the supergiant phase, and later the AGB phase are expected to be seen as color changes inside the ring. For reasonable ring expansion velocities and the assumption that the star formation is triggered simultaneously along the edge of the wave, it is shown that HST has enough resolution to resolve the supergiant "flash". Preliminary analysis suggests that a swing to the red is seen in the Cartwheel data inside the outer ring. However, the possibility that dust may complicate the picture is being explored by scheduled mid-IR observations of some of the rings with ISO. We refer the reader to the following review on ring galaxies[1].

References

1. Appleton, P. N. and Struck-Marcell, C. (1995), Collisonal Ring Galaxies, *Fundamentals in Cosmic Physics*, **In Press**

THE HOST GALAXIES OF $Z \approx 2$ RADIO-QUIET QSOS

ITZIAR ARETXAGA, B.J. BOYLE AND R.J. TERLEVICH
Royal Greenwich Observatory
Madingley Road. Cambridge CB3 0EZ. UK.

Based on an R-band imaging survey of high-redshift ($z \approx 2$) QSOs with the 4.2m WHT, we report the detection of extensions to the nuclear PSFs of two radio-quiet and one radio-loud QSO. The extensions are most likely host galaxies, with luminosities of at least $3 - 7\%$ of the QSO luminosity. The most likely values for the luminosities lie in the range $6 - 18\%$ of the QSO luminosity ($R \sim 19.8 - 20.9$ mag). Our observations show that, if the extensions we have detected are indeed galaxies, extraordinary massive and luminous galaxies are not only characteristic of radio-loud objects, but of QSOs as an entire class. For a detailed description of this work, see Aretxaga, Boyle & Terlevich (1995) MNRAS, 275, L27.

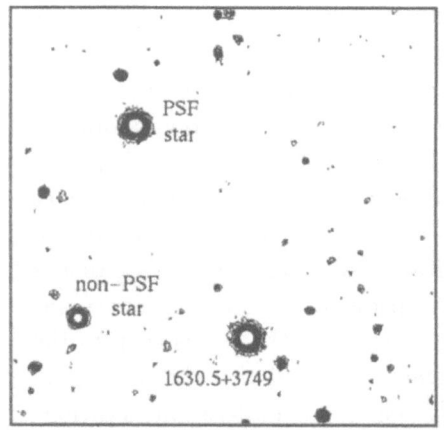 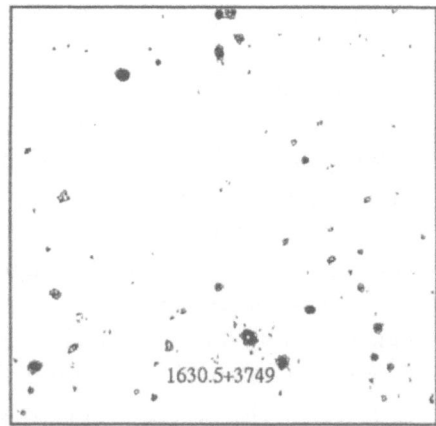

Figure 1: The left panel shows a $50'' \times 50''$ field of one of the radio-quiet QSOs (1630.5+3749) for which we have found and extension to its nuclear PSF. The right pannel shows the field after subtracting a 2D PSF profile to both QSO and stars. Note that the stars are subtracted while the QSO is left with significant residuals ($> 3\sigma$).

IA's work is supported by the EEC HCM fellowship ERBCHBICT941023. She acknowledges the organizers for partial support to attend the meeting.

A POSSIBLE EVOLUTIONARY SCENARIO FOR NGC 7217

E. ATHANASSOULA, A. BOSMA AND B. GUIVARCH
Observatoire de Marseille
2 Place Le Verrier, 13248 Marseille Cedex 4, FRANCE

AND

L. VERDES-MONTENEGRO
Instituto di Astrofisica di Andalucia
Apdo. Correos 3004, E-18080 Granada, SPAIN

NGC 7217 is an ordinary spiral galaxy with three rings whose size ratios are such that they can be associated with resonances, as for barred spirals. From 21-cm HI line data and BVRI CCD-images of this galaxy we find (cf. Verdes-Montenegro et al. 1995) : 1) a nuclear ring strong in $H\alpha$, 2) an inner ring seen clearly in a B - I colour map, and 3) an outer ring, with blue colours and strong HI-emission. After deprojection the disk has a mean ellipticity of 0.04 ± 0.01, while the position angle of the deprojected galaxy changes suddenly at 65" radius, where the minor axis becomes major axis. Thus a very mild oval distortion could exist, with the outer ring perpendicular to the oval. Merrifield and Kuijken (1994) find from the stellar kinematics that about 20 - 30 % of the stars are in retrograde orbits.

Did NGC 7217 have a bar which has since been destroyed ? This is possible : 1) From a simulation with sticky particles we find that rings can survive quite long after a bar has been turned off. 2) Orbit calculations show that a large fraction of the ergodic orbits present before a bar is switched off become retrograde. Thus if the bar had a substantial fraction of ergodic orbits, the presence of the retrograde orbits follows from a pendulum analogy.

There are several ways to decay the bar : 1) N-body simulations show that by hitting a barred galaxy with a companion the bar can be destroyed, and in such cases a substantial fraction of the stars is moving in retrograde orbits, as in NGC 7217. 2) A build up of a strong central mass concentration can scatter the x_1-orbits, while temporarily increasing the fraction of ergodic orbits.

ESTIMATION OF MG/FE RATIO IN EARLY-TYPE GALAXIES

B. BARBUY[1], J. A. DE FREITAS PACHECO[1,2] AND T. IDIART[1]
1 Instituto Astronômico Geofísico, USP
CP 9638, São Paulo, 01065-970, São Paulo, Brazil
2 Observatoire de la Côte d'Azur, BP 229,
06304 Nice Cedex 4, France

Through computed synthetic spectra for individual stars in the Mg_2 index region where different values of [Mg/Fe] were considered, and observed features Fe5270 and Fe5335 in a sample of 80 stars, we derived relations Mg_2 and $<Fe> = f(T_{eff}, \log g, [Fe/H], [Mg/Fe])$. These relations were used to compute integrated indices for single aged populations, where number counts in colour-magnitude diagrams were used. A multi-population model was then built through a convolution of the single stellar population indices with a metallicity and luminosity distribution for elliptical galaxies, where a chemical evolution model includes enrichment by type II and I supernovae. A grid of 80 multi-population models was computed, by varying the efficiency of star formation rate. In Figure 1 are shown the results for [Mg/Fe] = 0.0, 0.3 and 0.6 compared to observational data by Worthey et al. (1992). An average value of [Mg/Fe] = 0.3 is found for the sample ellipticals.

Worthey, G. Faber, S.M., Gonzalez, J.J.: 1992, ApJ, 398, 69

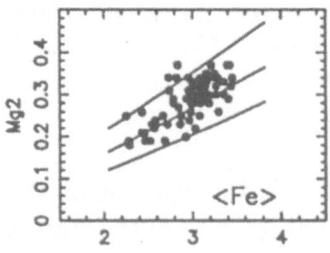

Figure 1. Mg_2 vs $<Fe>$ derived for multi-population models computed with [Mg/Fe] = 0.0, 0.3, 0.6 (solid lines) and sample galaxies by Worthey et al. (filled circles)

STARBURST CYCLE IN DISTANT CLUSTERS

A.J. BARGER, A.ARAGÓN-SALAMANCA AND R.S. ELLIS
Institute of Astronomy, Cambridge CB3 OHA England

W.J. COUCH
University of New South Wales, Sydney 2052 Australia

I. SMAIL
*The Observatories of the Carnegie Institution of Washington
Pasadena, CA 91101-1292 USA*

AND

R.M. SHARPLES
University of Durham, Durham DH1 3LE UK

A major puzzle in observational cosmology is the physical origin of a significant excess population of blue galaxies in the cores of distant rich galaxy clusters. This 'Butcher-Oemler' effect is now known to be a widespread starburst-related phenomenon. We test whether various spectral and photometrically-defined galaxy classes might represent different stages within a single cycle of star-formation. We compare the numbers of galaxies in various categories for three $z = 0.31$ clusters, AC103, AC114, and AC118, with evolutionary models generated according to the Bruzual & Charlot (1993) isochrone spectral synthesis code, assuming that some fraction of the model cluster population is viewed either before or during a secondary burst of star formation. We find good agreement between the model predictions and the number density of spectroscopically-confirmed members in the $H\delta$ versus $B - R$ plane for a cluster population in which 30 per cent of the member galaxies have undergone secondary bursts of star formation within the last ~ 2 Gyr prior to observation. As an additional check, we analyse a larger K_n-limited sample from newly-acquired infrared images and find good agreement between the models and the data in the $U - I$ versus $I - K_n$ plane for the same active cluster fraction. We conclude that the unusual galaxy population in distant clusters can be explained by a single cycle in which about 30 per cent of the cluster population experienced a secondary burst of star-formation within the last ~ 2 Gyr.

THE RADIO SPECTRUM OF SGR A*

T. BECKERT[1], W.J.DUSCHL[1,2], P.G.MEZGER[2], R. ZYLKA[2]
1: Institut für Theoretische Astrophysik, Heidelberg, FRG
2: Max-Planck-Institut für Radioastronomie, Bonn, FRG

Sgr A*, the enigmatic radio source located at the dynamical center of the Galaxy, is firmly detected in the frequency range of $\sim 1 -$ few 10^2 GHz. For $\sim 0.5 - 1$ GHz and in the MIR range only significant upper limits of the flux density are known. Between ~ 1.5 and 600 GHz the time averaged flux density S_ν is proportional to $\nu^{1/3}$ (ν: frequency). For frequencies higher than ~ 600 GHz as well as for those lower than ~ 1.5 GHz, S_ν drops sharply.

The radio spectrum can be understood as optically thin synchrotron radiation assuming a single homogeneous blob of relativistic electrons placed in a homogeneous magnetic field. The form of the spectrum imposes strict limits on the energy distribution of the electrons. The most important features of the distribution function $f(E)$ (E: energy) are high and low energy cut-offs. While the low energy cut-off must fulfil $f(E) \to 0$ for $E \to 0$, the cut-off at high energies must be steeper than exponential and thus excludes thermal relativistic electrons. The emitted spectrum of distributions meeting these conditions is independent of the detailed form between the low and high cut-off, thus establishing a class of quasi-monoenergetic electron distributions, all producing similiar synchrotron spectra. Assuming the source to be a spherical homogeneous blob, the turnover of the spectrum at ~ 1 GHz can be due to selfabsorption inside the source. Another explanation for the turnover is free-free-absorption in the surrounding hot gas of the central cavity. Both processes must occur. They provide an upper limit for the column density in Sgr A West of $N_H \sim 4.3 \cdot 10^{21} \mathrm{cm}^{-2}$ and a lower limit for the size (radius) of Sgr A* of $R \geq 1.2 \cdot 10^{13}$ cm.

References

Beckert T., Duschl W.J., 1995, Synchrotron radiation from quasi-monoenergetic electron distributions, to be submitted
Beckert T., Duschl W.J., Mezger P.G., Zylka R., 1995, A&A, in press
Duschl W.J., Lesch H., 1994, A&A 286, 431
Zylka R., Mezger P.G., Ward-Thompson D., Duschl W.J., Lesch H., 1995, A&A 297, 83

THE GALAXY POPULATION OF TWO DISTANT GALAXY CLUSTERS: 0016+16 (Z=0.54) AND 0939+47 (Z=0.41)

PAOLA BELLONI
Universitätssternwarte, Scheinerstraße 1, 81698 München

AND

HERMANN-JOSEF RÖSER
Max-Planck Institut Astronomie, Königstuhl 17, 69117 Heidelberg

We analyzed the galaxy population of two distant clusters of galaxies by means of optimized multi filter photometry. To this purpose, broad and narrow band filters have been chosen to detect features characteristic of elliptical (4000 Å break and the Mg_2 band) and E+A galaxies (H_β, H_γ and H_δ in absorption). Cluster membership and morphological types of the galaxies have been obtained by fitting the low-resolution spectral energy distributions with template spectra. Ten templates for E+A galaxies have been built up using Bruzual & Charlot (1993, ApJ 405, 538) population synthesis models with the assumption that these galaxies result from a strong star formation episode occuring in an early-type galaxy.

We obtained per cluster a sample of more than hundred galaxy members, suitable for studing evolutionary effects with a firm statistical basis. Furthermore, the already known E+A galaxies were successfully identified and new ones (29 in Cl0939+47 and 12 in Cl0016+16) were found (Belloni et al., 1995, *A&A* 297, 61). Their projected distribution in the cluster shows that they are more spread out over the field than the ellipticals. Our results confirm the Butcher-Oemler effect and the significance of the E+A galaxies in these clusters (about 20% of all member galaxies) with a large statistical improvement and a systematic approach.

Moreover, the large agreement between the morphological classification obtained with optimized multifilter photometry and that provided by Hubble Space Telescope images (Dressler et al.,1994, ApJ 430, 10) points out the effectiveness of the present approach in the recognition of morphological types of distant galaxies.

STAR FORMATION IN A MAGNITUDE LIMITED SAMPLE OF INTERACTING GALAXIES

NILS BERGVALL
Astronomical Observatory, Box 515, S-751 20 Uppsala, Sweden
EIJA LAURIKAINEN
Inst. de Astronomia, Obs. Astronomico National, Apartado Postal 70-264, Cd. Universitaria, C.P. 04510 Mexico, D.F
SUSANNE AALTO
Caltech, Astronomy Dept. 105-24, Pasadena CA 91125, USA
AND
LENNART JOHANSSON
Celsius Tech Systems AB, Järfälla, Sweden

We report on optical/near-IR spectroscopy and photometry of a magnitude limited sample of interacting pairs and merging galaxies and a control sample of apparently isolated galaxies (1,2). All observations were carried out at ESO, La Silla. When compared to the control sample, the interacting galaxies show only a moderate increase of star formation activity, in the central area typically a factor 2-3. Starburst activity seems to be very rare. Ongoing CO observations (Aalto, Horellou, Booth, Wiklind, Bergvall) indicate that these objects are not particularly rich in molecular gas. The interacting/merging galaxies have relatively high optical luminosities and high FIR luminosities and temperatures but these parameters are not correlated with other star formation signatures. We conclude that the interacting and merging galaxies in this sample, from the global star formation aspect, do not differ dramatically from scaled up versions of normal, isolated galaxies. This could suggest that many of the most lumino

components in interacting pairs could originate from multiple 'quiet' mergers.

References

1. Johansson, L., Bergvall, N. (1990), *Astron. Astrophys. Suppl. Ser.*, **86**, 167

SUPERGIANT SHELLS AND HOT GAS IN NGC 4449

D.J. BOMANS AND Y.-H. CHU
Univ. of Illinois, Dept. of Astronomy
1002 West Green St., Urbana, IL 61801, USA

AND

U. HOPP
Universitätssternwarte München
Scheiner Str.1, 81679 München, Germany

Supergiant shells are the largest interstellar structures in galaxies. They are outlined by long H II filaments enclosing an inner space containing very little warm or cold gas. In the LMC, two supergiant shells have been detected in X-rays indicating the existence of hot gas: LMC 2 (Wang & Helfand 1992) and LMC 4 (Bomans et al. 1994). It is not yet clear whether the hot gas in a galactic halo is supplied by such supergiant shells.

NGC 4449, at a distance of ~3 Mpc, is an irregular galaxy, quite similar to the LMC. A number of supergiant shells have been discovered in deep Hα images of NGC 4449 (e.g. Hunter & Gallagher 1992).

We have analyzed a 7850 sec ROSAT PSPC observation (RP600137) centered on NGC 4449, and compared it to our optical data. The ROSAT X-ray image shows three bright point sources and a widespread diffuse emission. In the west a large diffuse X-ray emission region is nicely delineated by long Hα filaments, indicating the existence of hot gas within a supergiant shell. This supergiant shell has an Hα morphology as spectacular as LMC 2 and a shell size and X-ray luminosity surpassing LMC 4 (the largest supergiant shell in the LMC). The X-ray luminosity of this supergiant shell is 9×10^{37} erg s^{-1}. It is remarkable that this supergiant shell extends hot gas to more than 2 kpc into the halo of NGC 4449!

References

Bomans, D.J., Dennerl, K., Kürster, M. (1994), *A&A L*, **283**, 21-25
Hunter, D.A., Gallagher, J.S. (1992), *ApJ L*, **391**, 9
Wang, Q., Helfand, D.J. (1991), *ApJ*, **379**, 327

HI IN HII–GALAXIES AND THEIR COMPANIONS

E. BRINKS, C. L. TAYLOR AND D. L. THOMAS
NRAO, P.O. Box O, Socorro, NM 87801, USA

AND

E. D. SKILLMAN
Univ. Minnesota, Minneapolis, MN 55455, USA

1. Introduction and Results

HII galaxies are dwarf galaxies which are currently actively forming stars. We speculated that an interaction with an optically faint, but gas–rich object might be responsible for their enhanced star formation. This prompted us to search for companions with the VLA in the 21–cm line of HI. This has several advantages over optical searches (Campos–Aguilar et al. 1991, AJ, 106, 1784; Telles & Terlevich 1995, MNRAS, 275, 1), e.g., the direct availability of redshifts.

The sample which we observed is volume limited ($V_\odot < 2500$ km s^{-1}), and is based on HII galaxies with $M_B > -19$ mag; it is drawn from the lists of Salzer et al. (1989, ApJS, 70, 447 & 479). We detected 20 out of 21 objects observed with the VLA in its D–array (45" resolution) configuration (Taylor et al. 1995, ApJS, 99, 427). We searched for HI companions within 30 arcmin, and within a velocity range of ± 250 km s^{-1}. Some 60% of the target galaxies have a companion (or are binary systems) which supports our working hypothesis that interactions *might* be the cause for the active star formation we see in the HII galaxies.

We have started a survey similar to the one described above, but on a sample of Low Surface Brightness (LSB) dwarf galaxies drawn from lists by Schombert et al. (1992, AJ, 103, 1107) and provided to us by Bothun (priv. comm.). These objects serve as a control sample for the HII galaxy observations. A first pass through our 18 LSB dwarf galaxies shows that only four of them have a companion, confirming our working hypothesis.

ISOPHOTAL EFFECTS IN FAINT GALAXY SAMPLES.

P. BRISTOW AND S. PHILLLIPPS
Department of Physics, University of Bristol, Bristol, UK

Using Monte Carlo style simulations of galaxy populations we create artificial faint galaxy samples which mimic those obtained by actual observational techniques. By comparison of samples selected according to total luminosity or luminosity within an isophote we are able to estimate the extent to which isophotal effects could cause number magnitude counts of faint galaxies to appear artificially steep (cf. McGaugh 1994, Phillipps 1993). We find that, if we assume a 'standard' non-evolving galaxy population (essentially that used by Broadhurst, Ellis & Shanks 1988 amongst others) then isophotal effects alone cannot account for the discrepancy between the observed steepness and no-evolution models, though they could significantly reduce the amount of evolution required and alter the median redshifts. Modifying the underlying galaxy population by the addition of a bivariate brightness dwarf component as observed in clusters (e.g. Irwin *et al* 1990) increases the significance of the isophotal effects, though only fractionally, despite the fact that such effects would be highly important for such a population considered on its own.

Further development of the simulation software allows us to produce simulated CCD frames, taking full account of observational conditions, which may be examined with the same image detection and photometry software as is applied to real observational data. This provides a route from intrinsic galaxy population models to apparent diagnostics (such as number counts) which comprehensively allows for selection effects.

References

Broadhurst, T. J., Ellis, R. S. & Shanks, T., 1988. *M.N.R.A.S.*, **235**, 827.
Irwin, M. J., Davies, J. I., Disney, M. J. & Phillipps, S., 1990. *MNRAS*, **245**, 289.
McGaugh, S, 1994. *Nature*, **367**, 538.
Phillipps, S., 1993. in Observational Cosmology., 308, eds Chincarini, G, Iovino, A, Maccacaro, T & Maccagni, D.

A SURVEY OF THE INTERSTELLAR MEDIUM
IN ELLIPTICAL GALAXIES: THE IONIZED GAS

N. CAON AND F. MACCHETTO
Space Telescope Science Institute,
3700 San Martin Drive, Baltimore, MD 21218, USA
Affiliated with the Astrophysics Division, Space Science Department,
ESA

AND

M. PASTORIZA
Instituto de Fisica, UFRGS,
Av. Bento Goncalves 9500, CP 15051, 91500 Porto Alegre RS,
Brazil

We have carried out an extensive program of observations of the ionized gas in 74 luminous elliptical and lenticular galaxies, selected to include a variety of properties in radio and X-ray emission, and in kinematical behavior. For each galaxy we have obtained broad-band R and V images and narrow-band images, centered at the Hα and [NII] emission lines, to derive the luminosity and distribution of the ionized gas. We found that a large fraction ($\simeq$ 70%) of E and S0 galaxies in our sample contain ionized gas. The gas morphology and size varies from small disks (mean diameter $1-4$ kpc) to large filamentary structures (extending up to 10 kpc from the galaxy center). Comparison with previous measurements shows reasonable agreement for a few galaxies, but considerable scatter for a large fraction, possibly due to differences in the limiting flux thresholds.

A significant correlation between Hα + NII and X-ray luminosities is found for those galaxies (38% of the sample) for which we have detected ionized gas and are listed as X-ray sources. Only weak correlations are present between infrared luminosities and the Hα + NII luminosity. A strong correlation is also found between the Hα + NII flux and the B-band flux within the region occupied by the line-emitting gas. This finding seems to support a scenario in which the ionization and excitation of the gas is in part or fully provided by UV photons produced by post-AGB stars.

SPECTROSCOPIC EVIDENCES FOR STAR FORMATION IN COOLING FLOW GALAXIES

N. CARDIEL AND J. GORGAS
Dpto. de Astrofísica, Universidad Complutense, Madrid

AND

A. ARAGON-SALAMANCA
Institute of Astronomy, Cambridge, UK

X-ray observations have led to the conclusion that many galaxy clusters are hosting *cooling flows*. The brightest cluster galaxies could have accreted masses of the order of 10^{11}–10^{12} $M_\odot$, but is still uncertain what the final fate of the accreted gas may be.

Trying to shed some light in the problem, we are carrying out spectroscopic observations of galaxies with and without cooling flows (see details in Cardiel, Gorgas & Aragón-Salamanca 1995, MNRAS in press). The analysis of Mg_2 and the 4000 Å break, good indicators of stellar populations in early-type galaxies, reveals that the nuclear measurements of both indices are well correlated with the mass accretion rate. We conclude that 5–17% of the mass flow is being converted into new stars with a "normal" IMF. This result is indicative that probably *all* the gas which is accreted *inside* the central regions of cooling flow galaxies is being converted into stars.

Our data show that line-strength gradients of galaxies inside large cooling flows flatten in the inner regions ($r \leq r_e/2$), which reveals that the star formation is concentrated toward the central parts. In the mean, Mg_2 and the 4000 Å break gradients in cooling flow galaxies are shallower than those found in giant ellipticals. Through the application of stellar population models in the galaxy regions where gradients have presumably been flattened, we derive that the fraction of V light that comes from the accretion population remains roughly constant with radius.

In addition, the indices attained in the outer regions ($r > r_e$) by some brightest cluster galaxies, both with and without cooling flows, are significantly lower than those usually found in normal ellipticals. Whether this low indices are indicative of recent star formation processes requires further observational data.

THE TOTAL MASS OF DDO 154

CLAUDE CARIGNAN
Département de physique, Université de Montréal,
C.P. 6128, succ "centre–ville", Montréal, Québec

The total mass and total extent of galaxies (including their dark halos) are fundamental parameters that are completely unknown for all galaxies. The best estimates we have for spiral and dwarf irregular galaxies come from detailed mass models using extended HI rotation curves. But, in every galaxy studied so far, such analysis only succeeded to derive lower limits of the total mass and total extent of their dark halo out to the last measured velocity point of the rotation curves which are still flat or even rising, implying that more dark mass is present at larger radii.

While it is conceivable that the total mass of galaxies could be determined by dynamical methods (satellite galaxies, galaxies in pairs or in groups), the main hope to derive the total extent is that, at least in some galaxies, the HI extends all the way to the edge of the halo (and still be detectable) so that the kinematics show the expected Keplerian decline if all the mass has been encountered.

Combining data from the DRAO interferometer array in Penticton, BC., with VLA observations, it was possible to recover all the single dish flux of DDO 154 of which $\sim 30\%$ was missing from the VLA data alone. The missing flux was found to be in an extended low surface brightness component in the outer parts of this dwarf irregular galaxy. The new combined data show that the HI disk extends to ~ 6 optical radii (R_{Ho}) or 21 optical scale lengths (α^{-1}) at a level of 1.0×10^{19} cm^{-2}. A total HI mass of 2.2×10^8 $\mathcal{M}_\odot$ is derived which gives an $(\mathcal{M}_{HI}/L_B) \simeq 7$. This allows to extend the rotation curve by more than 33% in radii.

The new combined data confirm that the rotation curve not only is declining in the outer parts, but that it runs parallel to the Keplerian decline expected if all the mass has been encountered. It thus seems that we have reached the edge of the mass distribution for DDO 154. This allows to derive a *total* mass of $\mathcal{M}_{tot} \simeq 3.0 \times 10^9$ $\mathcal{M}_\odot$, of which 90% is dark, and an upper limit to the radial extent of $R_{tot} \leq 8$ kpc.

AGE-METALLICITY RELATION: COMPARISON OF OPEN CLUSTERS' DATA WITH STELLAR POPULATIONS

Giovanni Carraro[1] & Yuen K. Ng[2]

[1]Department of Astronomy, Padova University, Vicolo dell'Osservatorio 5, 35122 Padova, Italy
[2]Leiden Observatory, P.O. Box 9513, 2300 RA Leiden. the Netherlands

The age–metallicity relation (AMR) from the Old Open Clusters population (Carraro & Chiosi, 1994; Friel & Janes, 1993) is compared with the disc stellar populations obtained from a recently developed model of the Milky Way by Ng 1994. A picture for the chemical evolution of the disc is presented in which the presence of a newly discovered Bar population (t = 8–9 Gyr, Z = 0.005–0.030) is taken into account. We suggest that the past history of the Galactic Disc has been significantly influenced by infall of metal poor gas from the halo and accretion events. The results are shown in Fig. 1.

Bibliography

Carraro, G. & Chiosi, C. 1994, *Astr. & Astrophys.* **287**, 761.
Friel, E. D. & Janes, K. A. 1993, *Astr. & Astrophys.* **267**, 75.
Ng, Y. K. 1994, *PhD. Thesis*, Leiden Observatory, the Netherlands.

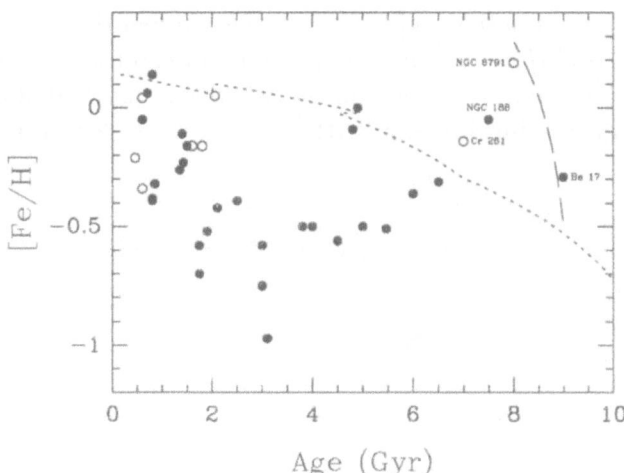

Figure 1: The Age–Metallicity Relation. The clusters are shown as filled and open circles. The AMR for the disc and bar population are shown respectively as short and dashed lines.

PHOTOIONIZATION EFFECTS ON GALAXY FORMATION

M. CHIBA
Astronomical Institute, Tohoku University, Sendai 980-77, Japan

AND

B.B. NATH
IUCAA, Post Bag 4, Ganeshkhind, Pune 411007, India

Abstract. We investigate the effects of the observed UV background radiation on galaxy formation. The biasing mechanism by photoionization is examined in detail, and its implication for galaxy number counts is discussed. Photoionization by UV radiation, J, decreases the cooling rate of the gas in halos, so that objects with only large density contrasts can self-shield against the radiation. In the context of the CDM model, we use the criterion that self-shielding is essential for star formation to calculate the mass function of galaxies (see Chiba & Nath 1994, ApJ, 436, 618 for details). We find that the cooling rate in the big (low-density) system, into which smaller objects are merged, is reduced by photoionization. This means that, in a merging dominated region, where the number density at the low-mass end ($M_b \le 10^{10} M_\odot$) is usually expected to decrease with time, the trend is *reversed* (the number of low-mass galaxies is more at lower redshifts z) due to the decreasing UV flux with time after $z \sim 2$ (see Figure 1).

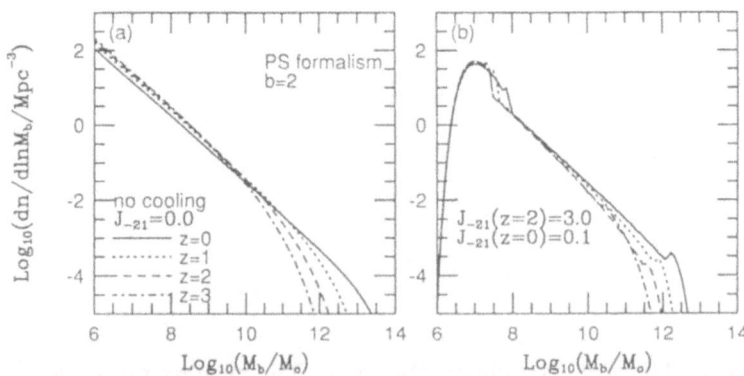

Figure 1. The Press-Schechter mass functions. (a) $J = 0$ (b) $dJ/dz > 0$

RECOVERING THE INTRINSIC METALLICITY DISTRIBUTION OF ELLIPTICAL GALAXIES

L. CIOTTI
Osservatorio Astronomico di Bologna
via Zamboni 33, I-40126 Bologna

M. STIAVELLI
Scuola Normale Superiore
P.zza de' Cavalieri 7, I-56126 Pisa

AND

A. BRACCESI
Dipartimento di Astronomia, Università di Bologna
via Zamboni 33, I-40126 Bologna

We present a simple recipe to derive the metallicity distribution of galaxies as a function of their integrals of the motion. Elliptical galaxies are known to possess metallicity gradients that frequently show variations of a factor of 2 from the center to the effective radius (Peletier, 1989). The observed gradients are the result of *projection* and *orbital mixing*. Orbital mixing arises because stars at a given radius may have their apocenters spanning a wide range of radii. Thus, the mean metallicity at any given point inside the galaxy is the result of the metallicity distribution in phase space weighted by the galaxian distribution function. We have proposed (Ciotti, Stiavelli & Braccesi, 1995) a simple inversion procedure allowing one to derive the dependence of metallicity on the integrals of motion for a spherical galaxy. The ideas of this paper can be generalized to oblate two-integrals models following the Hunter & Qian (1993) technique.

References

Ciotti, L., Stiavelli, M., and Braccesi, A., (1995), in press on *M.N.R.A.S.*
Hunter, C., Qian, E., (1993), *M.N.R.A.S.*, **262**, 401
Peletier, R., (1989), *PhD Thesis* (Groningen)

HOW MANY ULIRGS ARE MERGERS?

D.L. CLEMENTS
ESO, Karl-Schwarzschild-Strasse 2, D-85748 Garching b. Munchen

R.G. MCMAHON
IoA. Madingley Road, Cambridge. UK

AND

W.J. SUTHERLAND AND W. SAUNDERS
Physics Dept.. Oxford University. Keble Road, Oxford. UK

Ultraluminous IRAS Galaxies (ULIRGs) are unusual galaxies with luminosities $> 10^{12} L_\odot$ ($H_0 = 75$ kms^{-1}Mpc^{-1}), comparable to optically selected quasars, with up to 90% of this emitted in the 8-1000 μm region. Initial work on nearby ULIRGs suggested that all of these objects were in interacting or merging systems. Larger samples (e.g. Leech et al 1994) suggested that this might only be true for 2/3 of the objects. The triggering of ULIRG-type activity in the non-merging galaxies would then be difficult to understand.

We have recently completed a survey aimed at finding a large number of ULIRGs (Clements et al, 1995). This contains 91 ULIRGs. We here report R band imaging of the 56 ULIRGs in this sample brighter then Bj=19.5 using the ESO 2.2m telescope. These images show that 51 of the 56 ULIRGs are clearly disturbed systems. Four of the remaining five objects have nearby (in angular separation) companions. Only one object is isolated and undisturbed. Thus 91 % of the ULIRGs in this complete subsample are interactions/mergers, compared to 67 % in Leech et al (1994) or 61% in Zhenglong et al (1991). The origin of the discrepancy is unclear, but we believe that this is mainly due to faint signs of merging being missed in the earlier work.

References

Clements. D.L., et al., (1995), MNRAS, in press
Leech, K., et al. (1994), MNRAS, 267, 253
Zhenglong, Z.. et al, (1991), MNRAS, 252, 593

TOWARDS A COMPLETE LIBRARY OF STELLAR SPECTRA FOR EVOLUTIONARY SYNTHESIS

F. CUISINIER [1], TH. LEJEUNE [2], R. BUSER [1]
[1] *Astron. Inst. Univ. Basel, Switzerland*
[2] *Obs. Strasbourg, France*

The population and evolutionary synthesis of the integrated light of clusters and galaxies requires a good knowledge of the underlying stellar spectra. Libraries of observed stellar spectra can be used, but they have several disadvantages, e.g. uneven sampling — which causes problems in the integration phase. Furthermore, no comprehensive library of low- or high-metallicity stars does exist, which would be required to model chemical evolution. Libraries of synthetic spectra could — and should — solve these problems. Using Kurucz and more recent libraries dedicated to cool stars, we constituted a library covering the whole range of T_{eff}: 50 000 – 2500 K.

The comparaison between empirical and model computed temperature-colors relations shows differences that are too big to allow to use the cool synthetic spectra as such. We therefore developed a method to modify the synthetic spectra, in order to allow them to fit the observed temperature-colors relations.

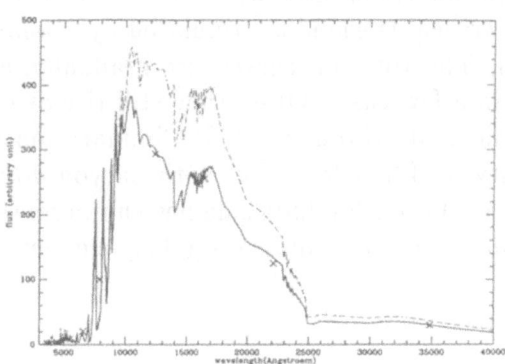

Figure 1. Original and modified spectrum at $T_{\text{eff}} = 3126$K. The crosses show the empirical fluxes in the UBVRIJHKL bands at this temperature.

PROPERTIES OF LOW SURFACE BRIGHTNESS GALAXIES

W.J.G. DE BLOK AND J.M. VAN DER HULST
Kapteyn Astronomical Institute, Groningen, The Netherlands
AND
S.S. MCGAUGH
Institute of Astronomy, Cambridge, United Kingdom

We have been working on multiband surface photometry, spectropho-tometry and HI synthesis data for 20 Low Surface Brightness (LSB) galaxies. LSB galaxies are well described by disks with an average central surface brightness of $\sim 23.4B$-mag arcsec^{-2}. They have scale lengths typical for high surface brightness (HSB) galaxies, though a large range of sizes is present. Their colours are blue, especially at the red side of the spectrum, where they are significantly bluer than HSB galaxies (de Blok et al. 1995a). Modelling and measurements of gas abundances (McGaugh 1994) suggests a low, stochastic star formation rate, and a lack of a large old population. The HI surface densities are a factor of three lower than those in HSB galaxies (de Blok et al 1995b). However the difference is not as large as in the optical. The HI disks are considerably larger, relative to the optical disks, than in HSB galaxies. The gas mass fraction is higher, indicating slow evolution. Star formation is inhibited by the low surface densities which are typically below the critical treshold as stipulated by Toomre's gravitational instability criterion. The rotation curves rise gradually, and are observed to flatten out only in a few cases. Often they still rise at the last measured point, or remain solid body through-out. Preliminary mass models suggest extended low density dark matter halos, with baryon dominated inner regions. The inferred evolution for LSB galaxies shows mass *and density* are fundamental parameters in determining a galaxy's evolutionary fate.

References

de Blok W.J.G., McGaugh S.S., van der Hulst J.M., 1995a, submitted
de Blok W.J.G., van der Hulst J.M., Bothun G.D., 1995b, MNRAS, 274, 235
McGaugh S.S., 1994, ApJ, 426, 135

BARS IN CUSPS

WALTER DEHNEN
Theoretical Physics
1 Keble Road, Oxford OX1 3NP, UK

In order to investigate the stability properties of galaxy models with central density cusps, N-body simulations of oblate models with density $\rho \propto m^{-1}$ $(m+a)^{-3}$ where $m^2 = R^2 + [z/q]^2$ and distribution functions $f(E, L_z)$ (computed as in Dehnen, 1995) have been performed with the following results.

1. An E7 model with identical amounts of stars of either sense of rotation was stable over 30 $t_{\mathrm{dyn}}(r=a)$. This is interesting for the bending instability has been argued to set in at about this flattening and be responsible for the absence of flatter elliptical galaxies (Merritt & Sellwood, 1994).

2. Rapidly rotating E$\gtrsim$E5 models quickly form weak bars inside the cusp, which are stronger for the more flattened, faster rotating initial configurations. The bars grow in a self similar fashion from inside out: the pattern speed decreases with increasing bar length and time. This process is initiated at the origin, where, because of finite N, the actual density no longer follows the power law, and stops when the edge of the cusp is reached. A typical example is given in the figure showing the x-y-coordinates of particles with $|z| < 0.1a$ after $\sim 20 t_{\mathrm{dyn}}(r=a)$ for an initially rapidly rotating E7-model. The bar has axis ratios of about 5:3:1, and extends almost to corotation. However, it has no sharp edge, but an inhomogenous density with a cusp steeper than the initial model. No sign of a buckling instability has been observerd.

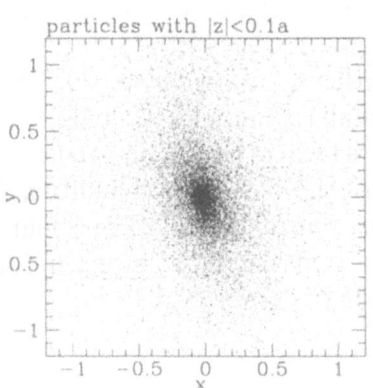

particles with |z|<0.1a

References

Dehnen W., 1995, MNRAS, 274, 919
Merritt D., Sellwood J., 1994, ApJ, 425, 551

COLOUR (≈AGE?) GRADIENTS IN SPIRAL GALAXIES

ROELOF S. DE JONG
Univ. of Durham, Dept. of Physics,
South Road, Durham DH1 3LE, UK

A sample of 86 galaxies was imaged in the B, V, R, I, H and K passbands to study their light and colour distribution as function of radius (de Jong & van der Kruit 1994). The radial colour gradients were compared with new dust models, which included both absorption and scattering, and with the stellar population synthesis models of Bruzual & Charlot (1993) and Worthey (1994). By requiring that the models had to fit all six passband photometry at the same time, the relative effects of dust, stellar age and stellar metallicity could be seperated (de Jong 1995a, 1995b). The main results from this investigation are:

— All galaxies become bluer with increasing radius. The colour at each radius correlates strongly with the average surface brightness at that radius, with Hubble type being an additional effect. Late type galaxies are bluer at the same surface brightness than early type galaxies.

— The reddening profiles predicted by the dust models are incompatible with the data when all colours have to be fitted at the same time. Dust cannot be the major cause of the colour gradients.

— The population synthesis models by Worthey (1994) indicate that the colour gradients cannot be caused by metallicity gradients alone.

— The best fit to the data is reached in a model where the colour gradients are mainly caused by an age gradient across the disk, with an additional metallicity gradient to explain the very red central colours. The colours of galaxies of type later than Sc indicate that they have in general a lower metallicity at all radii than the earlier types.

References

Bruzual G.A., Charlot S. 1993, ApJ 405, 538
de Jong R.S. 1995a, Ph.D. Thesis, Univ. of Groningen, The Netherlands
de Jong R.S. 1995b, in preperation
de Jong R.S., van der Kruit P.C. 1994, A&AS 106, 451
Worthey G. 1994, ApJS 95, 107

THE SHAPES AND AGES OF ELLIPTICAL GALAXIES

ROELOF S. DE JONG & ROGER L. DAVIES
University of Durham, Department of Physics,
South Road, Durham DH1 3LE, UK

Normally elliptical galaxies are thought to be old, evolved systems, but recently a controversy has arisen over the age of ellipticals. Measurements by Gonzáles (1993, Ph.D. thesis, UCSC) show that the $H\beta$ absorption indices of ellipticals span a range of values. Population synthesis models indicate that the $H\beta$ index is a good age indicator and hence, contrary to normal perception, the ages of ellipticals seem to span a range of values.

Here we investigate the hypothesis that the younger ages found in some elliptical galaxies are in fact due to an additional younger stellar population in a disk-like distribution superimposed on the old main body. In the figure we therefore examine the relation between $H\beta$ index (from Gonzáles' thesis and Davies et al. 1993, MNRAS 262, 650) and the shape of the isophotes as indicated by the C4 parameter (from Bender et al. 1989, A&A 217, 35; Peletier et al. 1990, AJ 100, 1091; Goudfrooij et al. 1994, A&AS 104, 179).

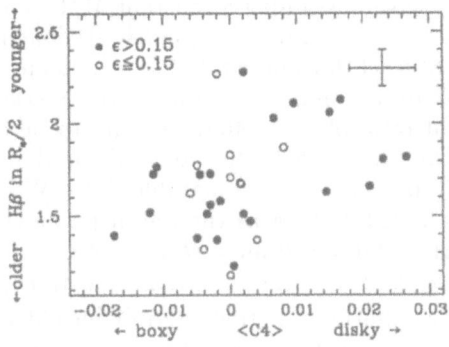

A weak trend can be seen in this diagram. There is an apparent lack of galaxies with boxy isophotes and high $H\beta$ index and of galaxies with disky isophotes and low $H\beta$ index. This could mean that galaxies with "young" centres (high $H\beta$ index) have an additional younger population distributed in a disk. A more detailed study is required to see whether higher $H\beta$ indices are indeed predominantly related to the light of the disky distribution.

ON THE DYNAMICS OF VERY FLATTENED ELLIPTICALS

H. DEJONGHE, V. DE BRUYNE AND P. VAUTERIN
*Sterrenkundig Observatorium, Universiteit Gent, Krijgslaan 281,
B–9000 Gent, Belgium*

AND

W.W. ZEILINGER
*Institut für Astronomie, Türkenschanzstraße 17, A–1180 Wien,
Austria*

Our aim is to study the generic phase-space structure of flattened ellipticals by investigating a few typical cases. Here we report on the E4 elliptical NGC 4697.

The construction of 2-integral and 3-integral distribution functions asks for photometry and kinematic data. In the course of the data analysis, we additionally detected a nuclear dust lane at 3.4″ or 0.4 kpc from the centre, which could be confirmed with HST data. This was put to good use to constrain the inclination.

A comparison of the Lucy-deprojection method with the multi-gaussian expansion method shows that both produce similar results, though the latter produces a smoother and analytically known spatial density.

The potential, obtained assuming a constant M/L, and the observed moments are the basis for quadratic programming models (Dejonghe, 1989). The 3-integral model is based on a Stäckel third integral. The original potential is retained where appropriate, in order to minimize errors due to the Stäckel approximation.

The mass-to-light ratio of NGC 4697 appears to be a fairly well determined $(M/L)_B = 4.8h_{50}$, and is much better constrained than would be the case for spherical anisotropic models with comparable data. We found that a 2-integral model cannot be excluded as long as the Satoh parameter k is allowed to vary. This qualifies an earlier claim of Binney *et al.*(1990). On the whole, the 3-integral model produces better and smoother fits, mainly with $\sigma_\varpi > \sigma_z$.

Our analytical distribution functions permit insight in the internal dynamics of the galaxy through detailed contour plots for all deprojected moments up to order 2. Because of the Stäckel approximation we can relatively easily construct a view of a flattened elliptical in action space.

References

Binney, J.J., Davies, R.L. & Illingworth, G.D., 1990, ApJ, **361**, 78
Dejonghe, H., 1989, ApJ, **343**, 113
Dejonghe, H., De Bruyne, V., Vauterin, P., Zeilinger, W.W., A&A, in press

SPECTRAL SYNTHESIS OF THE INTEGRATED SPECTRUM OF THE GALACTIC BULGE

SONYA DELISLE AND EDUARDO HARDY
Dép. de Physique, Université Laval and
Observatoire du mont Mégantic
Québec, Canada, G1K 7P4

We present preliminary results of the population synthesis of the Galactic Bulge using different techniques. The observations, taken with CTIO's 4m CS, cover the range 3600Å to 10000Å. The integrated spectrum includes light from stars in the line-of-sight. For the synthesis, we used first a program that we developed (Couture & Hardy (1993), and Hardy *et al.* (1994)) along with three libraries, one of composite population spectra, (Jablonka *et al.* (1990) and Bica (1988)), one of stellar spectra (Pickles (1985)), and one of equivalent widths measured on the integrated spectra of coeval populations (Bica and Alloin (1986a), (1986b) and (1987)). Results obtained with the spectra were inconclusive. With the equivalent widths, we found indication of the presence of a young and rich population (1 Myr and $[Z/Z_\odot] = +0.3$), but no old population or poor globular cluster-like population. Further tests are needed to evaluate the extent of the contamination from disk giants and its impact on the synthesis results. We also used Bica and Alloin's optimization program (Bica (1988)) with their library of equivalent witdhs. The contributions are spread over many groups but mainly intermediate-age, above-solar metallicity groups and old and poor groups. In all cases, no old metal-rich population was detected.

References

Bica, E. 1988, A&A **195**, 76.
Bica, E, and Alloin, D. 1986a, A&A **162**, 21.
Bica, E, and Alloin, D. 1986b, A&A Suppl. **66**, 171.
Bica, E, and Alloin, D. 1987, A&A **186**, 49.
Couture, J. and Hardy, E. 1993, ApJ **404**, 142.
Hardy, E., Couture, J., Couture, C. and Joncas, G. 1994, AJ **107**, 195.
Jablonka, P., Alloin, D. and Bica, E. 1990, A&A **235**, 22.
Pickles, A. J. 1985, ApJ. Suppl. **59**, 33.

NIR IMAGING OF THE BOX/PEANUT BULGE IN NGC 4302

R.-J. DETTMAR[1] AND A. FERRARA[2]

(1) Astronomisches Institut der Ruhr-Universität, D-44780 Bochum, Germany

(2) Osservatorio Astrofisico di Arcetri, I-50125 Firenze, Italy

Abstract. The central, spheroidal component of spiral galaxies is frequently found to be "box- or peanut" (b/p) shaped. More than 30% of edge-on galaxies may show this phenomenon (Dettmar 1995). To explain this large frequency a very common process for the origin of this specific shape is required. N-body simulations have shown that this can be attributed to instabilities and resonances caused by bars in spiral disks (e. g., Combes et al. 1990, Raha et al. 1991).

Previous statistics (Jarvis 1986, Shaw 1987, de Souza & dos Anjos 1987) are hampered by the low dynamic range of the photographic material used as well as by dust absorption in late type spirals and the difficulty to disentangle disk and bulge in such systems. We present K-band imaging of the very dusty edge-on galaxy NGC 4302 in the Virgo cluster supporting the presence of b/p bulges in late type spirals.

NIR Observations were obtained with the NICMOS-3 ARNICA camera at TIRGO[1] in spring 1994. The combined image of NGC 4302 clearly shows a flat-top "boxy" structure of the spheroid and an additional very thin component appears on the major axis cut. This latter component could be identified with a bar. The example of NGC 4302 demonstrates that b/p bulges are indeed present in disks of late type spirals and that bars can account for a significant fraction of these structures. As a consequence, the bulge/disk ratio of late type spirals has to be interpreted in an evolutionary scenario rather than as a primordial property.

References

Combes, F., Debbasch, F., Friedli, D., and Pfenniger, D. (1990) A&A 232, 82
de Souza, R. E., and dos Anjos, S. (1987) A&AS 70, 465
Dettmar, R.-J. (1995) in: IAU 169 "Unsolved Problems of the Milky Way" eds. L. Blitz and P. Teuben, Kluwer, in press
Jarvis, B. J. (1986) AJ 91, 65
Raha, N., Sellwood, J. A., James, R. A., and Kahn, F. D. (1991), Nature 352, 411
Shaw, M. A. (1987) MNRAS 229, 691

[1] TIRGO (Gornergrat, CH) is operated by CAISMI/CNR, Arcetri, Firenze, Italy

EFFECTS OF DUST EXTINCTION IN GALACTIC DISCS

A. DI BARTOLOMEO[1], G. BARBARO[1] AND M. PERINOTTO[2]

[1] *Dip. di Astronomia, Università di Padova, Padova, Italy*

[2] *Dip. di Astronomia, Università di Firenze, Firenze, Italy*

We have solved the radiative transfer equation taking into account both absorption and scattering into the line of sight to the observer to model the disc of spiral galaxies.

A dust model has been adopted, suitable for the diffuse interstellar medium of our Galaxy, to obtain, in all the considered spectral range ($1000 \div 10000$ Å), consistent quantities to describe both the absorption and the scattering properties of the dust.

New extinction curves, showing the internal absorption in a spiral galaxy, both if the scaleheight of stars and dust is the same (*case A*) or different (*case B*), have been derived (?). The most important results are the following:

1. The neglect of the scattering source is a hard approximation. In fact at $\tau_V = 0.72$ the overestimate of the internal absorption is of the order of 40% when the scaleheights of stars and dust are the same, and about $25 \div 30\%$ if the scaleheights are different. Moreover it increases with the optical depth.

2. The scattering deeply changes the shape of the observed mean interstellar galactic extinction curve. For instance the UV–bump, at $\lambda = 2200$ Å is lower and larger; the slope in the optical region is different.

3. The behaviour of the colour excesses, for *cases A* and *B*, as function of the optical depth, is different.

4. The use of the C parameter defined in studies of the surface brightness of differently inclined spirals, to infer the optical thickness is misleading, unless the C values are < 0.45. For $\tau > 2 \div 3$ in *case A*, C becomes negative.

References

Di Bartolomeo, A., Barbaro, G. & Perinotto, M., *MNRAS*, 1995, *accepted*

EMISSION LINE REGIONS IN M86

S. DÖBEREINER[1], J. HEIDT[2], S. J. WAGNER,[2]
[1]*Max-Planck Institut für Extraterrestrische Physik, Garching, Germany;* [2]*Landessternwarte Heidelberg, Germany*

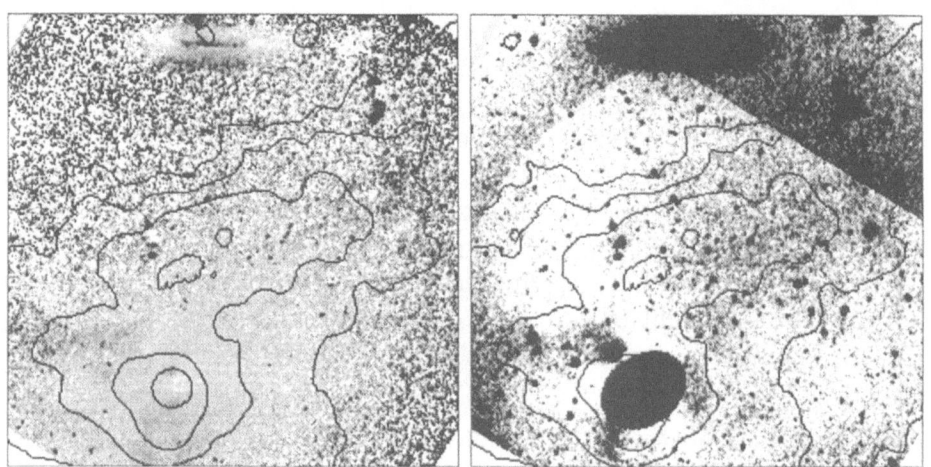

We present broad- and narrow-band (B, R, Hα+[NII] and Hα continuum) observations of M86 obtained with a focal reducer at the Calar Alto 1.23m telescope, to look for any signs of material cooling out from the hot X-ray emitting ISM. The above figures show a quotient image (13' x 13', north up) (Hα cont. - Hα) to the left, and a Hα image (right) from which the light of the stellar spheroid of M86 has been subtracted. Both are overlayed with ROSAT PSPC X-ray contours.

We found three regions of excess emission (dark in the figures): (A): Hα emission with complex morphology near and south of the center of M86, possibly related to cooling flow activity. (B): A broad fanshaped feature extending to the NE and E, visible as excess in the galaxy-subtracted images of all bands, although most prominent in the Hα band. The head of this feature is coincident with a small companion galaxy, so it is probably due to interaction. (C): A very narrow filament extending along the X-ray plume to the northwest. This filament may be due to cooling gas in the denser regions of the X-ray plume or to shock heating of yet unknown origin.

NGC 2188: A CASE STUDY FOR DISK-HALO INTERACTION

H. DOMGÖRGEN
Sternwarte der Universität Bonn, Germany

M. DAHLEM
Space Telescope Science Institute, USA

AND

R.-J. DETTMAR
Astronomisches Institut der Universität Bochum, Germany

Abstract. A deep Hα image of the nearly edge-on ($i \sim 86°$; Tully, 1988) irregular galaxy NGC 2188 shows the presence of spectacular filaments of ionized gas extending from a large star-forming complex several hundred pc into the halo of the galaxy (Dettmar, 1994). The origin and ionization mechanism of such extraplanar diffuse ionized features is not yet clear. One promising idea is that the extraplanar DIG is due to large scale mass circulation, i.e. a galactic "fountain" (Shapiro & Field, 1976) or "chimney" (Norman & Ikeuchi, 1989). Hα filaments could then be the photoionized walls of such chimneys or fountains.

Recently obtained VLA HI data also show peculiar HI filaments. East of the center of NGC 2188 a giant superbubble with a radius of 550 pc and energy requirement 1.6×10^{53} erg is found. However, a correlation between HI structures and Hα filaments cannot be found.

The HI observations reveal large-scale deviations from symmetry which are present in both, the gas distribution and in the velocity field. These suggest that NGC 2188 is a perturbed system, although it is not obviously an interacting galaxy.

References

Dettmar R.-J. 1994, in *The Physics of the Interstellar Medium and Intergalactic Medium*, PASP Conf. Series Vol. 80, Eds: Ferrara A. et al., p. 398
Norman C.A., Ikeuchi S. 1989, *ApJ* 345, 372
Shapiro P.R., Field G.B. 1976, *ApJ* 158, 279
Tully R.B. 1988, *Nearby Galaxy Catalog*, Cambridge University Press

THE $D_N - \sigma$ AND FUNDAMENTAL PLANE RELATIONS AS DISTANCE INDICATORS OF EARLY TYPE GALAXIES IN THE VIRGO AND FORNAX CLUSTERS

M. D'ONOFRIO AND S. ZAGGIA
Dipartimento di Astronomia dell'Università di Padova
Vicolo dell'Osservatorio 5, Padova, Italy

M. CAPACCIOLI AND G. LONGO
Osservatorio Astronomico di Capodimonte
Via Moiariello 16, Napoli, Italy

AND

N. CAON
Space Telescope Science Institute
Baltimore, USA

Abstract. We investigate the properties of early-type galaxies as distance indicators by applying the $D_n - \sigma$, Fundamental Plane (FP), and $\log(m) - \log r_e$ relations to a complete and fairly homogeneous sample of galaxies members of the Virgo and Fornax clusters. The relative mean distance of the two clusters can be derived with an accuracy up to $\sim 3\%$ provided that the internal kinematics of the galaxies is taken into account and a correct statistical approach is used.

The residuals of the $D_n - \sigma$ and FP relationships do not correlate with many structural and geometrical parameters of the galaxies: the mean effective surface brightness, the total luminosity, the average ellipticity and the Fourier coefficient a_4 of the isophotes, and, the exponent m of the $r^{1/m}$ law which parametrizes the shape of the light profiles. On the other hand, the kinematics of the galaxies affect both relations producing residuals that correlate with the maximum rotation velocity V_{max} and the (V/σ) ratio.

Once confirmed by future more accurate observations, this effect would introduce a sistematic bias in the distance determination of the clusters; a problem difficult to manage if the internal kinematics of the cluster members is unknown.

THE ULTRAVIOLET SPECTRA OF
OLD STELLAR POPULATIONS

BEN DORMAN
NASA/Goddard Space Flight Center

AND

ROBERT W. O'CONNELL
Astronomy Department, University of Virginia

We investigate the utility of UV spectra of old stellar populations as diagnostics of galaxy properties. For $\lambda < 3800$ Å, the integrated light of old stellar populations is dominated by two components: the hot stars which give rise to the UV upturn phenomenon at far-UV wavelengths, and the stars (i.e. main sequence, and subgiants) that lie closest to the turnoff. The mid-UV radiation from the turnoff varies strongly with metallicity, and less so with age; mid-UV light must be correct for the effect of the independent far-UV upturn component, but this is straightforward. Population synthesis models that account for the flux from the UV upturn can therefore determine the characteristics of the underlying stellar content from the mid-UV spectral region. The age & metallicity dependence of the far-UV 1550 Å is not well understood (see Dorman, O'Connell & Rood 1995, ApJ 442,105) since the mass loss mechanism on the RGB that produces very blue HB stars has no physical model. In contrast mid-UV indicators derive from the turnoff population which best represents the quantities we wish to measure, and are in addition visible at favourable optical wavelengths for $z \lesssim 1$.

We are investigating various spectral diagnostics with simple (single burst, single metallicity, Salpeter IMF) population models that include varying strengths of UV upturn. We find that a broadband index centered around 2200 Å measures a spectral region in which there are no strong lines and the continuum is very metal-sensitive. Models of metallicities [Fe/H] $= -0.47$, 0.0, & 0.3 occupy almost disjoint locations on a $15 - V$ vs $22 - V$ diagram, potentially helping to resolve the age/metallicity degeneracy. Higher precision data than is possible with the IUE satellite may be able to place (differential) ages on luminosity weighted galaxy populations.

POLARIZED RADIO EMISSION OF EDGE-ON GALAXIES:
OBSERVATIONAL RESULTS AND IMPLICATIONS

M. DUMKE [1], M. KRAUSE [1], R. WIELEBINSKI [1] AND U. KLEIN [2]
1 *Max-Planck-Institut für Radioastronomie*
Auf dem Hügel 69, D-53121 Bonn, Germany
2 *Radioastronomisches Institut der Universität Bonn*
Auf dem Hügel 71, D-53121 Bonn, Germany

We performed a small survey of the total and linearly polarized emission c edge-on galaxies at $\lambda2.8$ cm. The selected galaxies show very different levels in star forming activity, their environment (interacting or not), and many other $_{\text{l}}$ erties. In the following we present the main results of these observations.

Magnetic field orientation: No polarized emission could be detected in 1 5907. But the other galaxies in our sample do all show a large-scale magnetic predominantly parallel to the galactic plane, except NGC 4631: this strongly i acting starburst galaxy shows a vertical magnetic field structure in the central Thus it is obvious that dominant magnetic fields perpendicular to the disks of s galaxies are the exceptional rather than the normal case. But it is yet not which circumstances lead to such a field configuration. NGC 3628, also a starl galaxy, shows an ordered magnetic field parallel to the disk.

Radio spectra: From a combination of our measured flux densities with alr published values at other frequencies we found spectral indices α ranging from – to -0.86, consistent with the values found for normal disk galaxies. Nevertheles spectra seem to flatten with increasing nuclear star forming activity.

Galactic halos: We measured the distribution of the total intensity perpendi to the galactic disk. Although we detected emission above the disk in all gal. (except NGC 7331), only two of them show an extended radio halo. While NGC exhibits a galactic wind, NGC 891 is thought to be in a "chimney-mode", where material is ejected into the halo through chimneys.

Depolarization: The fractional polarization is found to be always lower thar theoretical value of $\sim 74\%$ and to increase with increasing distance from the gal. plane. Furthermore the amount and rate of this increase is different in all gala Since we can neglect Faraday effects at this wavelength, we modelled the degr polarization, based on beam depolarization. We found that the correlation leng the magnetic field increases with galactic height z and increases faster in gal. with lower star formation efficiency, supporting the idea that strong star form. leads to the transport of turbulences from the disk into the halo.

MAPPING THE GALACTIC CENTER

W.J. DUSCHL[1,2], S. VON LINDEN[3], T. WALTER[1], M. WITTKOWSKI[1,2]
1: Institut für Theoretische Astrophysik, Heidelberg, FRG
2: Max-Planck-Institut für Radioastronomie, Bonn, FRG
3: Landessternwarte, Heidelberg, FRG

Gas and dust in the inner region of the Galaxy are distributed in a flat, disk-like structure. We model the dynamics of this material in the framework of an accretion disk approach, and thus determine the efficiency of the radial transport of mass and angular momentum in the inner $\sim 200\,\mathrm{pc}$ of the Galactic Plane. Moreover, this allows us to establish the location (coordinates: galactic longitude l and depth normal to the celestial sphere) of molecular clouds from the observed positions (l) and radial velocities (currently, we neglect details of the vertical structure). Ultimately this will yield a map of the distribution of molecular clouds about Sgr A*.

In contrast to standard accretion disk models, we do not prescribe the viscosity by an α formulation but rather allow for additional external mechanisms (magnetic fields, supernova driven turbulence etc.) that may even dominate the transport of angular momentum. It turns out that the resulting viscosity is by about two orders of magnitude larger than what one would expect from an α accretion disk model.

We find that currently at radii of $\sim 100\,\mathrm{pc}$ from the Galactic Center $\sim 10^{-2}\,\mathrm{M_\odot/yr}$ are flowing towards Sgr A* and that highly non-axisymmetric processes, like bars, are *not* required to explain the transport of angular momentum and mass in the inner $\sim 200\,\mathrm{pc}$ of the Galaxy.

TIMESCALES FOR GALAXY FORMATION AND INTRA-CLUSTER MEDIUM ENRICHMENT

D. ELBAZ, L. VIGROUX
Service d'Astrophysique
CE Saclay-F91191 Gif-sur-Yvette Cedex-France

Recent observations with ASCA (Mushotzky 94, New Horizon of X-Ray Astron.-1st Results from ASCA) have confirmed the prediction (Arnaud et *al*. 92, A&A 254, 49; Elbaz et *al*. 95, to appear in A&A) that the Fe present in the intra-cluster medium (ICM) is mainly due to SNII and not SNIa: α-elements (O, Ne, Si, S) are overabundant with respect to Fe. To account for the large Fe mass present in galaxy clusters as well as these abundances ratios, one must consider that the IMF was different in the past, favouring high-mass stars. Moreover, the simplest way to eject the enriched ISM into the ICM is to consider that the same high-mass stars that were responsible for the Fe enrichment have also driven a galactic wind in young E/SO's.

Fe is nearly equally distributed between the galaxies and the ICM (Renzini 93, ApJ 419, 52) so galaxies must have ejected about half of their mass to enrich enough the ICM. A simple galactic wind model, where the wind happens when the thermal energy of the ISM is larger than its binding energy, shows that the starburst must extend at the scale of the whole galaxy in order to eject half of a galaxy mass with a large enough metallicity.

The timescale for such starbursts is typically a few 10^7 years, while the typical size of an L_* galaxy is of the order of 10 kpc. Hence, star formation should propagate at a speed of a few 100 km/s to account for such galactic starbursts, while nearby star forming regions show typical speeds close to 10 km/s, although in much different physical conditions (Elmegreen 92, in Star formation in stellar systems). Hence, the only way in such scenarii to avoid invoking extreme conditions would be that E/SO's result from the merging of dwarf galaxies of 10^9 $M_\odot$ which would have experienced a strong galactic starburst favouring high-mass stars, ejected this gas due to the low binding energy and merged afterwards. While the origin of a galactic starburst is difficult to understand for a whole L_* galaxy, it may be linked to galaxy interactions in the dwarf scenario.

SECULAR EVOLUTION OF THE SOMBRERO GALAXY.

ERIC EMSELLEM
Sterrewacht Leiden
Postbus 9513 2300 Leiden, The Netherlands

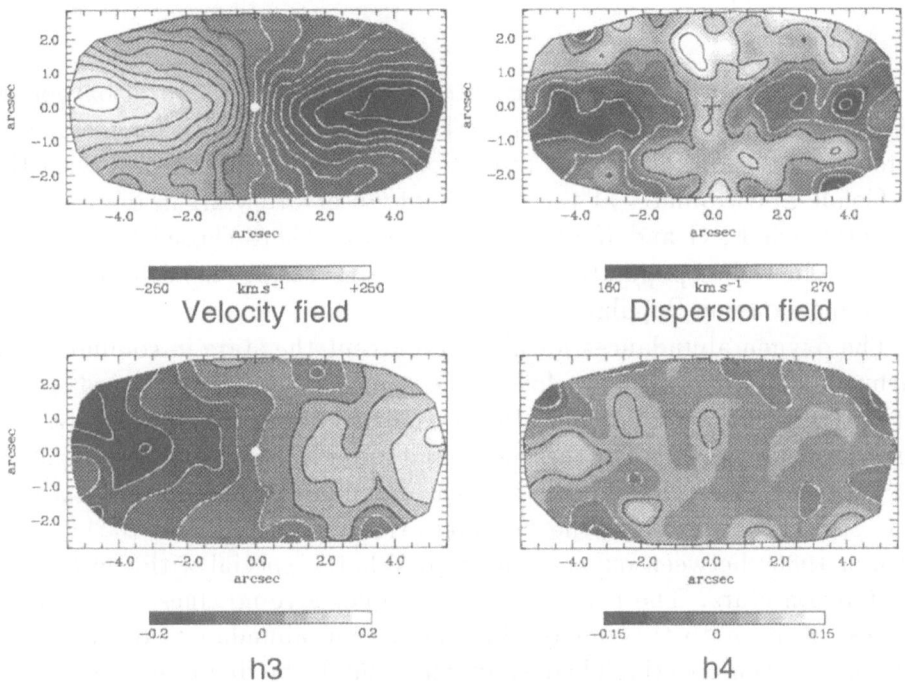

Figure 1. TIGER maps of the stellar velocities, velocity dispersions and the Gauss-Hermite coefficients (h_3, h_4) of the central region of M 104.

We present several pieces of evidence that there was an interaction between a bar, which is now dissolved (presumably by a small merger event - Emsellem 1995, A&A in press), and the interstellar medium (gas/dust) in the Sombrero Galaxy (M 104). This study has been achieved by combining new photometric and 2D spectroscopic data (HRCAM, TIGER - Fig. 1) with realistic models (a 3D spatial model of the dust distribution taking into account light scattering, and self consistent two-integral models). Such secular evolution of galaxies driven by bars may be a very common process (Friedli & Benz 1993, A&A 268, 65).

THE METAL-RICH DISK POPULATION

Oxygen abundances

SOFIA FELTZING AND BENGT GUSTAFSSON
Astronomiska observatoriet
Uppsala, Sweden

Edvardson et al 1993 demonstrated that rapid build up of oxygen in the inner Galaxy. Do these high [O/Fe] ratios prevail also for metal rich stars? We have obtained high resolution (R = 100 000) spectra with the 2D-Coude Spectrometer at the McDonald Observatory 2.7 m telescope for 50 metal rich F, G and K dwarfs. These stars where chosen to represent both an old stellar population which originates 'inside' the solar orbit and the young extreme Population I.

The oxygen abundances for all stars, except the stars in common with Barbuy and Grenon (1980), follow the same trends as in Edvardsson et al.(1993). Barbuy and Grenon (1980) studied 11 metal-rich dwarf stars on highly excentric orbits. For these we found lower iron abundances (typically 0.2 dex) and somewhat higher oxygen abundances. This means that the stars still stand out from the "normal" trend (see Fig. 1 in Barbuy and Grenon 1980) between oxygen and iron. What is special with the Barbuy and Grenon stars? The results for our samples give no clues to the origin of these stars or to the widely different oxygen abundances derived from different oxygen criteria. Their abundances might be due to erroneous T_{eff} or log(g). A change in T_{eff} by − 400 K or by − 0.3 in log(g), is however required in this case.

The absence of a gradient in [O/Fe] vs the V_{LSR} velocity suggests the existence of an efficient radial mixing in the gas disk as compared to the star formation rate. The full study, including abundances for $\alpha-$, s- and r−process elements, will be presented in Feltzing and Gustafsson (1995).

References

Edvardsson, B., Andersen, J., Gustafsson, B., Lambert, D.L., Nissen, P.E. and Tomkin, J. (1993) *A&A*, **275**, 101
Feltzing, S. and Gustafsson, B. (1995), in preparation
Barbuy, B. and Grenon, M. (1990) *in ESO/CTIO Workshop on "Bulges of Galaxies"*, ed. Jarvis, B.J. and Terndrup, D.M., p 83

UV TO NIR GALAXY EVOLUTION AND GALAXY COUNTS

MICHEL FIOC
Institut d'Astrophysique de Paris
98 bis, Bd. Arago, 75014 Paris, France

We have improved in the NIR the model of galaxy evolution developed by Guiderdoni & Rocca-Volmerange (A&A **186**, 1) in the UV and the visible, allowing thus a multispectral analysis of the evolution of galaxies and of faint galaxy counts. Since evolutionnary effects should be low in the near-infrared, cosmological ones might be put in evidence.

The recent M giants spectra and T_{eff} of Fluks et al. (A&Ass **105**, 311) are used and their bolometric corrections have been computed. More than 90 % of the K-luminosity of an ordinary galaxy comes from these stars. The infrared spectrum of other stars is a fit on observed colours. Stellar tracks for solar metallicity are from Bressan et al. (A&Ass **100**, 647) up to the EAGB and have been completed by the TPAGB phase as described in Groenwegen et al.(A&A **267**, 410).

The algorithm computs a burst of star formation and takes into account all the evolutionnary phases without loss of energy and avoids thus the oscillations of the colours. Synthetic spectra, including extinction, nebular emission and mass rejection are produced by the convolution of the bursts with star formation rates. The evolution of colours in (U-V,B-V) and (U-V,V-K) diagrams shows that different couples of age and star formation history may reproduce observed colours.

Predictions of number-magnitude counts in open and flat universes are compared to observations in the visible and the K for a classical luminosity function. For each cosmology and Hubble type, coherent z_{for} and SFR are choosen. Pure luminosity evolution in an open universe is compatible with K-counts but shows the well-known excess of faint blue galaxies in the blue. The normalisation of the LF on the K-counts is low and in agreement with bright blue counts of Maddox et al. (MNRAS **247**, 1p). An open universe in which the ellipticals and early-type spirals dominating the K-luminosity undergo a significant luminosity evolution is favoured by K counts but an additionnal population invisible in the blue bright counts and in the K seems however required to fit the faint blue counts.

S0 GALAXY LINE STRENGTHS AND GRADIENTS

DAVID FISHER, MARIJN FRANX
Kapteyn Astronomical Institute
University of Groningen
The Netherlands

AND

GARTH ILLINGWORTH
UCO/Lick Observatory
Board of Studies in Astronomy and Astrophysics
University of California, Santa Cruz

Line strengths and their gradients in Mg, Fe, and Hβ have been determined for a sample of 20 S0 galaxies in order to study the stellar populations of their bulges and disks and to investigate their relationship to elliptical galaxies. The data were obtained with the Lick Observatory 3m telescope in the long-slit mode over the region 4215−5615 Å with resolution 3.1 Å.

We infer bulge and disk gradients for the 9 most edge-on galaxies for which we have both major and minor axes profiles. The major and minor axes gradients in the bulge dominated inner regions are similar. At larger radii the disk profiles flatten considerably while the bulge Mg_2 strengths continue declining to lower values. Converted to [Fe/H] the bulges reach central values ~1-3 times solar and generally fall to [Fe/H] values lower than solar in our last measured points ($>1 \ r_e$). In contrast to the steep bulge Mg_2 gradients, the disk profiles are quite shallow. Based on our Mg_2 profiles, the average metal gradient found in the disks of our subsample is Δ [Fe/H] / $\Delta \ (r/h) = -0.06 \pm 0.05$, corresponding to a reduction in the mean metallicity of the disk stellar population by <15% per disk scale length (h).

Our findings do not support formation scenarios in which bulges formed either from heated disk material at late times after disk formation or through dissipationless stellar merging, as neither process includes mechanisms for producing the observed metallicity gradients. Our observations are better explained in terms of formation via dissipative collapse (or merging) at early-times.

KECK SPECTROSCOPY OF MODERATE REDSHIFT GALAXIES IMAGED BY HST

DUNCAN FORBES, A. PHILLIPS, D. KOO AND G. ILLINGWORTH
Lick Observatory, University of California, Santa Cruz, CA 95064 USA

Combining the results from Keck spectral and HST imaging data (Forbes *et al.* 1994), we have derived various quantitative parameters for 17 faint (I $\sim$ 21), distant (z $\sim$ 0.5) galaxies. Such redshifts correspond to a look-back time that is about half the age of the Universe and for which some scenarios predict significant galaxy evolution. We have measured disk scale lengths (with sizes ranging from 1–5 kpc) from fits to the surface brightness profiles and internal velocities with a rest frame resolution of $\sigma = 55$ to 80 km s^{-1} by fitting to the emission lines. The luminosity–disk size and luminosity–internal velocity relations for our moderate redshift galaxies are similar to the scaling relations seen for local galaxies, albeit with modest $\Delta M_B \sim 1^m$ brightening. We do not see evidence for a dominant population of starbursting dwarf galaxies, that have disappeared by the present epoch. Further details of this study can be found in Forbes *et al.* (1995). When large samples of kinematic data on distant galaxies are available, we will be able to trace galaxy evolution by mass as distinct from light.

References

Forbes, D. A., Elson, R. A. W., Phillips, A. C., Koo, D. C., & Illingworth, G. D. 1994, ApJ, 437, L17

Forbes, D. A., Phillips, A. C., Koo, D. C., & Illingworth, G. D., 1995, ApJ, submitted

STAR AND GLOBULAR CLUSTER FORMATION IN MERGERS

UTA FRITZE - V. ALVENSLEBEN
Universitäts-Sternwarte Göttingen
Geismarlandstr.11, 37083 Göttingen, Germany
ufritze@uni-sw.gwdg.de

The high burst strengths and star formation efficiencies found with spectrophotmetric and chemical evolutionary synthesis for mergers of gas-rich spirals led us to expect the formation of a secondary population of globular clusters (**GC**) with enhanced metallicity (F. – v. A. & Gerhard 1994, A&A **285**, 751 u.775). HST imaging of NGC 7252, NGC 4038/39 and NGC 1275 revealed rich populations of bright young star clusters (**YSC**).
We model the evolution of colours, luminosities, stellar absorption features, and spectra for star clusters of various initial metallicities and find that metallicity effects are of particular importance in early phases. Comparison with spectroscopy of the brightest GC candidates in NGC 7252 confirms their high metallicity $Z \sim 0.01$ (F. – v. A. & Burkert 1995, A&A **300**, 58). Once metallicity is known, precise age dating becomes possible. With ages of YSCs precisely known, our models follow their luminosity and colour evolution over a Hubble time. We derive mean ages for the YSCs of $1.37^{+1.6}_{-0.8}$ Gyr in NGC 7252, in agreement with the global starburst age, and of $\gtrsim 0.2$ Gyr in Antennae. With their present age and a mean effective radius $\langle R_{\mathrm{eff}} \rangle \sim 10$ pc (upper limit $\longrightarrow$ Meurer 1995, Nat **375**, 742), the YSCs in NGC 7252 can indeed be expected to be GCs. They will fade over a Hubble time to about the luminosity of ωCen. In Antennae, the LF of those YSCs with $R_{\mathrm{eff}} \leq 10$ pc, when evolved over a Hubble time, is found to look similar to the Galactic GCLF, except for an overpopulation of the faintest bins, at variance with van den Bergh's (1995, ApJ **450**, 27) results for the LF of **all** YSCs. Dynamical effects – strongest during the first Gyr and at the lowest mass (luminosity) bins – are expected to further change this LF and remain to be estimated (F. – v. A. & Burkert 1995, *in prep.*).
Acknowledgement This work is supported by the Deutsche Forschungsgemeinschaft, grant Fr 916/2-1.

DYNAMICAL STABILITY AND DYNAMICAL EVOLUTION OF THE DISKS OF SC GALAXIES

B. FUCHS, R. WIELEN
Astronomisches Rechen–Institut Heidelberg

AND

S. VON LINDEN
Landessternwarte Heidelberg–Königstuhl

We give a detailed description of the dynamical interaction of the stellar and gaseous components of a galactic disk. The stability of the two-component system against axisymmetric density perturbations is analyzed and the critical velocity dispersions of the stars and the gas, which control the instability, are determined. By comparison with the observed velocity dispersions it is shown that NGC 6946, a typical Sc galaxy, seems to be stable in the outer parts of the disk, but is dynamically unstable in the inner parts. The transition occurs exactly at the HII region disk boundary, so that the onset of gravitational instability appears to be related to the threshold of massive star formation (Kennicutt 1989). We study the dynamical evolution of the instability by numerical simulations using a code (cf. Casoli & Combes 1982), which implements both the stellar and gaseous components. It is shown that the gas forms after a short initial period of growing ring–like perturbations large cloud complexes, which induce by 'swing-amplification' (Toomre 1990) multi-armed spiral structures in the stellar and gas disks, exactly as observed.

Acknowledgements

We thank A. Toomre for enlightening discusions and F. Combes for letting use us her code.

References

Casoli F. and Combes F. (1982) *A&A* **110**, 287
Kennicutt R.C. (1989) *ApJ* **344**, 685
Toomre A. (1990) in *Dynamics and Interactions of Galaxies*, R. Wielen (ed.), Springer, p. 292

378

EVOLUTION OF GALAXIES THROUGH THEIR INTERACTION

Y. FUNATO AND J. MAKINO
University of Tokyo, 3-8-1, Komaba, Meguro-ku, Tokyo, 153, Japan

We investigated how the encounters between galaxies change their mass M and velocity dispersion σ. We carried out a series of direct N-body simulation of encounters of two spherical galaxies. In Figure 1, the relative change of energy $\Delta E/E$ are plotted against that of mass $\Delta M/M$ for various initial conditions. The filled and open symbols correspond to the cases of Plummer model and relaxed Hernquist model, respectively. Here $\beta \equiv (\Delta E/E)/(\Delta M/M)$.

We found a simple theory which explains the result. If the galaxies have halos with density $\rho \sim r^{-\alpha}$, the relation is expressed as $\Delta\sigma/\sigma = 0.25(\alpha - 3)\Delta M/M$ independendly of the collisoin parameters.

For the real galaxies, α is believed to be 4. Therefore the ratio β is 0.25. This implies that galaxies will assymptotically become to the state which is expressed as $\sigma \sim M^{0.25}$ through interactions. Assuming that M/L is constant, this relation is equivalent to the Faber–Jackson relation $\sigma \sim L^{0.25}$.

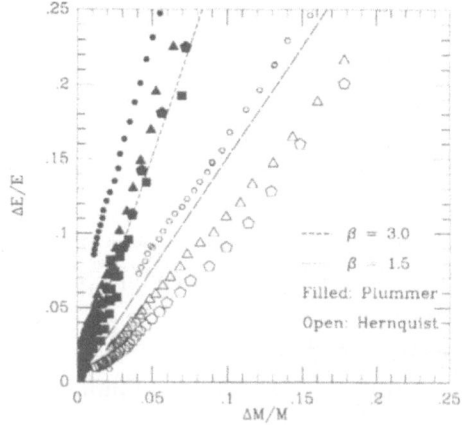

Figure 1. $\Delta E/E$ vs $\Delta M/M$

3D N-BODY SIMULATION OF THE MILKY WAY WITH GAS

R. FUX AND D. FRIEDLI

Geneva Observatory, CH-1290 Sauverny, Switzerland

We present the first self-consistent 3D barred model of the Milky Way including gas, as an evolved stage of an *unsymmetrised* N-body simulation with axisymmetric initial conditions (fig.1). The corotation lies at about 4 kpc and the central gas particles have condensed into a nuclear ring/disc. Distortions of the bar and other odd modes in the stellar distribution (a) affect the gas flow by producing one-side gas voids (b) and instabilities of the nuclear component like tilting, warping and off-centering (c-d). These peculiarities may help to reproduce the intriguing asymmetries and densely populated "forbidden" regions in the Galactic HI and CO $(l-v)$ maps (e-f). More details are given in the proc. of IAU Coll. 157 "Barred Galaxies".

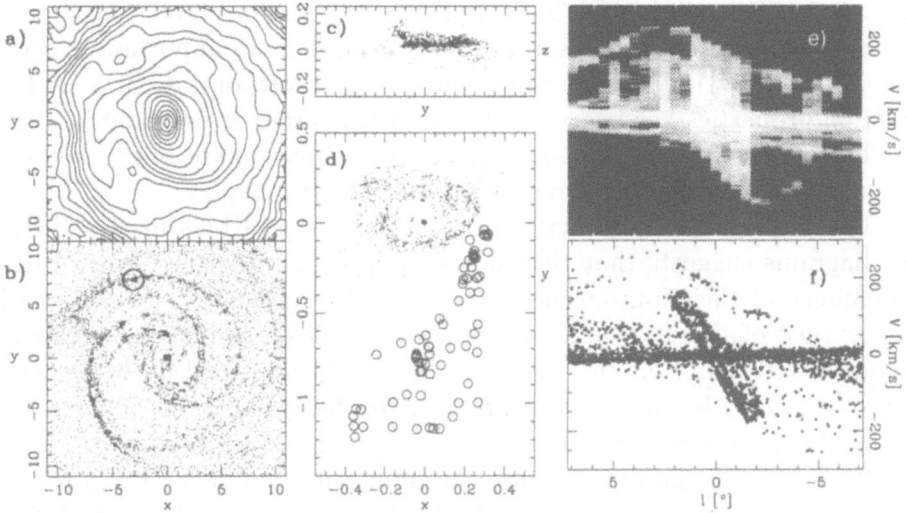

Figure 1. Outputs at $t = 1.5$ Gyr. a) Stellar isodensity curves, face-on. b) Gas particles, same scale. c) Nuclear gas ring, edge-on. d) Central zoom on frame b. The encircled dots are the particles responsible for the $V > 0$ and $l < 0$ strip in frame f. The length unit is 1 kpc. e) CO $(l-V)$ map of Dame et al. (1987). f) Model $(l-V)$ diagram for an observer located inside the circle of frame b, i.e. trailing the bar major axis by an angle of $20°$.

STUDY OF A COMPLETE SAMPLE OF Hα EMISSION–LINE GALAXIES FROM THE UCM SURVEY

J. GALLEGO, J. ZAMORANO, M. REGO, A.G. VITORES AND
O. ALONSO
Dpto. de Astrofísica, Universidad Complutense Madrid, Spain

The Universidad Complutense de Madrid survey is a long-term project with the aim of finding and analyzing star forming galaxies using the Hα line as the tracer for star formation processes. In order to obtain a representative and complete sample of the population detected, spectroscopic observations were carried out for the full sample of Hα emission–line galaxy (ELG) candidates of the UCM lists 1 and 2. The ELGs types most commonly found (47%) are intermediate to low-luminosity objects with a very intense star-formation region which dominates the optical energy output of the galaxy. This kind of ELGs is similar to the galaxy population detected in the blue objective-prism surveys. And what is more important, a second population (43%) of star-forming galaxies with low ionization or high extinction properties has been found. This ELGs group is detected neither in the blue (University of Michigan survey, Case survey) nor in other surveys (Kiso, IRAS, Markarian) using other selection techniques.

The position of the different natural groups of ELGs in the diagnostic diagrams suggests that they differ from one another in terms of metal abundance of their ionized gas, the ionization parameter and the relative importance of the starburst process in the galaxy, confirming a previous similar result found for the Michigan survey. Also a trend for lower metallicities at lower luminosities is present. No galaxy with metallicity lower than I Zw 18 has been found.

A study of the luminosity and spatial distribution of the UCM galaxies is presented. The two-point correlation function shows that UCM ELGs follow at great scale the distribution of the catalogue galaxies, although being less clustered in general. Using the UCM ELGs as SFR tracers a luminosity function for the SFR at the local Universe is computed. This result has important implications for our understanding of the origin of the star formation phenomena as well as the evolution of the galaxies.

THE ENRICHMENT OF THE INTRACLUSTER MEDIUM

Galactic Winds, Dwarf Galaxies, and Bimodal Star Formation in Ellipticals

B.K. GIBSON
Mt. Stromlo & Siding Spring Observatories
Australian National Univ., Weston Creek, Australia ACT 2611

AND

F. MATTEUCCI
Univ. of Trieste
Dept. of Astronomy, Via G.B. Tiepolo 11, Trieste, Italy 34131

Overview

Recent observational evidence for steep dwarf galaxy luminosity functions in several rich clusters has led to speculation that their precursors, via supernovae (SNe)-driven galactic winds, may be the source of the majority of gas and metals inferred from intracluster medium (ICM) x-ray observations (Trentham 1994). Utilising a fully self-consistent photo-chemical evolution package (Gibson 1995), and insisting that the post-galactic wind dwarfs obey the observed colour-luminosity-metallicity relations, we demonstrate that the bulk of the ICM gas does **not** originate within their precursors (Gibson & Matteucci 1995).

In a parallel study, we consider the present-day photo-chemical properties of elliptical galaxies, adopting the bimodal star formation scenario of Elbaz, Arnaud & Vangioni-Flam (1995). Based upon chemical evolution arguments *alone*, this scenario has been invoked by them to explain the observed metal mass, and their abundance ratios, in the ICM of galaxy clusters. Our fully self-consistent photo-chemical evolution analysis of their model highlights its failings: their predicted V-K colours are > 1 mag too red; their luminosity-weighted metallicities >0.7 dex too high; and their predicted metallicity dispersion is virtually non-existent (< 0.2 dex, versus the > 3 dex implied observationally) (Gibson 1995).

References

Elbaz, D., Arnaud, M., Vangioni-Flam, E., 1995, A&A, in press (and these proceedings)
Gibson, B.K., 1995, MNRAS, in press
Gibson, B.K., Matteucci, F., 1995, MNRAS, submitted
Trentham, N., 1994, Nature, 372, 157

STELLAR POPULATIONS IN DWARF ELLIPTICAL GALAXIES

J. GORGAS, S. PEDRAZ AND N. CARDIEL
Dpto. de Astrofísica, Universidad Complutense, Madrid

AND

J.J. GONZALEZ
Instituto de Astronomía, U.N.A.M., México

We present the results of a spectroscopic study in which we have measured line-strength indices in a sample of 5 dwarf and 2 compact elliptical galaxies (mostly from Virgo). Some conclusions about the stellar populations in dwarf E's are derived by comparing the sample with classical E galaxies. In the adjacent figure, we plot in the Hβ-[MgFe] plane the central indices of our sample of dwarf and compact E's together with data from González (1993, PhD Thesis). The grid represents the predictions of single-burst stellar population models from Worthey (1994, ApJS, 95, 107). The main conclusion is that, whilst the central regions of giant, intermediate and compact ellipticals span, according to stellar population models, a wide range in mean stellar ages, bright dwarf ellipticals are found to be old, compatible with a $\sim$ 10 Gyr old stellar population. This means that the dichotomy found in the Fundamental Plane between dwarf ellipticals and the gE-iE-cE sequence (Bender, Burstein & Faber 1992, ApJ, 399, 462) is also observed in the stellar populations.

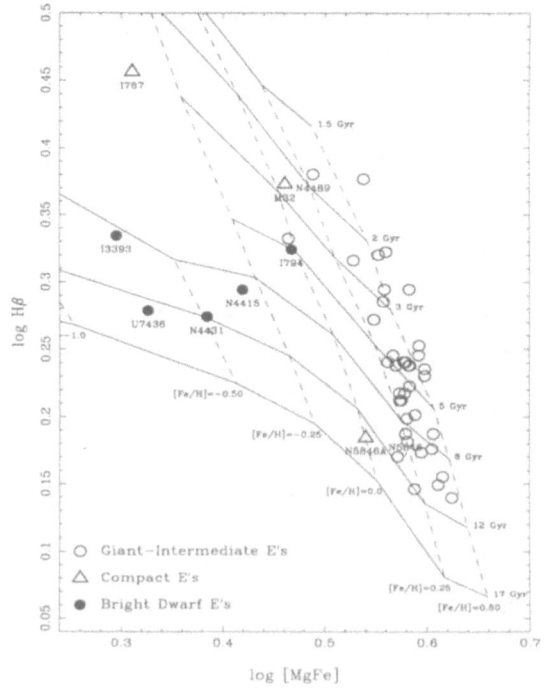

DUST VS. COLOR GRADIENTS IN ELLIPTICAL GALAXIES

PAUL GOUDFROOIJ

European Southern Observatory, Karl-Schwarzschild-Strasse 2,
D–85748 Garching bei München, Germany

Abstract. Color gradients in elliptical galaxies are commonly interpreted as being due to stellar population gradients (e.g., Davies et al. 1993, MN-RAS 262, 650). Here, I show that elliptical galaxies should generally contain a diffusely distributed component of dust, in *addition* to the "visible" component which is in the form of dust lanes or patches (see Fig. 1a). Employing a multiple scattering model for the dust, the presence of this newly postulated dust component is found to imply significant radial color gradients (see Fig. 1b). This should be taken seriously in the interpretation of color gradients in elliptical galaxies. This poster paper is based upon a (much!) more elaborate article by Goudfrooij & de Jong (1995, A&A 298, 784).

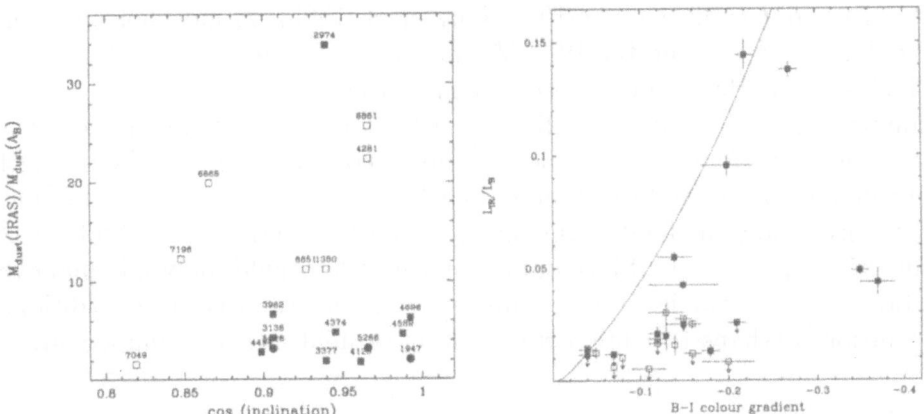

Figure 1. (**A, left**) The quotient of the dust mass derived from the IRAS data and the dust mass derived from optical extinction versus the cosine of the inclination angle of the dust lane in elliptical galaxies containing regular dust lanes (see Goudfrooij & de Jong 1995). The NGC designations of the galaxies involved are indicated. (**B, right**) The relation of L_{IR}/L_B with $\Delta\,(B-I)/\Delta\,(\log r)$ for elliptical galaxies in the "RSA sample" (cf. Goudfrooij et al. (1994, A&AS 104, 179)). Filled squares represent galaxies showing optical evidence for dust, and open squares represent galaxies without optical evidence for dust. Arrows indicate upper limits to L_{IR}/L_B. The dotted line represents the color gradient expected from differential extinction by a diffuse distribution of dust (see text).

CONSTRAINTS ON GALAXY EVOLUTION FROM FAINT REDSHIFT SURVEYS, KECK, AND HST

CARYL GRONWALL
UCO/Lick Observatory
Board of Studies in Astronomy and Astrophysics
University of California, Santa Cruz, CA 95064 USA

The nature of faint field galaxy evolution remains controversial. While many workers advocate exotic theories, such as rapid merging or disappearing populations, we have found that it is possible to explain the published counts and redshift data with traditional luminosity evolution models which derive an optimal set of *local luminosity functions* for different galaxy types (Gronwall & Koo 1995). Recently, there has been a tremendous amount of new data addressing this question, including 1) the measurement of the galaxy luminosity function vs. redshift from faint redshift surveys down to $B = 24$ and $I = 22$ (Colless 1995; Lilly et al. 1995), 2) morphological and angular size data from the HST Medium Deep Survey (Driver et al. 1995; Phillips et al. 1995), and 3) redshift measurements with Keck for a small sample of galaxies with $I > 22$ (Koo 1995). We have explored these new data and have found that while our model continues to provide an excellent match to the faintest observed redshift and angular size distributions, it underpredicts the faint counts and luminosity function evolution for very blue late-type galaxies. Since our current model includes only minimal evolution of these galaxies, the new observations suggest a need for additional evolution, perhaps through a starbursting or mild merging component.

References

Colless, M. 1995, *Wide Field Spectroscopy and the Distant Universe,* eds, S. J. Maddox & A. Aragón-Salamanca, (World Scientific), p. 263
Driver, S. P., Windhorst, R. A., Ostrander, E. J., Keel, W. C., Griffiths, R.E., & Ratnatunga, K. U. 1995, ApJ, 449, L23
Gronwall, C., & Koo, D. C. 1995, ApJ, 440, L1
Koo, D. C. 1995, this volume
Lilly, S. J., Tresse, L., Hammer, F., Crampton, D., & Le Fèvre, O. 1995, ApJ, in press
Phillips, A. C., Gronwall, C., Koo, D. C., Forbes, D. A., Illingworth, G. D., & Huchra, J. P. 1995, in preparation

SPATIALLY-RESOLVED INTERNAL KINEMATICS OF
⟨z⟩ ≈ 0.3 FIELD GALAXIES: EVIDENCE FOR ROTATION

P. GUHATHAKURTA AND K. ING
UCO/Lick Obs., U. of Calif., Santa Cruz, CA 95064, USA

H.-W. RIX
Max-Planck-Inst. für Astrophysik, 85740 Garching, Germany

M. COLLESS
Mt. Stromlo & Siding Springs Obs., Canberra, Australia

AND

T. WILLIAMS
Rutgers University, Piscataway, NJ 08855, USA

The nature of evolution in faint field galaxies remains a mystery. The Tully-Fisher relation, empirically relating the intrinsic luminosity of a spiral galaxy to its rotation speed, is an important tool for constraining the amount of luminosity evolution in distant field galaxies. Studying the luminosity-vs-linewidth relation for distant galaxies allows one to compare the luminosity of local and distant galaxies. The customary measure of a galaxy's rotation speed is the width of an emission line. It is important, however, to test whether the linewidth is a reliable measure of the galaxy's rotation speed or if it is dominated by turbulent motion within HII regions. In order to do this, we study the spatially-resolved kinematics and distribution of O[III] gas in about ten $\langle B \rangle \sim 21$ field galaxies at $\langle z \rangle \approx 0.3$.

We have used the CTIO 4-m telescope with the Rutgers Fabry-Perot imaging interferometer. The 2.5Å etalon provides a resolution of ≈ 120 km/s which is sufficient to measure the linewidth of most galaxies. For each galaxy, a series of images stepped by 1Å (~ 50 km/s) was obtained, centered around the (known) redshift of the O[III] line ($\lambda_{rest} = 5007$Å; $\lambda_{obs} \approx 6500$Å). We have well-resolved velocity fields for most of the galaxies in the sample (typical galaxy size is $3''$–$4''$; seeing was $\sim 1.3''$). Galaxy linewidths are significantly broader than the instrumental resolution. We compare the distribution of starlight and ionized gas, and study the O[III] velocity field. The largest galaxies show clear evidence of rotation.

SPECTROSCOPY OF THREE SBS GALAXIES

M.V. GYULZADIAN

Byurakan Astrophysical Observatory, 378433 Byurakan, Armenia

Three galaxies from Second Byurakan Sky Survey have been studied. The spectra of these galaxies were obtained at 6-m telescope of the Special Astrophysical Observatory of the Russian Academy of Sciences. For these objects line widths and emission line ratios were measured and ionization mechanism and chemical abundances were determined.

The FWHM for the all galaxies is in order of FWHM for narrow emission line galaxies. The excitation mechanism for them is photoionization.

Observed [SII]/Hα ratios are rather large in comparison to more conventional HII regions.

On the base of FWHM of emission lines, observed in the spectra of SBS 1122+610 and SBS 1139+601, these objects were placed on the diagrams of Baldwin et al. (1981) and Veilleux & Osterbrock (1987), and it was concluded that these galaxies are NELGs. SBS 1133+597 is HII galaxy.

References

Baldwin, J.A., Philips, M.M., Terlevich, R. 1981, PASP 93, 5
Veilleux, S., Osterbrock, D. 1987, ApJS 63, 295

LINE STRENGTH GRADIENTS IN LOW LUMINOSITY GALAXIES

C. HALLIDAY, G. BAGGLEY AND R. L. DAVIES
Department of Physics, University of Durham

M. BIRKINSHAW
Harvard Smithsonian Center for Astrophysics

AND

R. BENDER AND R. SAGLIA
University of Munich

Line strength gradients in luminous ellipticals show changes in metallicity of roughly a factor of two out to an effective radius (Worthey, Faber & Gonzalez, 1992; Davies, Sadler & Peletier (DSP), 1993; Gonzalez, 1993). The observed decline in Mg_2 line strength with increasing radius, while $H\beta$ remains roughly constant, has been interpreted to indicate an age gradient with the central parts of ellipticals being younger than the outer regions.

We have taken high quality long slit spectra of 14 low luminosity ellipticals (LLEs) ($-17 \geq M_B \geq -20$), and made preliminary measurements of Mg_2 and $H\beta$. We find measurements of Mg_2 in LLEs imply that the metallicity is lower by a factor of $\simeq 3$ than in the luminous galaxies but has a similar gradient. However the $H\beta$ values are similar to those in luminous ellipticals and remain constant as a function of radius.

Interpreted as single age stellar populations and using Worthey's models we would conclude that the ages of LLEs could be somewhat older than the more luminous systems. This conclusion may however be too naive. Mg is overabundant in luminous elliptical galaxies compared to Fe (Worthey, Faber & Gonzalez 1992, DSP) and this overabundance may not persist at lower luminosities. Accounting for this effect will increase age estimates for luminous ellipticals based on Worthey's models.

In the future we plan to measure line strengths for the weaker features in these spectra, and determine the rotation, line shapes and velocity dispersion with radius.

References

Binggeli, B., Sandage, A., and Tammann, G.A. (1985) *AJ*, **90**, 1681.
Davies, R.L., Sadler, E.M., Peletier, R.F., (1993) *MNRAS*, **262**, 650 (DSP).
Gonzalez, J.J., (1993) *PhD Thesis*, University of California, Santa Cruz.
Worthey, G., Faber, S., Gonzalez, J.J., (1992) *ApJ*, **398**, 69.

PECULIAR KINEMATICS IN THE CORE OF NGC 474

G. K. T. HAU[1], M. BALCELLS[2] & D. CARTER[3]

[1] *Inst. of Astron.* & [3] *Royal Greenwich Obs., Cambridge, UK*
[2] *Kapteyn Laboratorium, Groningen, The Netherlands*

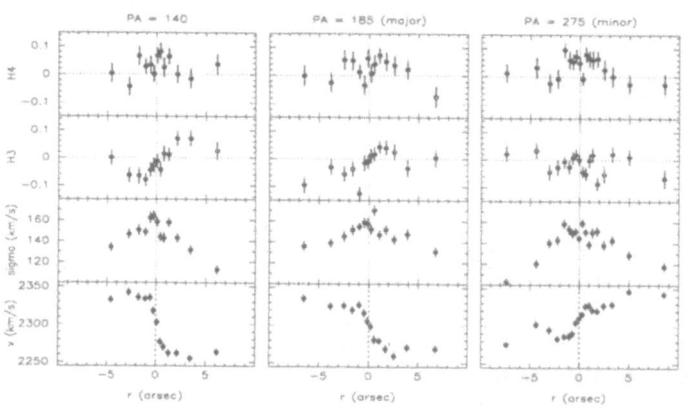

We present absorption line profile analysis of NGC 474, an elliptical with prominent, irregular shells. Profiles are parameterized with Gauss-Hermite polynomials (van der Marel & Franx ApJ 407 1993; Rix & White MNRAS 254 1992). The fastest rotation ($\sim 50\,\mathrm{km\,s^{-1}}$) and steepest central velocity gradient along the intermediate photometric axis rules out the possibility that NGC 474 is a face-on S0 (Schombert & Wallin AJ 94 1987) and suggests that it is triaxial. The asymmetry of the LOSVDs ($h3$ up to 0.08) indicates the presence of a subsystem with rapid, ordered rotation. The minor axis velocity curve shows a kinematic feature at 3-4" east of the nucleus, with no associated $h3$ or $h4$ features. Non-parametric LOSVD analysis (unresolved Gaussian decomposition, Kuijken & Merrifield MNRAS 264 1993) reveals a double-peaked profile at that location. In all position angles line-profiles are distinctly pointy for radii up to 2", and are consistent with zero further out. We have found similar central positive $h4$ terms in the shell galaxy NGC 2865 (Hau et al. MNRAS *in prep*). Cores with pointy LOSVDs are uncommon in ellipticals (Bender, Saglia & Gerhard MNRAS 269 1994). Positive $h4$ terms might contain important clues on the shell-formation mechanism in ellipticals.

HIGH-RESOLUTION IMAGING OF BL LAC HOST GALAXIES

J. HEIDT
Landessternwarte, Königstuhl, 69117 Heidelberg, Germany

AND

T. PURSIMO, A. SILLANPÄÄ, L.O. TAKALO, K. NILSSON
Tuorla Observatory, Väisäläntie 20, 21500 Piikkiö, Finland

We present high-resolution imaging of the BL Lac objects 1ES 0229+200 (z = 0.14), Markarian 421 (z = 0.031) and the prototype BL Lac (z = 0.069). Our goal is to study the properties of the host galaxies of this enigmatic sources and to compare them with their non-actice counterparts. The observations were carried out with the Nordical Optical Telescope (NOT) at La Palma under sub-arcsecond seeing conditions (FWHM < 0."8).

The surface brightness profiles of the host galaxies were analyzed as described in Bender and Möllenhoff (*A&A*, **177**, 71 (1987)). After masking the disturbing field stars/galaxies isophotes were fitted by ellipses down to 20.5 $mag/arccsec^2$ in Markarian 421, 23.5 $mag/arcsec^2$ in BL Lac and 25.5 $mag/arcsec^2$ in 1ES 0229+200. Additionally, we analyzed the surface brightness profile of the companion of Markarian 421.

All BL Lac objects are embedded in giant elliptical galaxies. Their host galaxies show ellipticities in the range from 0.2 to 0.4 and constant position angles. The α_4-Fourier coefficient is consistent with zero in 1ES 0229+200 and Markarian 421 and is positive in BL Lac. This is already the third BL Lac object, where "disky" isophotes have been found in their host galaxies. The companion of Markarian 421 shows a strong bulge and very high positive α_4-Fourier coefficients. Hence it is likely that this galaxy is an early-type spiral, contradictionary to previous claims in the literature.

In order to estimate an absolute magnitude for the host galaxies of the three BL Lac objects we fitted a de Vaucouleurs profile to the observed surface brightness profile excluding the inner 3 × FWHM arcsec from the fit. We derived M_R = -24.1 for 1ES 0229+200, M_R = -25.1 for Markarian 421 and M_R = -24.5 for BL Lac (H_0 = 50 km/s/Mpc, q_0 = 0). They are in the same range as determined in other studies.

PHYSICAL PROPERTIES OF SUBCRITICAL ACCRETION DISKS

O. M. HEINRICH
Institut f. Theoretische Astrophysik
Tiergartenstr. 15, 69121 Heidelberg

Despite of the great interest in accretion disks around supermassive black holes there are still numerous controversies about their structure and their physical properties. Therefore we have calculated the structure of such disks for a wide range of parameters. For the viscosity we use a standard α-description with the main component of the stress tensor $t_{r\phi}$ proportional to the gas pressure. The disk structure depends sensitively on material properties such as opacities and specific heats. In our calculations we have used an equation of state and mean opacities taking into account a list of the most important ionization processes as well as the radiation contributions to thermodynamic quantities.

The most interesting features of our models can be summarized as follows:

• Accretion disks around supermassive black holes are optically very thick due to a rather high surface density ($\Sigma \approx 10^5...10^7 g/cm^2$). This means that the radial inflow velocity is almost everywhere (except a very narrow boundary layer near the inner edge) very small compared with the velocity of sound and thus advection effects are negligible. Furthermore, the timescale for the viscous evolution is very long.

• Since these disks are very massive, selfgravity becomes important for $r \approx 10^3 r_g$ and the disk is subject to a gravitational instability. This might indicate that the outer part of the disk is fragmented and a region of intensive star formation. The wind from such a surrounding stellar torus may considerably contribute to the feeding of the inner accretion region.

• In the inner part of the disk radiation pressure dominates over gas pressure. Nevertheless the dominant energy transport mechanism is not radiative transfer but turbulent heat transport.

Heinrich, O.M. (1994) Astron. Astrophys. 286, pp. 338-343
Falcke, H. and Heinrich, O. M. (1994) Astron. Astrophys. 292, pp. 430-438

STATISTICS OF HI EXTENTS OF IRREGULAR GALAXIES

G. LYLE HOFFMAN
Lafayette College
Easton, PA 18042-1782, USA

Using the $3.'2$ Arecibo beam, we have mapped the HI envelopes of a sample of 70 irregular galaxies (Sdm, Sm, Im and BCD), including 45 from the Virgo Cluster Catalog, 14 Field galaxies at similar redshifts, and 11 Nearby dwarfs chosen to give a complete sample out to 6 Mpc, within the Arecibo declination range. To explore correlations among variables characterizing the size and dynamics of these galaxies, we have combined this sample with all other mappings (multiple single beams and synthesis array) of irregulars and of spiral galaxies spanning the same redshift range (out to about 20 Mpc) available in the literature. In all, there are 128 mapped dwarfs and 119 mapped spirals in this dataset. We obtain the following correlations: $r_{H,max} = (2.75 \pm .15)r_{25}^{(0.890\pm.034)}$ where both HI and optical radii are in kpc; $L_B = (3.59\pm.80)\times10^6 r_{gm}^{(2.70\pm.11)}$ where L_B is in solar luminosities and r_{gm} is the geometric mean of $r_{H,max}$ and r_{25}; $L_B = (125\pm36)V_c^{(3.62\pm.19)}$ where the rotation speed is in km s^{-1}; $M_H = (9.3 \pm 1.1) \times 10^6 r_{gm}^{(1.977\pm.056)}$ where the HI mass is in solar units; and $L_B = (1.9^{+2.7}_{-1.1}) \times 10^{-3} M_{dyn}^{(1.164\pm.043)}$ where $M_{dyn} = V_c^2 r_{gm}/G$.

Arguably the most important result of our survey is the uniqueness of the Giovanelli and Haynes "protogalaxy" HI 1225+01 in any plot of HI radius vs. optical radius or luminosity. DDO 154 and DDO 137 are seen in such plots to be on the tail of a continuous distribution while HI 1225+01 sits alone. Objects with HI radius more than 10 times the optical extent (r_{25}) are evidently quite rare. The scatter for dwarf irregulars is larger than that for spirals in all cases, apparently in part intrinsically so, and increases as the systems become smaller. Nevertheless, it is apparent that the relationship for dwarfs fits smoothly onto the extension of that for spirals with no pronounced dislocation or change in slope. Further implications will be discussed in a forthcoming paper.

A SUPERNOVA REMNANT IN THE DWARF HO IX

ULRICH HOPP
Universitätssternwarte München, D 81679 München, FRG

H.U. ZIMMERMANN
MPI für Extraterrestrische Physik, D 85740 Garching, FRG

MANFRED STICKEL
MPI für Astronomie, D 61117 Heidelberg, FRG

AND

CHRISTIAN HENKEL
MPI für Radioastronomie, D 53121 Bonn, FRG

The dwarf irregular galaxy Ho IX (DDO 66) is a satellite of the giant spiral M81. Triggered by an interaction with M81 (Hopp & Schulte-Ladbeck, 1987 AA 187, 5), Ho IX may have recently undergone a burst of star formation. On very deep Calar Alto 3.5m telescope R images, we detected a ring-like nebula in the northeastern part of Ho IX, which is situated near to the giant CO molecular cloud recently discussed by Henkel et al. (1993 AA 273, L15). The structure of the nebula resembles supernova remnants like CTB 80. No HI-features can be detected at its location, especially no hole. It shows a blue central point source (R=21.7). A Lucy deconvolution of the 0.9" seeing images indicates further faint point sources, perhaps a small stellar cluster. An optical long slit spectrum of the nebula shows emission lines with ratios as in old supernova remnants while it excludes normal HII regions, planetary nebula, or the LMC X-1 type nebula. The position of the central object coincides with a bright and variable ROSAT point source which was already detected by the Einstein spacecraft. The X-ray spectrum is rather steep and shows strong intrinsic absorption. The nature of the X-ray source and its possible connection to the optical nebula is not yet understood. Miller (1995, ApJ 446, L75) who independetly found the nebula interpretated his spectra as a supershell surrounding an OB association. Our optical nebula line ratios severely deviates from Miller's. Our ratios as the X-ray spectrum point more to a SNR at the M81 distance, most probably belonging to Ho IX.

THE COLD GAS CONTENT OF ELLIPTICAL GALAXIES

W.K. HUCHTMEIER
Max-Planck-Institut für Radioastronomie
Auf dem Hügel 69, D-53121 Bonn, Germany

L.J. SAGE
Department of Astronomy, University of Maryland
College Park MD 20742, USA

AND

C. HENKEL
Max-Planck-Institut für Radioastronomie
Auf dem Hügel 69, D-53121 Bonn, Germany

The 100m radiotelescope at Effelsberg has been used to observe two samples of elliptical galaxies in the 21cm line of neutral hydrogen. One sample is defined by the elliptical galaxies in the Revised-Shapely-Ames catalog (RSA) (Huchtmeier 1994, *Astron.Astrophys*286, p.389); the other sample is defined by all elliptical galaxies with IRAS $100\,\mu$ fluxes $\geq 500mJy$ north of declination -31^0 (Huchtmeier, Sage, Henkel 1995 *Astron.Astrophys.* in press). Among the detected galaxies there are 23 (RSA) and 24 (IRAS) isolated elliptical galaxies free of confusion by nearby galaxies with similar radial velocities. Global properties of these two samples of elliptical galaxies are discussed: their HI-properties, optical and IR luminosities, their optical colors, their masses of dust and of molecular hydrogen.

Elliptical galaxies from the RSA and most elliptical galaxies from the IRAS sample have the same mean M_{HI}/L_B ratio : 0.030 ± 0.026; only a small group of objects from the IRAS sample is several times richer in HI ($M_{HI}/L_B = 0.206 \pm 0.105$). These "HI-rich" elliptical galaxies have blue colors like spiral galaxies and have a tendency towards higher average dust temperatures. The large number of elliptical galaxies in compact groups (in this sample) suggests that gravitational interactions and mergers may be an important source of interstellar matter for elliptical galaxies.

THE CLUSTERING OF FAINT GALAXIES:
AND THE EVOLUTION OF $\xi(R)$

J.D. HUDON AND S.J. LILLY
Dept. of Astronomy
60 St. George St.
University of Toronto
Toronto. Ontario
Canada. M5S 1A7

Abstract. The two-point angular correlation function, $\omega(\theta)$, is constructed from a catalog of 13,000 objects in 24 fields distributed over an area of 2 degrees square and complete to a limit of $R = 23.5$. The amplitude and slope of our correlation function on arcminute scales are in broad agreement with recent CCD results in the literature and decreases with depth. No evidence is found for a flattening of the slope of the correlation function away from $\delta \sim 0.8$. Using the redshift distribution from the recent I-band selected Canada-France Redshift Survey, the observed $\omega(\theta)$ implies a value of the clustering length $r_0 = 1.86 \pm 0.42h^{-1}$ Mpc ($q_0 = 0.5$) at $z = 0.48$. This is consistent with the clustering of optically selected local galaxies, if clustering has developed with epoch. Specifically, clustering evolution to a CfA-like sample (with $r_0 = 5.5h^{-1}$ Mpc) would require growth in clustering stronger than the type seen in CDM-like models. Evolution to a less clustered IRAS-like sample (with $r_0 = 4h^{-1}$ Mpc) would require evolution represented by stable clusters of fixed physical size. If there was no growth in clustering (i.e galaxies are fixed in comoving space) then a local sample that is very weakly clustered (with $r_0 = 3h^{-1}$ Mpc) would result. These possibilities are discussed in the context of our understanding of the nature of the faint galaxy population.

THE OVERABUNDANCE OF MAGNESIUM OVER IRON IN BULGES OF SPIRAL GALAXIES

P. JABLONKA
DAEC,Observatoire de Paris-Meudon, F-92195 Meudon

N. ARIMOTO
Institute of Astronomy, University of Tokyo, Mitaka, Japan

AND

P. MARTIN
Steward Observatory, Tucson, Arizona 85721, USA

We have collected integrated light spectra of bulges of 28 spiral galaxies. Our data sample the Hubble sequence uniformly from S0 to Sd types, and cover a large range in magnitude ,viz, about −16 to −22 mag in r-Gunn band. In short, disks do not contribute to more than 14% of the total integrated light in our spectra, and all galaxies are analyzed under the same conditions. More to read in a forthcoming paper.

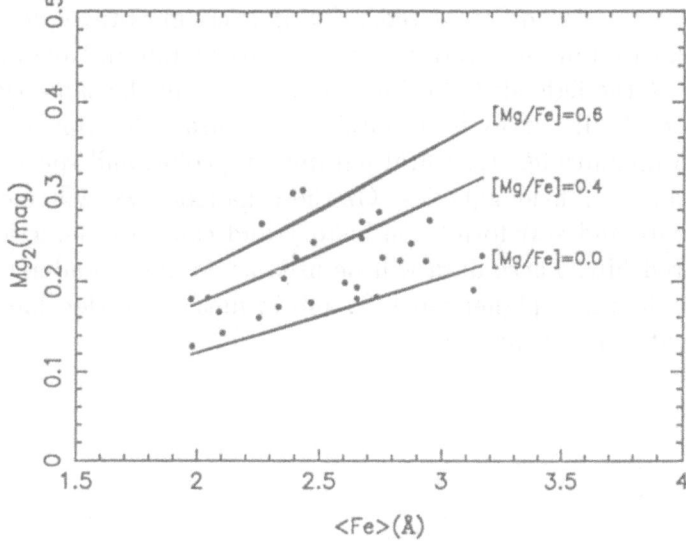

Figure 1. Mg₂ vs the mean of the equivalent widths of the iron lines at 5270Å and 5335Å. The overplotted lines correspond to the predictions of Barbuy et al. model (This volume).

SPECTROSCOPIC AND PHOTOMETRIC NEARBY FIELD GALAXY SURVEY

R.A. JANSEN AND M. FRANX
Kapteyn Astronomical Institute, Landleven 12, P.O.Box 800, NL-9700 AV Groningen, The Netherlands

D.G. FABRICANT
Harvard-Smithsonian Center for Astrophysics, 60 Garden Street, Cambridge MA 02138, USA

AND

N. CALDWELL
F.L. Whipple Observatory, Smithsonian Institution, P.O.Box 97, Amado AZ 85645, USA

Galaxy evolution is one of the key questions in current astronomy. Observations of strong and recent galaxy evolution conflict with previous ideas of orderly and early evolution of galaxies.

The galaxy evolution theories can be tested by comparing the images and spectra of galaxies at different redshifts, *if* account is taken for the biases in the comparison of an "average" nearby galaxy and an "average" distant galaxy and *if* the light distributions are sampled in the same way.

The purpose of our survey is to obtain an accurate description of the distribution of magnitude, structural parameters, color, and spectral type for a large number of field galaxies. Of these galaxies we will measure star formation rate and star formation history and calculate detection rates at increasing redshift. These data will be used as an aid in understanding the spectra of galaxies at higher redshift, and in measuring the changes in star formation rates over time.

ANISOTROPY AND MASS IN ELLIPTICAL GALAXIES

G. JESKE[1], O.E. GERHARD[2], R.P. SAGLIA[3], R. BENDER[3]

[1] *Landessternwarte Heidelberg, Germany*

[2] *Astronomisches Institut, Univ. Basel, Switzerland*

[3] *Institut für Astronomie, München, Germany*

Changes in the anisotropy of the stars and changes in the galaxy potential due to dark matter manifest themselves differently in the observable line (velocity) profile shapes (VPs) of elliptical galaxies (Gerhard 1993, MNRAS 265, 213). In Fig. 1, this is illustrated for a set of realistic spherical models (Jeske *et al.* 1995, preprint).

Radially (tangentially) anisotropic distribution functions lead to more peaked (more flat-topped) VPs than in the isotropic case; i.e., to $h_4 > (h_4)_{iso}$ and $h_4 < (h_4)_{iso}$, respectively. In an inhomogeneous stellar system, an increase in radial (tangential) anisotropy at intrinsic radius r is accompanied by an increase (decrease) of h_4 at projected radius $R \simeq r$. As the mass of the model at large r is increased, at constant anisotropy both the projected dispersion and h_4 increase. Increasing β at constant potential, on the other hand, lowers σ and increases h_4. Thus by modelling σ and h_4 both $M(r)$ and β can in principle be found.

We are currently applying these ideas to several E0 galaxies for which line profile measurements to $\sim 2R_e$ have been obtained by the techniques described in Bender, Saglia & Gerhard 1994 (MNRAS **269**, 785).

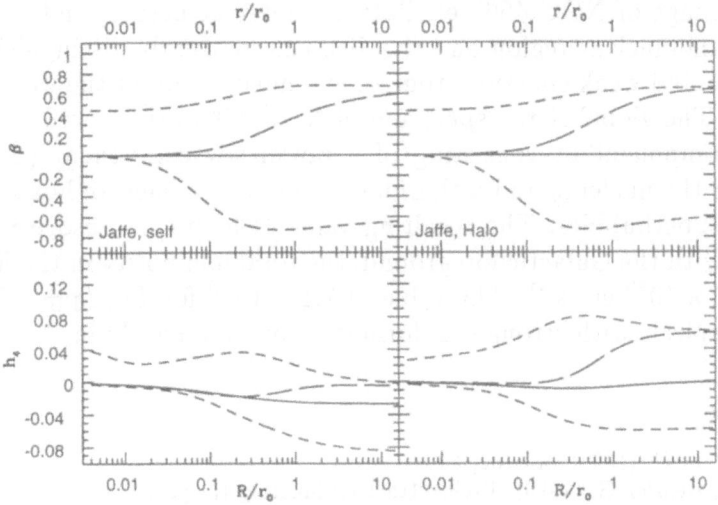

Figure 1. Anisotropy parameter β and VP-parameter h_4 for representative Jaffe models in self-consistent (left) and halo potential (right). Shown are the isotropic model (solid) and several quasi-separable, radially and tangentially anisotropic models (dashed).

NGC 4569: X-RAY OBSERVATION OF A SPIRAL GALAXY WITH NUCLEAR STARBURST ACTIVITY

N. JUNKES, G. HENSLER
Institut für Astronomie und Astrophysik
Universität Kiel, D-24098 Kiel, Germany

We investigate the distribution of soft X-rays and their spectral characteristics for a sample of nearby nuclear starburst galaxies in order to probe their evolution. NGC 4569 is a bright early-type spiral in the Virgo cluster, one of the few blue-shifted galaxies outside the local group. It is gas-deficient in the outer spiral arms, the neutral hydrogen strongly concentrated in the inner region [1]. The bright nucleus, embedded in a normal stellar bulge, is probably the result of a recent star formation episode [4]. Based upon optical spectroscopy of its nucleus [5], the galaxy has been classified as a LINER. The results on NGC 2903 will be presented separately [2].

Our target was observed with the ROSAT PSPC for $\approx$ 18 ks. In contrast to NGC 1808, where the bulk of X-ray emission arises from the nucleus [3], the X-ray image of NGC 4569 exhibits extended structure, and contributions from the nuclear region and the disk can be clearly distinguished. In addition, we find weak emission from an extended region at the west side of the galaxy. The global X-ray spectrum of NGC 4569 can be well described by a two-component fit consisting of a power-law component (predominantly from the nucleus) and a thermal plasma component of hot gas from SNRs and superbubbles. The resulting value of hydrogen absorption is in agreement with the Galactic foreground, the total luminosity in the ROSAT band is $\approx 2 \times 10^{40}$ erg s^{-1}. The value of 1.1×10^{-3} for L_X/L_{FIR} is typical for normal spirals with strong star formation predominantly in the nucleus.

References

[1] Cayatte V. et al. (1990), AJ 100, 604
[2] Junkes N., Hensler G. (1996), Proc. 11th IAP Meeting (in press)
[3] Junkes N., Zinnecker H., Hensler G., Dahlem M. & Pietsch W. (1995), A&A 294, 8
[4] Staufer J.R., Kenney J.D. & Young J.S. (1986), AJ 91, 1286
[5] Willner S.P. et al. (1985), ApJ 299, 443

THE SIGNATURES OF GALACTIC DISK EVOLUTION IN VERTICAL COLOUR PROFILES OF EDGE-ON GALAXIES

A. JUST, B. FUCHS AND R. WIELEN
Astron. Rechen-Inst.
Mönchhofstr. 12-14, D-69120 Heidelberg, Germany

AND

C. SCORZA
Landessternwarte
Königstuhl, D-69117 Heidelberg, Germany

Galactic disks are heated by some gravitational scattering process. The velocity dispersion and scale height increases with the age of the stellar populations as can be observed directly in the solar neighbourhood. This evolutionary effect can also be derived from the vertical structure of spiral galaxies seen edge-on. We use a physical model of a selfgravitating disk composed of isothermal subpopulations with increasing age and velocity dispersion and an exponential dust distribution (Just et al. 1995). The emissivity of the stars is computed from the luminosities in the different bands of single age populations which are computed with the method of photometric evolutionary synthesis (Einsel et al. 1995).

We apply these models to profiles of the edge-on galaxy IC 2531 in U,B,V,R,I,K bands and on CCD data of NGC 891 in V,I, and K. We find that the general features of the luminosity and colour profiles are well reproduced by the models. For IC 2531 we find the same stellar composition in age and metallicity as in the solar cylinder but with a different heating rate. In NGC 891 the central blue disk is much stronger consistent with a constant SFR instead of a strongly decreasing one.

References

Einsel, C., Fritze-v. Alvensleben, U., Krüger, H., and Fricke, K.J. (1995) *A&A* **296**, 347
Just, A., Fuchs, B., and Wielen, R. (1995) submitted to *A&A*

ABSORPTION FEATURES OF "L$_\alpha$ DAMPED" TYPE IN SPECTRA OF VERY DISTANT QSO AS AN INDICATOR OF YOUNG CLUSTERS OF GALAXIES

B.V.KOMBERG
Astro Space Center of P.N.Lebedev Physical Institute
117810 Profsoyuznaya 84/32, Moscow, Russia

It is shown that at present one can not exclude a possibility that rare, wide and deep absorption feature of L_α damped type could be formed in a rich gas medium of distant clusters of galaxies, not verialized yet. In future massive galaxies are likely to be formed in the central regions of these clusters. Nuclei of these galaxies become very active during a short time, i.e. they pass a stage of QSO. Narrower absorption lines of metals and/or 21 cm associated with L_α damped systems can be formed in the clouds belonging to separate galaxies of these clusters.

On the basis of the process described above one can make some observational conclusions concerning characteristics of absorption lines of "L_α-damped" type:

1) scales of absorbing region can reach hundreds of kiloparsecs;

2) "L_α-damped" features can not be observed at low Z, because of strengthening of X-ray radiation emitted by hot gas in the central regions of near rich clusters.

3) velocities of metal absorption lines associated with "L_α-damped" features can differ up to hundreds of km/s from Z_{damp};

4) widths of "L_α-damped" absorption lines can reach hundreds of km/s and correspond to the dispersion of velocities in young systems of galaxies;

5) under assumption that $T_s^{gas} \gg T_{background} = 2.7(1 + Z_{damp})$, gas in a young system of galaxies can emit a flux about $1\,mJy/\Box^o$ in 21 cm line with $\lambda = 21$ cm $(1 + Z_{damp})$.

SELF-REGULATED STAR-FORMATION IN GALAXIES

J.KÖPPEN, CH.THEIS AND G.HENSLER
Institut für Astronomie und Astrophysik,
Universität Kiel, Germany

In the chemodynamical models of galaxies the energy input from massive stars into the ambient medium results in a self-regulation of the star formation rate (SFR). A thorough analytical and numerical study of this model shows that there is always a strong and negative feed-back, and the SFR becomes almost independent of the assumed stellar birth function (SBF). The time-scale to reach this equilibrium is much shorter than the gas consumption time-scale, hence the models evolve along this solution for most of the time. This mechanism provides a physical explanation for a quadratic dependence of the SFR on gas density. For more details cf. Köppen et al. (1995), A&A **296**, 99 and in preparation.

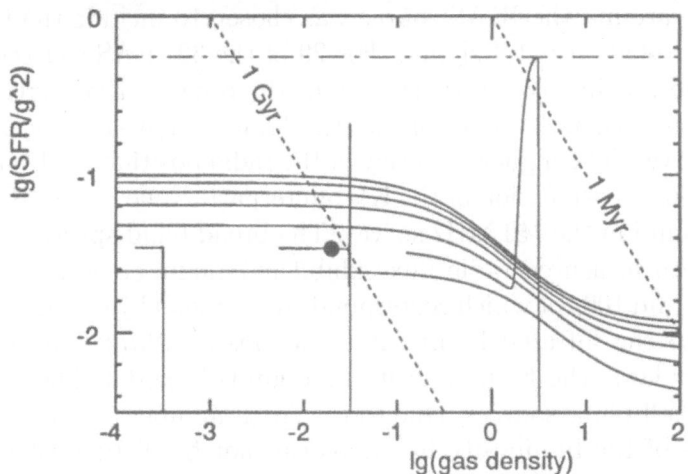

Figure 1. The ratio SFR/g^2 of the star-formation rate and gas density as a function of the gas density g (in $M_\odot/pc^3$): shown are the analytical equilibrium solutions for various values of the constant of proportionality in the SBF (log C = 9, 7, ... , −1 from top to bottom). Also depicted are results from three numerical models with initial gas densities of log g = 0.5, −1.5, and −3.5. Lines of equal star-formation timescale are short-dashed. The filled dot marks the equilibrium for our 'standard' value C = 0.55 at the density of the local interstellar medium (n_{gas} = 1 cm^{-3}).

IDENTIFICATIONS OF FAINT IRAS SOURCES

M.W. KÜMMEL, S.J. WAGNER
Landessternwarte Heidelberg
Königsstuhl 12
69117 Heidelberg
Germany

From overlapping scans in the IRAS all-sky survey and additional pointed observations the deepest far infrared survey before ISO exists in the region around the North Ecliptic Pole (NEP) (Hacking P. and Houck J.R., ApJS **63** p. 311). This survey contains detections up to 10 and fluxes up to 100 times fainter than the IRAS survey. In the central square degree around the NEP we combine the far IR-survey with deep radio data at 151 MHz and 1.5 GHz (Visser, A.E. et al., A&AS **110** p. 419, Kollgaard, R.I. et al., ApJS **93** p. 145) and own observation at $2.2\mu m$ (K') and $435nm$ (B). The error circle around the IRAS source was chosen to include the true source with 85% probability (1.4 sigma). For 29 of the 32 IRAS sources we found at least one possible counterpart. Ten of the objects have multiple (up to four) counterparts in K'. Four of the IRAS sources have counterparts in the 1.5 GHz survey. The higher accuracy of the radio position ($\sim 1''$) allowed an unambiguous identification of the K' counterpart. None of the IRAS sources could be found in the 151 MHz survey. The broad band spectra of the three galaxies with measured radio flux exhibit maximum emission between the radio band and $100\mu m$ which corresponds to emission by cool dust ($< 50\,\mathrm{K}$). Contrary to the infrared luminosity functions at $12\mu m$ and $60\mu m$ which show power laws, the K' luminosity function is bimodal. The brightest K' objects are all point sources. Due to the small number statistics the power law indices of the luminosity functions can not be distinguished. We find a linear relationship between the K' flux and the flux at $60\mu m$ and $12\mu m$ over at least one decade. The large deviations by individual sources make an identification of the correct counterpart through this relation impossible. The spectral energy distributions of unambiguously identified sources span only one decade in energy (νS_ν), i.e. they have flat energy distributions. This suggests an identification of K' objects with flat energy distribution in case of multiple counterparts.

THE TILT OF THE FUNDAMENTAL PLANE OF ELLIPTICAL GALAXIES: DYNAMICAL AND STRUCTURAL EFFECTS

B. LANZONI
Dipartimento di Astronomia, Università di Bologna
Via Zamboni 33, 40126 Bologna, Italy

L. CIOTTI
Osservatorio Astronomico di Bologna
Via Zamboni 33, 40126 Bologna, Italy

AND

A. RENZINI
European Southern Observatory,
Garching b. München, Germany

We explore several structural and dynamical effects on the projected velocity dispersion as possible causes of the fundamental plane (FP) tilt of elliptical galaxies (Ciotti, Lanzoni & Renzini, 1995). Specifically, we determine the size of the systematic trend along the FP in the orbital radial anisotropy, in the dark matter (DM) content and distribution relative to the bright matter, and in the shape of the light profile that would be needed to produce the tilt, under the assumption of a constant stellar mass to light ratio. Spherical, non rotating, two–components models are constructed, where the light profiles resemble the $R^{1/4}$ law. For these we can exclude orbital anisotropy as the origin of the tilt, while a systematic increase in the DM content and/or concentration may formally produce it. Also a suitable variation of the light profile can produce the desired effect, and there may be some observational hints supporting this possibility. However, fine tuning is always required in order to reproduce the tilt, while preserving the tightness of the galaxies distribution about the FP.

References

Ciotti L., Lanzoni B., and Renzini A., 1995, submitted to *M.N.R.A.S.*

EVIDENCE FOR A SIZABLE AGE SPREAD AMONG GALAXIES FROM THE UV UPTURN PHENOMENON IN EARLY-TYPE SYSTEMS

YOUNG-WOOK LEE
Yonsei University Observatory, Shinchon 134, Seoul 120-749, Korea

JANG-HYUN PARK
Korea Astronomy Observatory, Taejeon 305-348, Korea

The suggestion of Lee (1994, ApJ, 430, L113) that the age spread among galaxies is responsible for the systematic variation of ultraviolet (UV) upturn among the early-type systems is confirmed here with the detailed population synthesis models. Our models suggest that the far-UV spectra of these systems are composite of hot metal-poor horizontal-branch (HB) stars (and their post-HB progeny) and metal-rich post-asymptotic giant-branch (PAGB) stars. The systematic variation of UV upturn, however, depends on the contribution from hot metal-poor HB stars and their progeny, which in turn depends on the ages of oldest stellar populations in galaxies. Consequently, our models predict that the strength of absorption features, such as C IV and Si IV, is anticorrelated with the strength of UV upturn. This is consistent with the far-UV spectra for NGC1399 and M31 obtained by the Hopkins Ultraviolet Telescope aboard the Astro-1 space shuttle mission. We note that the opposite trend is expected in other's models that favor metal-rich solution for the UV upturn phenomenon without age spread among galaxies. Our result implies a prolonged epoch of galaxy formation, in the sense that more massive galaxies (in denser environments) formed first. With the assumption that the UV upturn phenomenon is solely due to the age variations among galaxies, we tentatively estimate the difference in age between the giant ellipticals and our Galaxy to be 4 billion years or more. This suggests that the best estimate for the lower limit of the age of the Universe is close to 20 Gyrs, which of course would be in conflict with the current estimate of the H_0 together with the standard cosmological models with zero cosmological constant. The reader is referred to Park & Lee (1995, ApJL, in press) for details of this work.

SIMULATION OF MASS TRANSPORT IN DISK GALAXIES

S. VON LINDEN AND J. HEIDT
Landessternwarte, Königstuhl, D-69117 Heidelberg

AND

H.P. REUTER AND R. WIELEBINSKI
MPI für Radioastronomie, Auf dem Hügel 69, D-53121 Bonn

The large-scale dynamics and evolution of disk galaxies is controlled by the angular-momentum transport provided by non-axisymmetric pertur-bances through their gravity torques. To continuously maintain such gravitational instabilities, the presence of the gas component and its dissipative character are essential.

By using 2D N-body simulations of a self-gravitating disk, composed of stars and gas, we investigate quantitatively the efficiency of the mass and angular-momentum transport and its relevance for a large-scale mass accretion rate in disk galaxies. Clouds are simulated by spheres of constant density and they have masses between 10^4 und 10^7 $M_\odot$. The hydrodynamics is based on the cloud-cloud collision model of Combes & Gerin (1985, *A&A*, **150**, 327-338) and Casoli & Combes (1982, *A&A*, **110**, 287-294)

We verify this theoretical considerations with observations. We present as an example NGC 7331. Using optical and radio observations we compare and discuss the secular redistribution of matter in the disk of this galaxy. The CO observations show an extended, torus-like accumulation of molecular gas at a radius of 4-8 kpc. The central 4 kpc of the galaxy is partly depleated from molecular gas. The position-velocity cuts along the major axis hints at the presence of an inner disk/ring structure. The optical I-band image shows deformation of the isophotes in the inner 10 arcsec which may be due to a stellar bar.

The simulation reproduce a small nuclear bar as well as the locations of the ring-like enhancement of molecular material around the turnover points. We are able to describe a possible development of the stellar and molecular disk.

We thank F. Combes for providing us her numerical code and for helpful discussions about the code.

FIELD GALAXY EVOLUTION STUDIES WITH AN OPTICAL MULTICOLOR DEEP-SKY SURVEY

Charles T. Liu (U. of Arizona), Richard F. Green (NOAO),
Patrick B. Hall (U. of Arizona) and Patrick S. Osmer (Ohio State U.)

We are investigating field galaxy evolution with the optical multiband survey of Osmer et al. (1995, in preparation; Hall et al. 1992, Bull. A.A.S., 24, 1136), which covers some 3000 sq. arcmin of sky with CCD photometry in six broad-band filters from 3000-10000Å, complete down to R~23. The sample contains some 9000 cataloged galaxies, of which 7000 have data in at least three colors.

Our approach has been to identify both the star formation rate (SFR) and redshift of the field galaxies in the survey from their multicolors. Model galaxy colors in the U/B/V/R/I(7500Å)/I(8600Å) passbands were produced using spectral energy distributions of typical E/S0, Sbc, Scd and Irr galaxies created with the spectrophotometric data of Kennicutt (1992, ApJS, 79, 255) and Coleman et al. (1980, ApJS, 43, 393). In the spirit of Koo (1985, AJ, 90, 418) and others, it is possible to distinguish the spectroscopic Hubble types and redshifts of each galaxy type using different projections in multi-dimensional color space (Fig. 1). Since our I-band filters are relatively narrow (1000Å FWHM), we have additional leverage for high-redshift objects near z~1.

We have applied these models to a complete subsample of 438 red galaxies from the sample. Red galaxies were selected for several reasons: **(1)** the 4000A break is generally strong, aiding in more accurate redshift determinations; **(2)** degeneracies caused by such factors as reddening are less of an obstacle; **(3)** luminosity evolution in the red galaxy population can be well quantified using no-evolution models such as these. Our preliminary analysis appears to confirm the effectiveness of our technique for redshift identifications of these data. With the overall galaxy sample, we hope to compute luminosity functions of faint field galaxies with respect to magnitude, redshift and spectral type. With that information, we will estimate the total SFR in field galaxies with respect to cosmic time and the fractional SFR contributions from galaxy sub-types.

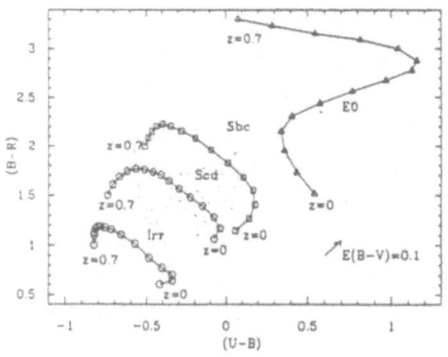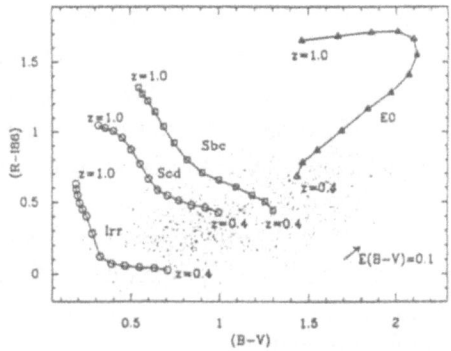

Fig. 1 Slices in multicolor space for blue *(left)* and red *(right)* colors in the Osmer et al. (1995) dataset, showing the color-vs.-redshift evolutionary tracks for typical E0, Sbc, Scd and Irr galaxies. A representative subsample of galaxies in the dataset are superposed.

NEAR-INFRARED SPECTROSCOPY OF NGC 253:
STARBURST AND SURROUNDINGS

D.LUTZ AND F.PRADA
Max-Planck-Institut für extraterrestrische Physik
Postfach 1603, 85740 Garching, Germany

Near-infrared longslit spectra of NGC 253 obtained with IRSPEC at the ESO NTT are presented. By analysis of the ^{12}CO 2.29μm bandhead we find that the stellar population in the central starburst region (r ~ 150 pc) rotates more slowly than the gas, but has a velocity dispersion of 128 km/s, about twice the value found for emission lines from the gas in this region. This implies an about five times higher dynamical mass than previously derived (Rieke et al. 1980), removing the need to invoke a lower mass cutoff in the starburst initial mass function. The peak of near-infrared emission is displaced from the dynamical center.

We discuss extinction values derived from line ratios of [Fe II], H, and H_2. While part of the differences can be explained by the wavelength dependency of derived extinction in case of mixed emitters and absorbers, there are also significant differences between the extinction for H and H_2. This requires differences in local extinction and calls for caution in the interpretation of H vs. H_2 line ratios in other starburst galaxies.

Along the major axis of the galaxy, the H and [FeII] emission are similarly distributed and trace the starburst region of radius about 150 pc. The H_2 emission is more extended, reaching out to at least 300 pc. This halo of H_2 emission which can be interpreted in terms shocks, which would have to be slow to avoid excitation of [FeII], or of a huge photon dominated region. The required properties of such a PDR - dense molecular clouds in a hot thin medium are consistent with properties inferred from Far-infrared observations (Carral et al. 1994). In none of the near-infrared lines do we find evidence for linesplits due to the NGC 253 superwind.

Carral,P., et al. 1994, *Astrophys. J.* **423**,223
Rieke,G.H., et al. 1980, *Astrophys. J.* **238**,40

A LINKUP BETWEEN SEYFERT'S AND THEIR SURROUNDINGS

A.P. MAHTESSIAN[1], E.YE. KHACHIKIAN[1] AND H. TIERSCH[2]
[1] *Byurakan Astrophysical Observatory, 378433 Byurakan, Armenia*
[2] *Potsdam University, 14482 Potsdam, An der Sternwarte 16, Germany*

The observed physical characteristics of galaxies are (at least partly) determined by the influence of the surroundings after the formation of the galaxies. The examination of the frequency of the occurence of active galaxies, i.e. Seyfert galaxies, in different galaxy systems compared to a sample of field galaxies gives possibly an insight into this problem.

The galaxy groups were taken from the CfA redshift survey, the list of them is published in Mahtessian (1992, Soob. Byurakan. Obs. 65).

The investigated correlations are:

1. *The occurence of Seyfert's in groups with different number of members:* In the sequence going from single galaxies (i.e. "groups" with one member) to member-rich groups the relative number of Sy1 and Sy2 does not change, but the Sy3 are more often in galaxy groups than among the single galaxies.

2. *The correlation between the morphological content of a group and the occurence of Seyfert's:* The galaxy groups with and without Seyfert's do not distinguish from each other, i.e. the morphological content of a group has obviously no influence on the existence of Seyfert's within the groups.

3. *The correlation between the density of galaxies in the groups and the occurence of Seyfert's:* As a parameter of the density the mean pairwise distance between the galaxies is used. We could not find that the number of Seyfert's depends on the density of the galaxy groups.

4. *The correlation between the velocity dispersion within galaxy groups and the occurence of Seyfert's:* The groups with Seyfert's have in the mean a larger velocity dispersion than the groups without Seyfert's. This difference is statistically more significant for groups with Sy1 and Sy2.

5. *The correlation between the crossing time τ of the galaxy group and the availability of the Seyfert's:* Groups with Seyfert's have in the mean a smaller crossing time than groups without Seyfert's.

A MODULAR TOOL FOR THE EVOLUTIONARY SYNTHESIS OF STELLAR POPULATIONS

CLAUDIA MARASTON
Dipartimento di Astronomia, Universitdi Bologna
Via Zamboni 33, 40126 Bologna, Italy

An innovative tool for the construction of **Evolutionary Synthesis models of Stellar Populations** is presented. It is based on three independent matrices giving respectively 1) the fuel consumption during each evolutionary phase as a function of stellar mass, 2) the typical temperatures and gravities during such phases, and 3) colors and bolometric corrections as a function of gravity and temperature. The first matrix allows to calculate the relative contribution of each phase to the bolometric light of the population, thanks to the so-called **Fuel Consumption Theorem** (Renzini & Buzzoni 1986, in *Spectral Evolution of Galaxies*, ed. C. Chiosi and A. Renzini (Dordrecht:Reidel),p.195)). The other two matrices allow to distribute such contributions in the various passbands, and thus to compute the synthetic colors of the population as a function of age. The modular structure of the code allows to easily assess the impact on the synthetic spectral energy distribution of various model ingredients (e.g. uncertainties in the stellar evolutionary models, mixing length, temperature distribution of the horizontal branch (HB) stars, AGB mass loss, color-temperature transformations, etc.) Models have been computed with Y=0.27, Z=0.02 and for three choices of a double-slope IMF ($s = 1 + x = 2.35, 3.5, 1.5$ for $M \geq 0.6 M_\odot$ and $s' = s - 2.5$ for M $< 0.6 M_\odot$). Among the main results, it is found that: i) the appeerence of an extended AGB has a strong impact on the integrated $(V - K)_0$ color, which increases by $\simeq 1.5$ mag in $\Delta t \simeq 400$ Myr; ii) the inclusion of a blue-HB developing at ages around 10 Gyr produces a hook-shape in the $(U - B)_0$ vs $(B - V)_0$ locus described by aging stellar populations, in agreement with what observed for Magellanic Clouds Clusters.

A detailed description of the variuos results will be given in a forthcoming paper.

ENVIRONMENTAL EFFECTS ON THE PROPERTIES OF SPIRAL GALAXIES: ISOLATED PAIRS OF SPIRALS

I. MÁRQUEZ

Institut d'Astrophysique de Paris

98 bis, Bd Arago, 75014 Paris (France)

We characterized **isolated spiral galaxies** as a reference for the properties of interacting ones: we selected all the spirals in CfA catalog with $m_B \leq 13.0, \delta \geq 0$, a$\leq$ 4' and $73^o > i > 32^o$. The isolated ones were those having no companions in $\Delta D \leq 0.5$Mpc and $\Delta V \leq 500$ km/s and no satellites galaxies in their neighborhood (from Nilson catalog and Palomar charts). To select the **isolated pairs of spirals** we used Karachentsev's catalog applying similar conditions for m_B, δ, a, i and the same isolation criteria. We used broad band CCD images (Johnson B,V e I), narrow band CCD images ($H\alpha$) and long slit spectra in $H\alpha$ region. The BVI images analysis consisted on applying sharp-divided methods, obtaining simulated images (from bulge/disk decomposition and galaxy orientation in the sky) and Fourier analysis. From $H\alpha$ images we determined total emission, size, distribution and flux of HII regions. We also obtained the rotation curves.

Isolated Galaxies: they present uniform disks, smooth color gradients and flat rotation curves. From the Principal Component Analysis applied to the analyzed properties we find that two eigen-vectors explain the 95%: (1) Mass, luminosity, size or specific angular momentum; (2) B/D or G (inner gradient of the rotation curve).

Isolated Pairs: they show distorted morphologies (warps, X isophotes, prominent outer rings, tails, plumes, bridges). Interacting spirals are more luminous, with higher surface brightness, bluer total colors (greater dispersions) with stronger color gradients and redder central colors. Star forming processes are more efficient. Some of the interacting galaxies show **TYPE II PROFILES** (but not even an isolated one). Some of the interacting galaxies show declining rotation curves. These results are explained as due to the mass redistribution provoked by the interaction, which implies local inestabilities and mass transport to the center (**Márquez, I., 1994, PhD Thesis**, Universidad de Granada, Spain, and references therein).

PROBING THE HALO OF CENTAURUS A:
A MERGER DYNAMICAL MODEL FOR THE PN POPULATION

A. MATHIEU AND H. DEJONGHE
Sterrenkundig Observatorium, Krijgslaan 281, Gent, Belgium

AND

X. HUI
Astronomy Department, Boston University, Boston, MA 02215

We use planetary nebulae observations (Hui *et al.* 1995) to build dynamical models of the dust-lane elliptical galaxy NGC 5128 (Centaurus A). The PN photometric and kinematical data extend out to 20 kpc ($\sim 4r_e$) along the major axis and 10 kpc along the minor axis. Our models are built using a Quadratic Programming technique (Dejonghe 1989). The method produces fits to the data set, which consists of the photometry field (E2, well fitted by a $r^{1/4}$-law) together with the major- and minor- axis rotation curves and velocity dispersion profiles. Assuming the merger hypothesis for Cen A, we describe its kinematics in a spherical potential by two sub-systems, one rotating about the intrinsic short axis and the other about the intrinsic long axis of the galaxy.

We show that no self-consistent model can match both photometry and kinematics of Cen A; the model fails to reproduce the high values of the major axis velocity dispersion at large radii, clearly indicating the presence of a dark halo. On the other hand, models including a dark halo can produce satisfactory fits to the complete data set. Our best fit model consists of 50% of dark matter for a total mass of $4 \times 10^{11} M_\odot$. The mass-to-light ratio increases from 5 at 5 kpc to 12 at 50 kpc. Different dark matter halos are compatible with our data set and the corresponding total masses interior to 50 kpc range from $3 \times 10^{11} M_\odot$ to $6 \times 10^{11} M_\odot$. For our QP best fit model, 75% of stars are rotating about the short axis and 25% about the long axis.

References

Dejonghe, H. 1989, ApJ, 343, 113
Hui, X., Ford, H.C., Freeman, K.C. & Dopita, M.A. 1995, ApJ in press

X-RAY OBSERVATIONS OF ELLIPTICAL GALAXIES BY ASCA SATELLITE

K. MATSUSHITA AND K. MAKISHIMA
Department of Physics, University of Tokyo
7-3-1,Hongo,Bunkyo-ku,Tokyo,Japan

Using ASCA, we have confirmed that the ISM of X-ray bright elliptical galaxies are surprisingly metal poor, as compared to the theoretical predictions. In fact the exact values of the derived metallicity depend considerably on the plasma emission codes. However ,the overall metallicity cannot be larger than ~ 1 solar. For low L_X/L_B galaxies, all the available plasma codes suggest abundances less than half a solar. The ASCA spectra may be compatible with somewhat higher metallicity if we assume there is an additional low-temperature component (e.g. $kTe \sim 0.3$ keV). However, the derived abundance can not be over 1 solar. In particular, the Si abundance turns out to be < 1.5 solar, confirming the metal-poor nature of the ISM. These ASCA results are in severe contradiction with most of the SN Ia rate, particularly that of Tammann (1982). Considering further that a fairly long time (10^{9-10}yr) is needed for the stellar mass loss to accumulates into the ISM, it is suggested that the SN Ia rate has remained quite low throughout Hubble time.

Are there plausible mechanisms that reduce the apparent metallicity? Hiding heavy elements in dust is one possibility. However the dust sputtering time is rather short. Dilution of the ISM by the ICM would do the job, but it is inconsistent with the ASCA data since relatively isolated ellipticals also exhibit low abundances. Finally, presence of an unknown continuum source would produce a similar effect, but such a component is not seen in the lowest L_X/L_B galaxies (Matsushita et al. 1994).

In conclusion, the ISM abundance remains a big puzzle that must be challenged by future investigations, from both theory and observation.

References

Matsushita, et al. (1994), ApJL, **436**, L41

ANISOTROPIC DISTRIBUTION FUNCTIONS FOR THE ELLIPTICAL GALAXY NGC 1600

MICHAEL MATTHIAS AND ORTWIN GERHARD
Astronomisches Institut der Universität Basel
Venusstrasse 7, CH-4102 Binningen, Switzerland

Three-integral (3I) dynamical models for NGC 1600 were constructed as follows: (i) Lucy-inversion of CCD photometry and gravitational potential as in Binney, Davies, Illingworth (ApJ 361, 78, 1990), assuming axisymmetry. (ii) Third integral by perturbation theory as in Gerhard & Saha (MN 261, 311, 1991). (iii) Two- and three-integral distribution functions as in Dehnen & Gerhard (MN 261, 311, 1993), assuming various anisotropy patterns. The kinematic results from these models are presented in Fig. 1. The best-fitting 3I model (solid line, right panels) has outward-increasing radial anisotropy on the major axis and is nearly isotropic on the minor axis. The M/L of the various 3I-models varies only slightly around M/L=6.2.

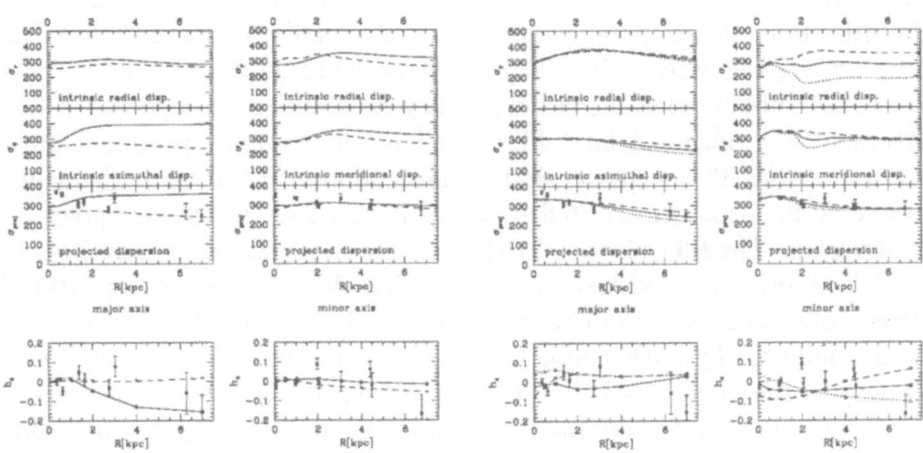

Figure 1. Kinematics of NGC 1600 with models superposed. Observed kinematics σ_{proj} and h_4 are taken from Bender, Saglia & Gerhard (MN 269, 785, 1994). Left two panels show two-integral models $f(E, L_z)$ (solid line) and $f(E, S_m)$ (see Dehnen & Gerhard 1993). Right two panels show some three-integral models. Top to bottom: intrinsic velocity disperions σ_r, σ_ϕ (major axis) resp. σ_θ (minor axis), observable quantities σ_{proj}, h_4.

OPTICAL AND IR PROPERTIES OF RADIO GALAXIES AS A FUNCTION OF THEIR RADIO POWER

L. MAXFIELD, S.G. DJORGOVSKI, D. THOMPSON

M.A. PAHRE, R.R. DE CARVALHO
Palomar Observatory, Caltech, Pasadena, CA 91125, USA

AND

M. VIGOTTI, G. GRUEFF
Istituto di Radioastronomia C.N.R., Bologna I-40100, Italy

We compare optical and infrared photometric and spectroscopic properties of high-redshift radio galaxies from the 3CR and B3 surveys. At a given redshift and a fixed restframe frequency, the two samples differ on average by an order of magnitude in radio power, thus providing a fair baseline in radio powerfor a range of redshifts. We present new optical and IR photometry and spectrosopy for a number of B3 sources. We combine these data with the existing corresponding information on B3 and 3CR sources, in order to explore different correlations of source properties with redshift, and among themselves. B3 sources follow the same trend as 3CR's in the K band Hubble diagram, although they do seem to be slightly fainter on average at a given redshift. This trend is slightly more prominent in the Gunn r band. This suggests that some fraction of the observed light in the r and K bands is contributed by an active nucleus, which also powers the radio lobes. The B3's also tend to have lower emission line luminosities than 3CR's at any given redshift, suggesting that there may be a correlation between line luminosity and radio power. Such a correlation is clearly seen and is followed by both samples. It suggests that the UV emission lines are largely powered by the active nucleus, ostensibly a hidden quasar, which is also responsible for the radio emission. We also examine the behavior of the optical and radio PA alignments for the combined B3+3CR data set. We find that high-power and high-redshift subsamples for both B3's and 3CR's show the alignments more prominently, but we still cannot tell which of these variables dominates this effect. This work was supported in part by the NSF PYI award AST-9157412, and the Bressler Foundation.

SPATIAL DISTRIBUTION OF FAR INFRARED AND RADIO CONTINUUM EMISSION IN SPIRAL GALAXIES

Y.D. MAYYA AND T.N. RENGARAJAN
Tata Institute of Fundamental Research, Colaba, Bombay-5, INDIA

A study of 8 nearby spiral galaxies (NGC 2903, 3079, 3198, 3628, 4303, 4321, 4656 and 6946) is carried out using the radio continuum (RC) and far infrared (FIR) images at 1′ resolution. These images are used to study the radial gradients in the ratios of FIR to RC (Q_{60} and Q_{100}), warm dust temperature ($Td(60/100)$) *etc.* The main results are illustrated with NGC 2903 as an example in Fig. 1, where azimuthally averaged quantities are plotted. $Td(60/100)$ decreases away from the center (45–25 K), increasing again by ~ 5 K in outer galaxies. Typically Q_{60} decreases by a factor of three away from the center in a given galaxy, but has an order of magnitude spread in the pixel values over all the galaxies. In contrast, Q_{100} shows flatter gradient, which is expected from the observed temperature gradient. 20 cm RC emission profile is also shown in Fig. 1. The RC and FIR profiles can be fitted by a combination of central gaussian and exponential disk components. In general RC and FIR have about the same fraction of exponential component with the exception of NGC 3628, in which the FIR is dominated by the gaussian while the RC is mostly disk component (see Fig. 2). In 5 of the remaining 7 galaxies, the exponential component contributes $> 50\%$ of the total. In general RC scale lengths are larger than the FIR.

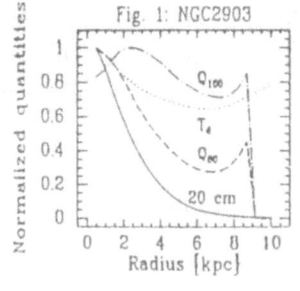 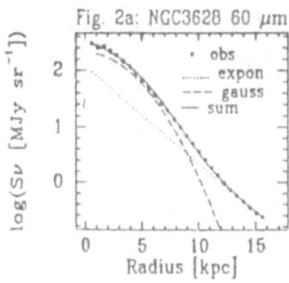 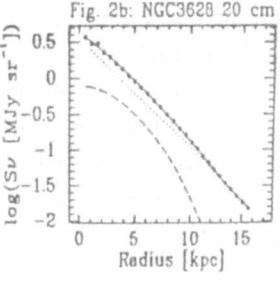

ARP 142:

Another Interacting Galaxy with Very Large Internal Motions?

C.M. MCCAIN, K.C. FREEMAN AND P.J.QUINN
Mt. Stromlo & Siding Spring Observatories
Private Bag, Weston Creek PO, Weston, ACT 2611, Australia

1. Introduction

We present our study of Arp 142, an interacting system which consists of irregular (NGC 2936) and spheroidal (NGC 2937) components. We discovered that NGC 2936 has very high internal motions, much higher than what is expected from the dynamics of interaction of such galaxies.

2. Discussion

NGC 2936 has a long dust lane extending towards its companion, however in K-band images the dust lane disappears and the chaotic irregular looks more like a double-armed face-on distorted spiral. Still, one can see matter being stripped off from the spiral due to its interaction with NGC 2937.

Velocities calculated from the $H\alpha$ and [NII] emission lines give a total internal velocity of about 1000 kms^{-1} contained inside the chaotic component. This velocity spread is much larger than one might expect from the dynamics of interaction of such galaxies. Our K-band magnitude and the Tully-Fisher law indicate that the expected internal velocity spread for NGC 2936 should be $\lesssim 500$ kms^{-1}.

Hardly any $H\alpha$ and [NII] emission is seen from the nucleus of NGC 2936 but outside about 5″ (2.5 kpc), the emission lines are double- or triple-peaked with velocity differences of up to 400 kms^{-1}, and with unequal strengths. This may suggest the existence of an ionisation cone around the nucleus, although the [NII]λ6583/$H\alpha$ ratio does not indicate that shocks are important here, which is usually the case for these cones. The velocity contours averaged over the multiple peaks are like those of a distorted but rapidly rotating disk.

CFM is grateful for the travel support from the Astronomical Society of Australia.

SEARCHING THE CONTINUUM FOR PRIMEVAL GALAXIES

MARSHALL L. MCCALL AND MICHAEL M. DEROBERTIS

York University, Department of Physics and Astronomy
4700 Keele St., North York, Ontario, Canada M3J 1P3

A population of primeval galaxies (PG's) should be detectable by directly imaging with two intermediate-band filters tuned to either side of the Lyman break (DeRobertis, M. M., and McCall, M. L. 1995, A.J., **109**, 1947). In the figure below, the solid and short-dashed curves show the flux (left scale) as a function of redshift from a PG 0.7 Gyr old with a total stellar mass of $5 \times 10^{10} \mathcal{M}_\odot$ as seen through filters with rest-frame passbands 890 ± 30 Å ('β') and 1010 ± 30 Å ('ρ'), respectively, moved to redshift 5. The upper curves depict the colour $\beta - \rho$ (right scale); the dotted line is for the 0.7 Gyr population, and the dot-dashed line is for a 7.5 Gyr model. A source can be identified as a PG if it can be clearly detected in the ρ filter *and* if it has a colour greater than +0.75 mag. Confusion with any old stellar systems at lower redshifts can be eliminated by supplementing observations with Gunn r and i. The colour condition selects Lyman break objects between redshifts 4.7 and 5.4, a range over an order of magnitude greater than is achievable through an emission line survey. The discriminatory power of the technique is not affected by internal dust.

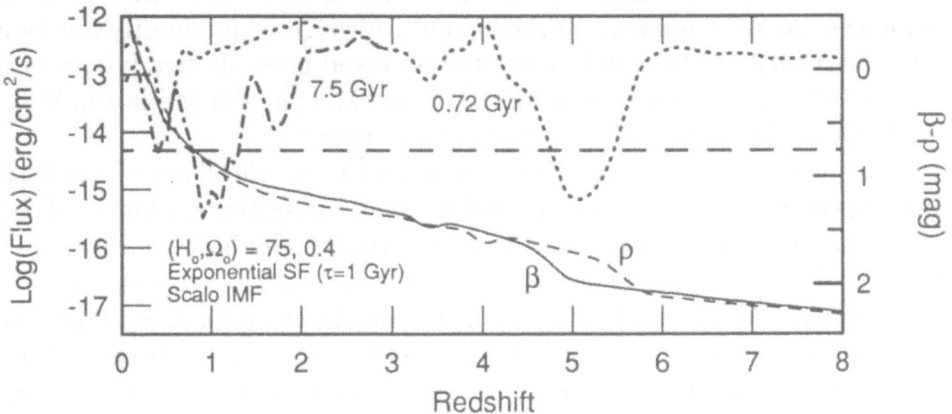

A MODEL FOR THE GLOBULAR CLUSTER LUMINOSITY FUNCTION

DEAN E. MCLAUGHLIN AND RALPH E. PUDRITZ

Dept. of Physics and Astronomy, McMaster University

Hamilton, Ontario L8S 4M1 Canada

We develop the idea (Harris & Pudritz 1994, ApJ, 429, 177) that, like currently forming star clusters and associations, globular clusters (mean mass $\simeq 3 \times 10^5 M_\odot$) were born in the 'cores' of much larger ($\sim 10^8 - 10^9 M_\odot$) star-forming complexes which we call 'supergiant molecular clouds,' or SGMCs. The number $N(m)$ of protoclusters at mass m is then determined by a steady-state balance between their growth by core–core collisions, and their self-destruction via the side effects of star formation. This mass spectrum is ultimately passed on to the globular cluster system (GCS) itself, by virtue of the very high star-formation efficiency required to produce a bound stellar cluster from a gaseous core.

The major influence on the shape of the GCS mass spectrum is the ratio β of fiducial core disruption and collision timescales. Our models are further characterized by a *mass-dependent* core lifetime: below a critical mass m_*, star formation is too passive to disrupt a core; but above this limit, cores will self-destruct in a finite amount of time. We identify m_* with the peak magnitude of the globular cluster luminosity function $[\phi \sim m N(m)]$. Its value and the peak mass m_1 of the luminosity-weighted luminosity function $[\psi \sim m^2 N(m)]$ are then used to fit the observed mass spectra (above m_*) of the Milky Way, M31, and M87 GCSs (see McLaughlin & Pudritz 1996, ApJ, 456, in press; also Harris, these proceedings).

Our main results are: (1) The ratio β, and hence the shape of the GCS mass spectrum, is expected to be independent of position within a galaxy. (2) m_1 varies among GCSs, and is roughly that mass above which a core's collisional growth time is longer than its lifetime. (3) More massive cores in a given SGMC must be shorter-lived; specifically, the data imply that core disruption times scale as $m^{-0.6}$ above m_*. (4) β is significantly larger, and the GCS mass spectrum shallower, in M87 than in the Local Group spirals. This is likely an effect more of environment than of Hubble type alone.

PROPERTIES OF E & S0 GALAXIES IN THE COMA CLUSTER

DÖRTE MEHLERT, RALF BENDER & ROBERTO SAGLIA
Universitätssternwarte München, D-81698 München
GARY WEGNER
Dartmouth College, Hanover, NH 03755, USA
INGER JØRGENSEN
The University of Texas at Austin, Austin, TX 78712, USA

As one of the richest nearby clusters, Coma is the ideal place to study the structure of galaxies as a function of environmental density, thus to constrain the theories of galaxy formation and evolution. For a magnitude limited sample of ≈ 40 E and S0 galaxies we want to obtain spectra with sufficient S/N and spatial resolution, that we can derive the rotation curves, the velocity dispersions profiles and the radial gradients of the line indices of Mg, Fe and Hβ. Following questions will be addressed:

• Are the radial velocity dispersion profiles and the rotation of galaxies in high density environments similar to those in low density environments? Data for galaxies in low density environment are available from Bender et al. (1994, *MNRAS*, **269**, 785). Are the centrally measured velocity dispersions representative for the mean kinetic energy of the galaxy?

• Can the scatter in the Fundamental Plane (FP) - which tightly correlates the radii, surface brightnesses and (central) velocity dispersions (Djorgovski & Davis, 1987, *ApJ*, **313**, 59; Dressler et al. 1987, *ApJ*, **313**, 42) - for the Coma cluster be reduced if the mean kinetic energy is used instead of the central velocity dispersion? Can we derive stronger constraint on the variations in the M/L ratio than already implied by the FP?

• The radial gradients of the line indices can be used to test the hypothesis that the metallicity gradient depends on the so-called "escape velocity" of the stars introduced by Franx & Illingworth (1990, *ApJ*, **359**, L41) . Also we can check whether the age of the stellar population varies with radius. Ages and metallicities can be estimated from the data with the use of stellar population models (Worthey 1994, *ApJS*, **95**, 105; Bruzual & Charlot 1993, *ApJ*, **405**, 538).

• How does the radial variation of stellar populations and kinematics within the galaxies vary as a function of the clusters density profile?

We already obtained spectra for 19 galaxies with total magnitudes $B_T = 11.^m - 13.7^m$ at the MDM 2.4m telescope, the McDonald 2.7m telescope and the 3.5m telescope at Calar Alto. First rotation curves and velocity dispersion profiles show the typical shapes for the different types of galaxiesx (E, S0, E/S0).

FLARE ACTIVITY IN THE NGC 1275 NUCLEUS.

N.I. MERKULOVA, L.P. METIK
Crimean Astrophysical Observatory
334413, p/o Nauchny, Crimea, Ukraine

The simultaneous UBVRI – observations of some Seyfert galaxies have been carried out at Crimean Astrophysical Observatory using the 1.25m telescope. Photometric errors are less than 0.01 mag. By analysis of the NGC 1275 light curves within one night we are discovered at least two types of variability: **1)** The flares with the maximal amplitudes of 5 – 30 % in the U-band and durations of 15 – 30 minutes. **2)** For the first time we have found **rapid red flares**. For instance, the red flare on the light curve 22-23.10.1992 lasted ∼ 65 minutes (see Figure. Vertical axes are in magnitudes). The amplitude in the filter I is ∼ 25 % (top panel). The light curve in the U-band is on the bottom panel. Fluxes in the U,B,V-bands were almost constant. One can see the flux decreasing in some filters before this flare. We conclude, that light curve of NGC 1275 nucleus could be represented by a superposition of rapid flares of different types.

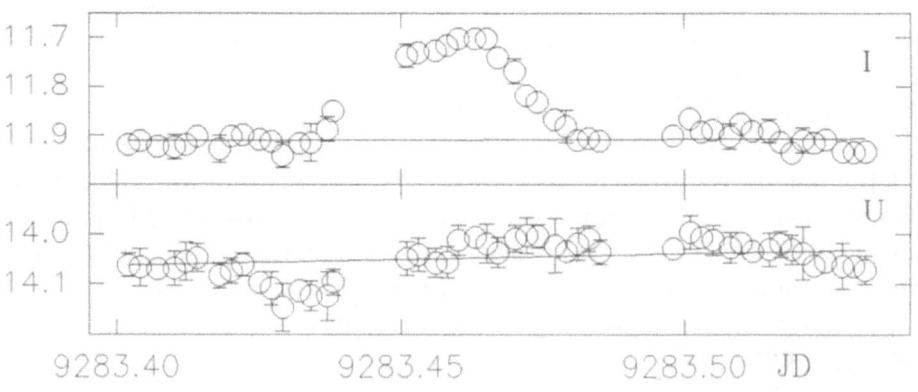

Figure: Rapid red flare on the light curve of NGC 1275

THE ANATOMY OF STARBURSTS IN THE ULTRAVIOLET

G.R. MEURER
The Johns Hopkins University
Department of Physics and Astronomy
Baltimore, MD 21218, U.S.A.

We have obtained *Hubble Space Telescope* ultraviolet (UV; $\lambda_c = 2320$ Å) images of nine starburst galaxies using the FOC, with the aim of characterizing the anatomy of starbursts with $\sim$pc resolution at a wavelength where hot massive stars dominate the luminosity output. The images have been analysed in detail, and full results can be found in Meurer et al. (1995).

Starbursts have a highly irregular UV morphology both in terms of isophote shape and surface brightness profile. This is true even if the numerous clusters they contain are removed from the images. Despite this irregularity, most (7/9) have similar effective surface brightnesses, which corresponds to a star formation rate of 0.7 $M_\odot$ Kpc^{-2} yr^{-1} in stars with mass 5 – 100 $M_\odot$, (those we detect directly). This similarity suggests that a negative feedback mechanism places an upper limit on the star formation intensity. The lower limit may be set by our UV brightness selection.

On average about 20% of the UV light comes from compact clusters. The rest comes from diffusely distributed high mass stars. Thus starbursts manufacture star clusters at a high efficiency, but clusters are not the building blocks that starbursts are made out of. The brightest clusters are preferentially found near the center of starbursts, suggesting that cluster formation may be related to the mechanism that limits their starburst intensity. The clusters have luminosities and sizes consistent with the hypothesis that they are young globular cluster. The cluster luminosity function is a power law of index ~-2. Although very different in form from that of globular clusters, it does not rule out the proto-globular cluster interpretation since the clusters in starbursts need not be coeval.

References

Meurer, G.R., Heckman, T.M., Leitherer, C., Kinney, A., Robert, C., and Garnett D.R. (1995) *Astron. J.*, accepted.

THE CENTRAL BAR IN M 94

C. MÖLLENHOFF
Landessternwarte, Königstuhl, Heidelberg, Germany

AND

M. MATTHIAS AND O.E. GERHARD
Astronomisches Institut, Universität Basel, Switzerland

Surface photometry in I, J, K of the oval disk galaxy M 94 (NGC 4736) reveal a weak central stellar bar of 0.7 kpc semi-major axis length, comprising $\approx 14\%$ of the total light within 20". By stellar kinematics the existence of a small spheroidal bulge with $v/\sigma \approx 0.8$ was discovered. The ionized gas (H_α) in this region shows global and local deviations from the stellar kinematics. Model calculations of closed orbits for the cold gas in the combined potential of bar, disk, and bulge predict large non-circular motions in equilibrium flow. However, these do not fit the observed gas kinematics; obviously hydrodynamical forces play a role in the central region of M 94.

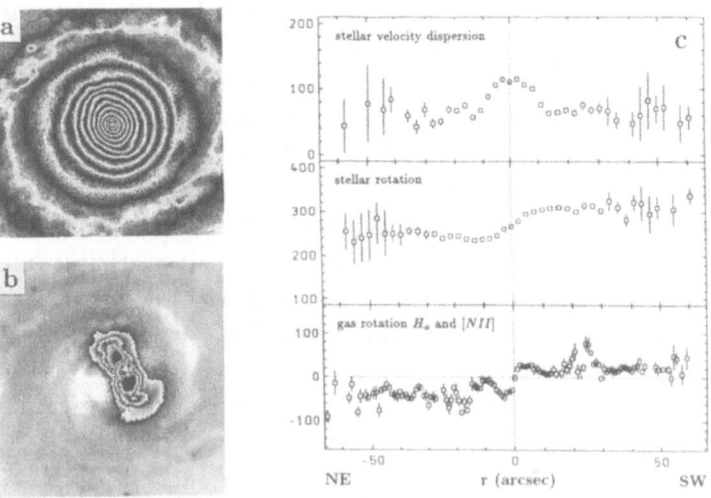

Figure 1. a) I-image of M 94, 75"×75". b) Residual bar after subtraction of an axisymmetric model for disk and bulge. c) Stellar velocity dispersion, stellar rotation curve, and perturbed gas rotation curve along $P.A. = 45°$ (in km/sec).

AN AUTOMATED SEARCH FOR LOW SURFACE BRIGHTNESS DWARF ELLIPTICAL GALAXIES

Z. MORSHIDI, R.M. SMITH AND J.I. DAVIES
Department of Physics & Astronomy, University of Wales, College of Cardiff, P.O.Box 913, Cardiff CF2 3YB, UK

A new automated technique applied to APM scans of UKSTU photographic plates has been used to search for LSBGs in the Fornax cluster. 92 galaxies, almost all of which were classified by Ferguson (1989) as dwarf ellipticals, were found. Fields to the East and West of the cluster were found to have 46 and 13 LSBGs, respectively, as shown in the figure below (□ are LSBGs, ▲ are brighter galaxies).

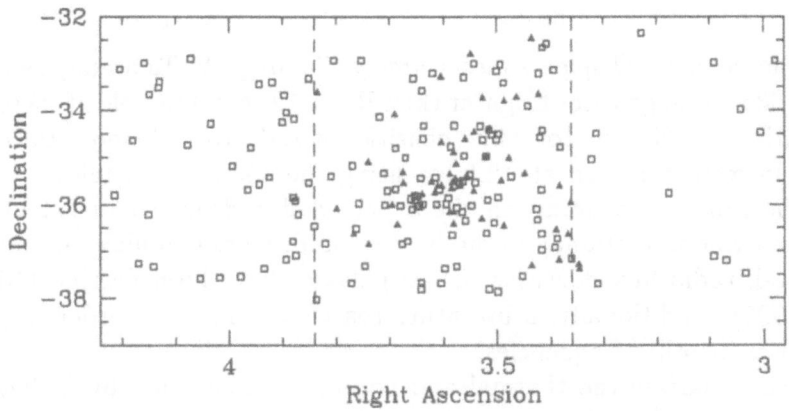

The LSBGs are clustered weakly towards the Fornax plate, in confirmation of the suggestion of Davies (1989). An estimate of the clustering size of the LSBGs indicates that they have a scale length approximately 5 times that of the brighter galaxies.

References

Davies, J.I., 1989, Ph.D. Thesis, University of Wales College of Cardiff
Ferguson, H.C., 1989, Astron. J., **98**, 367

RESULTS OF A RADIO CONTINUUM SURVEY OF SPIRAL GALAXIES AT 10.55 GHz

S. NIKLAS AND R. WIELEBINSKI
*Max-Planck-Institut für Radioastronomie, Auf dem Hügel 69,
D-53121 Bonn, Germany*

U. KLEIN
*Radioastronomisches Institut, Universität Bonn, Auf dem Hügel 71,
D-53121 Bonn, Germany*

AND

J. BRAINE
*IRAM, 300 Rue de la Piscine, F-38406 St. Martin d'Hères,
Franc*

The *Revised Shapley-Ames Catalog* (Sandage & Tamman, 1981) contains 335 spiral galaxies brighter than $B_T = 12$. A subsample of these galaxies with $\delta \geq -25°$ and total flux densities at 1.49 GHz (Condon, 1987) above 10 mJy were observed at $\lambda 2.8$ cm using the 100-m radio telescope of the MPIfR Bonn. Depending on the expected flux density and extent of the source two observational methods were used: cross-scanning and mapping. In total, radio flux densities of 192 galaxies have been derived (Niklas *et al.*, 1995). Additionally, a literature search was made in order to get flux densities at other frequencies.

One can separate the thermal and non-thermal emission by fitting to the radio spectra of a galaxy an optically thin thermal component with a spectral index of -0.1 and a synchrotron component with a spectral index α_{nth}. The fit parameters are the thermal fraction f_{th}^{1GHz} at 1 GHz and the non-thermal spectral index α_{nth}. The mean values of the derived distributions are: $\overline{f_{th}^{1GHz}} = 0.07 \pm 0.01$, $\sigma = 0.05$ and $\overline{\alpha_{nth}} = 0.85 \pm 0.02$, $\sigma = 0.12$. The thermal fraction seems to be independent of morphological type. The left diagram of Fig. 1 shows the distribution of α_{nth}. Sa and Sab galaxies and some irregular galaxies tend to have flatter spectra than Sb/Sc galaxies.

A test of the separation of thermal and non-thermal emission was made. We have calculated the flux density of the thermal radio emission at the corre-

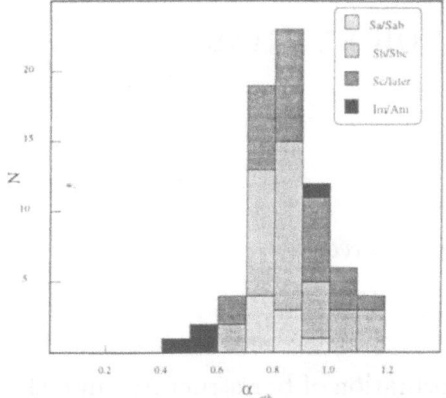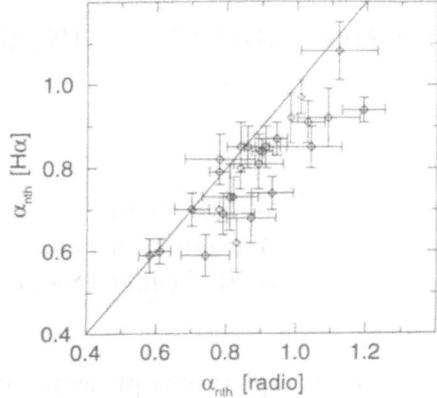

Figure 1. The left diagram shows the distribution of the derived α_{nth}. The different grey scales represents the morphological types. The right diagrams shows the plot of $\alpha_{nth}^{H\alpha}$ versus α_{nth}^{radio}. The solid line corresponds to perfect agreement.

sponding frequencies using the Hα data of Kennicutt & Kent (1983). These thermal fluxes were subtracted from the composite radio spectra. We fitted a power law to the residual spectrum in order to derive the non-themal spectral index $\alpha_{nth}^{H\alpha}$. The right diagram of Fig. 1 shows $\alpha_{nth}^{H\alpha}$ versus α_{nth}^{radio}. There exists a good correlation between the derived quantities. The shift towards flatter $\alpha_{nth}^{H\alpha}$ may be due to optical absorption in the host galaxies.

References

Sandage, A., Tammann, G.A. 1981, A Revised Shapley-Ames Catalog of Bright Galaxies (Washington, D.C.: Carnegie Institution of Washington)(RSA)

Condon, J. 1987, *ApJS*, **65**, 465

Niklas, S., Klein, U., Braine, J., Wielebinski, R. 1995, *A&AS*, **113**

Kennicutt, R.C., Kent, S.M. 1983, *ApJ*, **88**, 1094

426

BARRED GALAXIES: INTRINSIC OR EXTRINSIC?

M. NOGUCHI
Astronomical Institute, Tohoku University
Aoba, Sendai 980-77, Japan

A unified picture is presented of the formation of bar structures in disk galaxies of various morphological types. In order to discuss bar formation in the context of galactic disk formation, a simple analytic model is constructed of the growth of galactic disks by infall of primordial gas from haloes and subsequent star formation in the disks. It is monitored during the course of disk growth whether or not the condition for spontaneous bar formation (i.e., bar instability) is fulfilled for the stellar disk component.

It is found that the infall timescale is a key parameter which controls the dynamical property of the resulting stellar disk. Disks which grow fast by rapid infall experience gas-rich phases, in which massive gas clumps arising from gravitational instability in the gas disk dynamically heat the stellar disk component. When the disk has fully grown and becomes mostly stellar, it has already acquired enough random motions to suppress bar instability. On the other hand, when the gas infall from the halo region proceeds slowly, star formation (though less intense than in rapid infall cases) keeps gas mass in the disk low, leading to a dynamically cold stellar component due to lack of strong heating by massive gas clumps. Therefore the stellar disk becomes unstable and forms a bar once its mass fraction relative to the total galaxy mass reaches a critical value.

Based on this result, we propose that late-type barred galaxies, the disks of which are considered to have formed by slow accretion of the halo gas, have intrinsic origin whereas the bars in early-type galaxies, whose disks are likely to have grown quickly, have been formed in tidal interactions with another galaxies. This formation scenario, coupled with the numerical result which reveals a striking difference in structure between tidal and spontaneous bars, can explain the observed dichotomy that early-type galaxies generally have a flat bar while late-type galaxies have a bar of exponential type.

Bar dissolution and bulge formation

C.A. NORMAN

*Space Telescope Science Institute and Johns Hopkins University, Baltimore, MD 21218,
U.S.A.*

H. HASAN**

Space Telescope Science Institute, 3700 San Martin Dr., Baltimore, MD 21218, U.S.A.

and

J.A. SELLWOOD

*Dept. of Physics and Astronomy, Rutgers University , Piscataway, NJ 08855-0849,
U.S.A.*

We discuss the general classification of secular evolution in galaxies into terms of stellar dynamics. We present two-dimensional N-body simulations of a disk galaxy in which a central mass concentration is imposed after the formation of a strong bar. We show that the bar dissolves almost completely if the central mass concentration exceeds approximately 5% of the combined disk and bulge mass. This behavior can be understood in terms of previous work on single particle orbits (Hasan & Norman 1990, Hasan et al. 1993); the sustaining orbits aligned with the bar become stochastic as the Inner Lindblad resonance moves out past the minor axis of the bar. We present arguments that bar formation and subsequent thickening and dissolution will create a bulge-like stellar distribution from the central part of the disk. We discuss the predictions of such a model including the point that barred Scs with sufficient central mass concentrations should be building bulges now. We emphasize that bulges can come from a number of different mechanisms and we discuss the current evidence at both high and low redshift.

Currently we do not know from either observational or theoretical studies whether the scenario presented by the bar-central mass dissolution model actually leads to significant late time evolution along the Hubble-sequence. Observations of galaxy morphology at high redshift with the Hubble Space Telescope will help answer this question when the high redshift distribution of galaxies along the Hubble sequence is compared to the normal low redshift distribution. Catching a galaxy in the act of bulge formation by observing a suitable barred Sc with a sufficiently large central mass concentration and with a Population I bulge and disk kinematics may also be very productive. For recently heated bulge components the kinematics of any young A-type stars in the bulge should be found to have cylindrical rotation.

We plan to pursue these studies by performing 3D N-body simulations.

References

Hasan, H. and Norman, C., 1990, *ApJ* **361**, 69.

Hasan, H., Pfenniger, D. and Norman, C., 1993, *ApJ*, **409**, 91.

** PRESENT ADDRESS: ASTROPHYSICS DIVISION, NASA HEADQUARTERS, WASH-
INGTON D.C., U.S.A.

OLD STARBURSTS IN ACTIVE NUCLEI:
THE CONNECTION BETWEEN STARBURST
AND SEYFERT ACTIVITY.

L. ORIGLIA[1], A.F.M. MOORWOOD[2], E. OLIVA[3]
[1] *Torino Observatory, I–10025 Pino Torinese*
[2] *European Southern Observatory, D–85748 Garching*
[3] *Arcetri Observatory, I–50125 Firenze*

Infrared spectra of selected stellar absorption features (CO 2.29, 1.62 and Si 1.59 μm) of a sample of Seyferts and many calibrators (stellar clusters, normal galaxies and well known starbursters) are presented. For HII and Seyferts we also report Brγ measurements.

The measured equivalent widths in old stellar systems and HII galaxies are remarkably similar and do not provide reliable diagnostics for distiguishing young from old/metallic systems. However, the features are much broader (i.e. the dynamical mass is larger) in ellipticals (200-300 Km/s) than in spirals and HII galaxies (100-150 Km/s) of similar luminosity.

A more sensitive age indicator is therefore the light to mass ratio (L_H/M) one infers from the observed H ($1.65\mu m$) stellar luminosities and line velocity dispersions. Compared to optical lines (e.g. the CaII triplet) our method has the advantage of providing a direct measurement of the non–stellar contribution to the observed luminosity (which is important in Sy1's), and of minimizing the reddeding correction.

All HII galaxies and several Sy2's exhibit $L_H/M>3$ and ~ 5 times those of ellipticals and spirals, Brγ spectra indicate that the starbursts in Sy2's are older than in HII galaxies. The L_H/M of Sy1's is similar to that of spirals and typical of old disk/bulge populations.

These results are therofore compatible with an HII$\rightarrow$Sy2$\rightarrow$Sy1 evolutionary scheme where the two Seyfert types are intrinsically different objects.

THE INFLUENCE OF SPIRAL ARMS AND BAR ON THE LARGE-SCALE GALACTIC MAGNETIC FIELD EVOLUTION

K. OTMIANOWSKA-MAZUR AND S. VON LINDEN
Astron. Observatory, Jagiellonian Univ., Kraków, Poland,
Landessternwarte, Heidelberg, Germany

AND

H. LESCH
Max-Planck-Institut für Radioastronomie, Bonn, Germany

Recent observations of radio polarization from nearby galaxies show that the large-scale galactic magnetic field is aligned with spiral arms and bars and the magnetic field vectors in the interarm regions possess a spiral structure which has the same pitch angle as that in spiral arms. Our present project is going to address the following questions: **What is the structure and evolution of the large-scale galactic magnetic field under the influence of spiral and bar structure in a galactic disk? To which extent could the resulting magnetic field account for the observed spiral pattern of magnetic field in nearby galaxies?** The model is based on the particle-particle numerical scheme (SPH) involving two components: stars and molecular gas. The magnetic field is connected with the latter one. The magnetic field computations were performed first in two dimensions for 100 velocity fields: from 10^7 to 10^9 yrs. The resultant magnetic field is strongly affected by spiral arms, however at the given evolutionary stage its structure is different from the velocity field at the same time. The magnetic pitch angle distribution shows that the magnetic field "remembers" all the past velocity steps. The magnetic pitch angle distribution resulting after beam smoothing could quite well fit observations. The present model with fully 3D velocity field of interstellar gas should clear the problem if the magnetic field under the realistic velocity evolution of gas could explain the observed structure of large-scale magnetic field with constant pitch angle in the whole disk.

THE ENERGETICS OF FLAT AND ROTATING EARLY-TYPE GALAXIES AND THEIR X-RAY LUMINOSITY

S. PELLEGRINI
Dipartimento di Astronomia
via Zamboni 33, I-40126 Bologna

AND

L. CIOTTI
Osservatorio Astronomico di Bologna
via Zamboni 33, I-40126 Bologna

1. The problem and the results

A multivariate statistical analysis of data measuring the optical and X-ray properties of the *Einstein* sample of early-type galaxies (Eskridge *et al.* 1995) showed that: 1) on average S0s have lower X-ray luminosity L_X at fixed optical luminosity L_B than do Es; 2) at fixed L_B the X-ray brightest galaxies are also the roundest; this correlation holds for both morphological subsets of Es and S0s. We investigate whether a flat and partially rotationally supported galaxy is expected to host a different gas flow phase (and so a largely different amount of hot gas) with respect to a spherical pressure supported galaxy of the same L_B. This is accomplished using the global energetic balance of the hot gas flows, and axisymmetric two-component galaxy models (Ciotti and Pellegrini 1995).

It results that, for a general stellar system, the critical variations in the energy budget can be produced only by a change in the galaxy structure, not by rotation, independently of the problem of the unknown amount of thermalization of the ordered stellar motions. Reasonable flattenings at fixed L_B can make the gas less bound, even when the central stellar velocity dispersion is comparable to or higher than that of the round galaxy.

References

Eskridge, P., Fabbiano, G., and Kim, D.W., (1995), *Ap. J. Suppl. Ser.*, **442**, 523
Ciotti, L., and Pellegrini, S., 1995, submitted to *M.N.R.A.S.*

NUMBER COUNTS AND EVOLVING DWARFS

S. PHILLLIPPS[1] AND S.P. DRIVER[2]

[1] *Department of Physics, University of Bristol, Bristol, UK*

[2] *Department of Physics & Astronomy, Arizona State University*

Recent redshift surveys (eg. Colless et al. 1993) have shown that the excess galaxies seen in faint B band number counts are not evolved giants at high redshift, but low to moderate luminosity objects at modest redshifts. This led to the suggestion (eg. Cowie et al. 1991) that there was once an extra population of dwarfs which has now disappeared, ie. there is non-conservation of galaxy number. We have investigated a picture in which the dwarfs have merely faded to become very low surface brightness galaxies like those now turning up in nearby clusters (eg. Turner et al. 1993).

The key features of our model (see Phillipps & Driver 1995) are (1) a separate steeply rising ($\alpha = -1.5$) LF for the dwarfs in addition to the standard ($\alpha = -1.0$) Schechter function at brighter magnitudes ($M^*_{dwarfs} = M^*_{giants} + 3$), (2) a rate of fading of these dwarfs consistent with models of simple 1 Gyr duration starbursts, and (3) due allowance for selection effects which limit the detectability of low surface brightness objects. By varying the relative numbers of dwarfs and the rate of fading we have been able to produce a model with conservation of galaxy numbers which does indeed fit the steep number counts, the redshift distributions *and* the local LF. This model has a large population of dwarfs ($\phi^*_{dwarfs} \simeq 2\phi^*_{giants}$) which fade by 1.8^m since $z \simeq 0.5$ *and* has a wide range of surface brightnesses so that only about 60% of the dwarfs are detectable in the type of local surveys used to determine the LF.

References

Colless M.M. et al., 1993, MNRAS, 261, 19

Cowie L.L., Songaila A., Hu E.M., 1991, Nature, 354, 400

Phillipps S., Driver S.P., 1995, MNRAS, 274, 832

Turner J.I., Phillipps S., Davies J.I., Disney M.J., 1993, MNRAS, 261, 39

MULTIPLE GASEOUS DISKS IN NGC 1052

H. PLANA AND J. BOULESTEIX
Observatoire de Marseille 2 place Le Verrier F 13248

We present detailed velocity field of NGC 1052 in [NII]6583Å line obtained with scanning Perot-Fabry. Extended ionized gas was studied by Davies et al. (1986) showing a clear counter-rotation between ionized gas and stars. It is an E4 galaxy type with 'Boxy' isophotes (Nieto et al. 1991). Taking benefit from high spectral resolution and 2D mapping of Scanning PF, we have discovered several components in our profiles (Plana 1995). Their study shows two main gaseous components: a first one with a Major Axis at PA=45° and a second one at PA=30°. The ionized gas extension is approximately 40" for the first component and 20" for the second. The main component contains 100 times more gas than the second one.

The two gaseous components are counter-rotating with respect to one another with approximately the same velocity amplitude ($\sim$ 150 km/s). There is strong evidence for a triaxial shape. In order to derive the intrinsic galaxy shape and viewing angles of gas disks, we use the model described by de Zeeuw and Franx (1989) based on Stäckel potential. Using Bertola's representation (1991), we are able to determine possible values of viewing angles θ and ϕ and possible values of axial ratios $\frac{b}{a}$ and $\frac{c}{a}$ for both components. We can consider component 1 in XY plane and component 2 in ZY plane with $5° < \phi < 15°$, $\theta = 56°$ (plane XZ is forbiden by theory). But computation of axial ratios does not show consistent values for this hypothesis.

In conclusion, NGC 1052 shows 2 counter-rotating gaseous components, one is a disk lying in the perpendicular longest axis plane and the second seems to be a polar ring precessing in the perpendicular shortest axis plane.

References

Bertola F. et al. 1991, ApJ, 373,369.
Davis R. et al. 1986, ApJ, 302, 234.
Nieto J.L. et al. 1991, A&AS, 88, 559.
Plana H. et al. 1995, A&A, in press.
de Zeeuw T. et al. 1989, ApJ, 343, 617.

GALAXY EVOLUTION IN DISTANT CLUSTERS

B.M. POGGIANTI
Kapteyn Instituut
P.O. Box 800, 9700 Groningen, The Netherlands

AND

G. BARBARO
Dipartimento di Astronomia
vicolo dell'Osservatorio 5, 35122 Padova, Italy

A significant evolution has been detected in intermediate redshift clusters ($z < 0.9$), first by photometric studies ([1], [2]), which showed an excess of blue objects; subsequent spectroscopic studies revealed anomalies in most of the galaxies, mainly consisting of excessively strong Balmer lines. In order to explain the spectroscopic observations, bursts of star formation superimposed to the traditional scenario of galactic evolution are needed. The analysis of spectral lines and colours by means of an evolutionary synthesis model ([3]), including both the stellar contribution and the emission of the ionized gas, allows in most of the cases the determination of the time elapsed since the end of the burst and the fraction of galactic mass involved in it. In the clusters considered (AC103, AC114, AC118 at z = 0.31, [4]), the theoretical analysis demonstrates that the bursts affect substantial galactic mass fractions, typically 30 % or more. The observations can be equally well reproduced by either elliptical+burst models or by spiral+burst models in which the star formation is truncated at the end of the burst. The analysis of an UV colour such as (1550-V) is proposed as a valid method to distinguish between the two cases for Hδ strong red galaxies.

References

1. Butcher H.R., Oemler A.Jr, 1978, ApJ 219, 18
2. Butcher H.R., Oemler A.Jr, 1984, ApJ 285, 426
3. Poggianti B.M., 1995, PhD Thesis, University of Padova
4. Couch W.J., Sharples R.M., 1987, MNRAS 229, 423

SEARCH FOR EMISSION LINE GALAXIES TOWARDS NEARBY VOIDS

C.C. POPESCU [1], U. HOPP [2] AND H. ELSÄSSER [1]

[1] *Max Planck Institut f. Astronomie, D–69117 Heidelberg, FRG*
[2] *Universitätssternwarte München, D–81679 München, FRG*

Recent redshift surveys of galaxies (e.g. the Center for Astrophysics Survey (CfA)), revealed that bright galaxies are distributed in sheet-like structures which surround large voids. Still under debate is the question whether all galaxies follow such a distribution or if less luminous galaxies are more equally distributed. Given the limitation of the actual surveys, the observed emptiness of the voids may be a result of observational bias.

Several studies of the spatial distribution of galaxies, especially of dwarf galaxies, were carried out to overcome some of these biases. Unfortunately the results were contradictory and no definitive conclusions were drawn. We therefore started a project to search for emission-line galaxies (ELG) towards nearby voids in order to study their large scale spatial distribution. We use for this purpose the IIIa-J objective prism plates from the Hamburg QSO Survey. The plates are digitized and an automatic procedure was applied to select the candidates. All the selected objects were observed with follow-up spectroscopy, by means of the 2.2 m and 3.5 m telescopes of the Calar Alto Observatory.

We have obtained a final sample of 203 objects, of which 196 are ELG, four are galaxies with absorption lines and three are QSOs. Almost half of our objects are newly discovered ones and three quarters of the given redshifts are new. The apparent magnitudes, as derived from the objective prism plates, range between $15.0 \leq B \leq 19.5$. The sample is dominated by nearby galaxies, with a peak in the redshift distribution at 0.015.

The first results indicate that there are a few very isolated ELG. Most of these are distributed along the rim of the voids.

THE LARGE RANGE OF DARK MATTER CONTENT IN DWARF GALAXIES AND ITS IMPLICATIONS

S.A. PUSTILNIK AND V.A. LIPOVETSKY
Special Astrophysical Observatory, Russian AS

J.-M. MARTIN
Observatoire de Paris-Meudon, ARPEGES

AND

T.X. THUAN
Astronomy Department, University of Virginia

We present the analysis of a new set of radio and optical observations of a large sample of **Byurakan** Blue Compact Galaxies. HI spectra were obtained with the **Nançay** 300-m and **Green Bank** 43-m radio telescopes. CCD-images were taken with the **KPNO** 0.9-m and **Whipple Observatory** 1.2-m telescopes. Dark Matter (DM) to luminous mass ratios in these BCGs were found to vary from about less than 0.5 up to 14. Recent data taken from the literature indicate this same range. This result has important consequences on models of dwarf galaxy formation, indicating possibly different formation mechanisms. The standard CDM model of dwarfs formation requires large DM halos. However the formation of dwarfs as tidal debris resulting from strong interactions of massive spirals leads naturally to dwarfs with low content of DM. On Fig.1 we show DM to luminous mass ratio versus rotational velocity for our BCGs and some other galaxies.

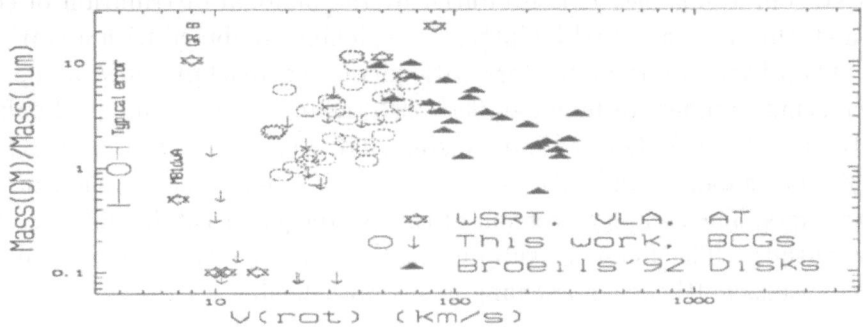

GALAXIES AND PHOTOMETRIC SIGNATURE OF STARBURST

K.D. RAKOS AND T.I. MAINDL

Institut für Astronomie der Universität Wien, Austria

AND

J.M. SCHOMBERT

NASA HQ, Washington D.C., USA

In a series of papers (see Rakos & Schombert 1995, ApJ 439, 47), we have carried out CCD photometry of galaxies in rich clusters at various redshifts. We use a modified Strömgren system, modified in the sense that the filter set is "redshifted" to the cluster of galaxies in consideration (i.e., $\lambda_{obs} = \lambda(1 + z)$, therefore no k-correction). We call it uz, vz, bz, yz to distinguish it from the original zero redshift system.

Our observations confirm a strong, rest frame Butcher-Oemler effect where the fraction of blue galaxies increases from 20% at $z = 0.4$ to 80% at $z = 0.9$. We believe that the majority of blue galaxies in clusters are triggered by interactions partly with a population of low surface brightness galaxies which fade and are then destroyed by the cluster tidal field and partly by merging processes so that spirals have been converted to SO galaxies. This view is also supported by the local distribution of blue galaxies within a cluster. The maximum number of blue galaxies is placed in the distance between 0.5 and 1 Mpc from the center.

In the diagram $bz - yz$ vs. mz the starburst galaxies are well seperated below $mz = -0.2$. We see that the mz index delivers a very good signature for starburst galaxies. It is produced by the bimodal distribution of colors in a starburst galaxy and by intrinsic reddening combined with a starburst.

The observations in clusters with increasing redshift show, along with increasing numbers of blue galaxies, also the increasing number of galaxies with $mz < -0.2$. One of our clusters, CL 317+1521 at $z = 0.583$, has 60% blue galaxies and 42% have $mz < -0.2$, photometric signature for starbursts. The complete version of this poster-paper can be obtained from the authors. The financial support of the Austrian Fonds zur Förderung der wissenschaftlichen Forschung is acknowledged.

ISOLATING RED GIANTS IN M31'S OUTER SPHEROID: THE METALLICITY GRADIENT

D. REITZEL AND P. GUHATHAKURTA
UCO/Lick Observatory, Univ. of California, Santa Cruz

AND

A. GOULD
Dept. of Astronomy, Ohio State Univ.

The aim of our project is to measure the metallicity gradient in the spheroid of the Andromeda galaxy (M31). Deep KPNO 4-m *UBRI* images of a $15' \times 15'$ field in the outer spheroid of M31 are being used to isolate a sample of red giant branch (RGB) stars. These stars are distinguished from the more numerous distant field galaxies on the basis of broadband colors and image morphology. The color technique uses *UBRI* photometry for isolating faint stars while rejecting 97% of the background galaxies (Gould et al. 1992). This yields a sample of candidate RGB stars in M31's spheroid, located about 20 kpc from its center (in projection). The shape of the M31 RGB will be compared to those of Galactic globular clusters spanning a wide range of (known) metallicities. The metallicity of the M31 outer spheroid will be derived by interpolation. Combined with the metallicity measurement at 8.6 kpc in M31's spheroid (Pritchet & van den Bergh 1988), our new measurement will permit determination of the metallicity gradient of M31. This quantity is important for understanding galactic evolution as it provides a means to distinguish between the dissipational collapse model (Larson 1974) and the accretion model (Searle & Zinn 1978).

References

Gould, A., Guhathakurta, P., Richstone, D. and Flynn, C. (1992) Evidence for Dwarf Stars at D∼100 kpc Near the Sextans Dwarf Spheroidal Galaxy, *Ap.J.*, **388**, 345-353.

Larson, R.B. (1974) Dynamical Models for the Formation and Evolution of Spherical Galaxies, *MNRAS*, **166**, 585-616.

Pritchet, C.J. and van den Bergh, S. (1988) Stellar Populations in the Inner Halo of M31, *Ap.J.*, **331**, 135-144.

Searle, L. and Zinn, R. (1978) Composition of Halo Clusters and the Formation of the Galactic Halo, *Ap.J.*, **225**, 357-379.

GLOBULAR CLUSTER SYSTEMS IN FORNAX

M. KISSLER-PATIG
ESO, Garching

S. KOHLE, M. HILKER AND T. RICHTLER
Sternwarte der Universität Bonn

AND

L. INFANTE AND H. QUINTANA
Grupo de Astrofísica, Universidad Católica de Chile

We present V and I photometry of the globular cluster systems of the early-type galaxies NGC 1374, NGC 1379, NGC 1387, NGC 1427, and NGC 1399, obtained with the 100 inch telescope of Las Campanas Observatory.

The widths of the color distributions of the four E galaxies are compatible with the errors of our photometry, pointing to a single globular cluster population. In NGC 1399 the color distribution is much broader, confirming previous results of the wide range of age/metallicity of the globular clusters. No gradient in color was seen in any galaxy, rising again the question of the reality of such a gradient in NGC 1399.
The individual distances obtained from the globular cluster luminosity function are in good agreement within our sample, as expected for the compact Fornax cluster, and lead to a turn-over magnitude of $V = 23.7 \pm 0.2$ mag, or a distance of $(m - M) = 31.1 \pm 0.2$, assuming an absolute turn-over magnitude of $V = -7.4$ (following W.E.Harris, 1991, ARAA 29, 543). This confirms results from the PNLF and SBF methods, that Fornax and the core of Virgo are at equal distance from us.

We also investigate clustering properties, colours and surface brightnesses of dwarf and LSB galaxies in Fornax. Within a few arcmin of NGC 1399 the faint galaxy density rises significantly towards the center, which makes an accretion scenario and a correlation with the extraordinary rich GCS of NGC 1399 likely.

OXYGEN ABUNDANCES IN DIFFUSE ELLIPTICALS AND THE METALLICITY-LUMINOSITY RELATION FOR DWARF GALAXIES

M. G. RICHER
DAEC, Observatoire de Paris – Section de Meudon
5 Place Jules Janssen, F-92195 Meudon Cedex, France

M. L. MCCALL
Dept. of Physics & Astronomy, York University
4700 Keele Street, North York, Ontario, Canada M3J 1P3

AND

N. ARIMOTO
Institute of Astronomy, University of Tokyo
Mitaka, Tokyo 181, Japan

Using theoretical models of the planetary nebula populations in galaxies, we investigate whether the current oxygen abundances in bright planetary nebulae can be used to predict the oxygen abundance that persisted in the interstellar medium when star formation stopped. In all galaxies, these models predict that a gap develops between the abundances observed in bright planetary nebulae and those that persisted in the interstellar medium when star formation stopped. This abundance gap depends primarily upon the oxygen abundance achieved in the interstellar medium when star formation stopped, though it also has some sensitivity to the history of star formation. The gap is always less than 0.5dex in these models. For the Milky Way, the predicted abundance gap, 0.14dex, is identical to that observed. The abundance gap magnifies the abundance-related differences between diffuse ellipticals and dwarf irregulars found by Richer & McCall (1995, ApJ, 445, 642). Diffuse ellipticals are confirmed to have larger oxygen abundances than similarly luminous dwarf irregulars, and to have larger [O/Fe] ratios than dwarf irregulars with the same oxygen abundance. The simplest explanation for both of these observations is that diffuse ellipticals formed on shorter time scales than dwarf irregulars. Given this difference in the history of star formation, diffuse ellipticals cannot be the faded remnants of dwarf irregulars.

NEAR-IR IMAGING SPECTROSCOPY OF CD GALAXY NGC1275

B. J. SAMS, R. GENZEL, A. KRABBE, N. THATTE AND H. KROKER
Max Planck Institut für extraterrestrische Physik
Giessenbachstrasse 85748 Garching, GERMANY

1. The Continuum and Molecular Line Emission

H and K band imaging spectroscopy of the central 12″ (4.2 kpc) of NGC1275 using the Max-Planck-Institut für extraterrestrische Physik imaging spectrometer "3D" maps the gas density and temperature in the core and separates the contribution of Seyfert emission to the core light.

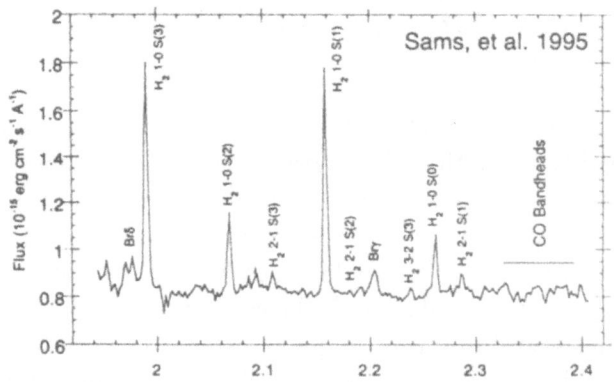

Figure 1: Integrated K-band spectrum of central 3″.25

The K band spectrum integrated over the central 3″.25 shows many H_2 lines whose excitation is 1750 K, consistent both with shock heating and excitation by a central source. The derived H_2 column density is $N(H_2) = 7.6 \times 10^{15} cm^{-2}$. The continuum slope varies from $I(\lambda) \propto \lambda^{-2.7}$ within the central 1″, to a nearly black body slope of $I(\lambda) \propto \lambda^{-3.6}$ at a radius of 3″, showing that The central source dominates the continuum emission. We fit the K-band continuum emission from the central 1″ with a combination of K5III stars which dominate the light in the NIR, emission from hot dust with emissivity $\propto \lambda^{-1}$, and Seyfert emission $\propto \lambda^{-1}$. We find a a stellar contribution of 0.075, a Seyfert contribution of 0.8, a dust contribution of 0.125 at 475 K, and a K band optical depth of 0.75. This causes dilution of the CO_{sp} index by a factor of roughly 13 in the center.

THE NEAR INFRARED FEH LINES AS INDICATORS OF SURFACE GRAVITY OF M STARS

RICARDO PIORNO SCHIAVON AND BEATRIZ BARBUY
Instituto Astronômico Geofísico - USP
CP 9638, São Paulo, 01065-970, SP, Brazil
ripisc@astro1.iagusp.usp.br, barbuy@vax.iagusp.usp.br

We compute synthetic spectra in the region around 1 μm, including the Wing-Ford band (WFB) of Iron Hydride (FeH) in the calculations. This band is known to be a good indicator of surface gravities of M stars. Employing Kurucz model atmospheres, we study the response of the intensity of the WFB to atmospheric parameters and check our results against observations of M dwarfs. This study is part of an ongoing project which aims to investigate the M dwarf-to-giant ratio in galaxies, through a population synthesis method, exploring a number of spectral indicators in the near infrared, such as the WFB, the NaI, CaII and CO near infrared features.

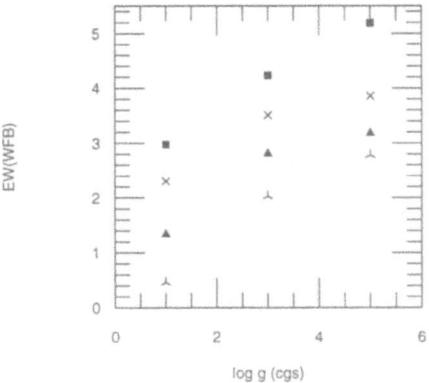 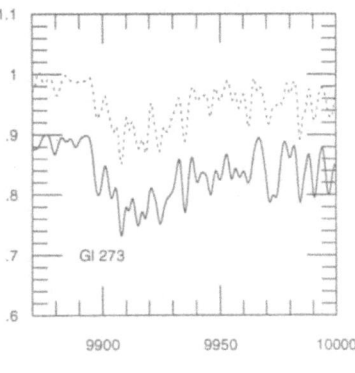

Figure 1. Left: the behaviour of the Equivalent Width of FeH lines in the spectral interval 9890-9970Å as a function of stellar surface gravity (log$g$) for effective temperatures of Teff = 3500, 3700, 3900 and 4200 K. Right: comparison between synthetic and observed spectra of an M dwarf star of (Teff, log g, [Fe/H]) = 3500, 5.1, 0.0. This work was based upon observations collected at Laboratório Nacional de Astrofísica, Brazópolis, Brazil.

EVOLUTION AND STATE OF THE LOCAL ISM

T. SCHMUTZLER[1] AND D. BREITSCHWERDT[2]
[1] *Institut für Theoretische Astrophysik,*
Im Neuenheimer Feld 561, D-69120 Heidelberg, Germany
[2] *Max-Planck-Institut für Kernphysik,*
Postfach 10 39 80, D-69029 Heidelberg, Germany

The most puzzling observations concerning the LISM (distance < 100 pc) can be explained by a fast adiabatically cooled gas in the cavity of an old superbubble. The ultrasoft X-ray background and contributions to the C- and M-bands are due to the continuum emission of delayed recombination [1]. In contrast to collisional ionization equilibrium (CIE) models, but consistent with recent observations [2], our model predicts a lack of emission lines and a low emissivity in the EUV range. In the figure below we compare the emissivities resulting from CIE at $T = 10^6$ K and those from our model at $T = 4.2 \times 10^4$ K. The basic feature of our model is a thermally self-consistent approach of the time-dependent evolution.

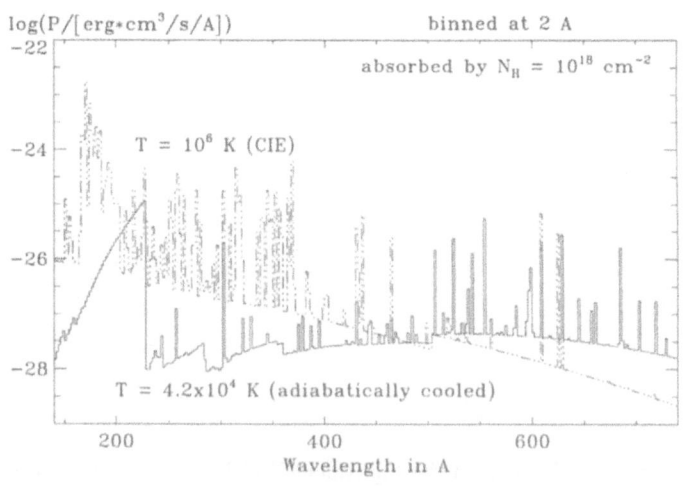

References

1. Breitschwerdt, D., Schmutzler, T., 1994, *Nature*, **371**, 774–777
2. Jelinsky, P., Vallerga, J.V., Edelstein, J., 1995, *Astrophys. J.*, **442**, 653–661

STELLAR POPULATIONS OF DWARF IRREGULAR GALAXIES

R.E. SCHULTE-LADBECK
Dept. of Physics & Astronomy, University of Pittsburgh, Pittsburgh, PA 15260, USA

AND

ULRICH HOPP
Universitätssternwarte München, D 81679 München, FRG

We recently completed two-color CCD photometry of resolved stars in 11 dwarf irregular galaxies (Hopp & Schulte-Ladbeck, 1995, A&AS 111, 527), selected because of their relative isolation from massive galaxies in the Kran-Korteweg – Tammann sample (1979, AN 300, 181). The B-R color magnitude diagrams (CMD) show that all galaxies studied had star forma-tion activity in the last $\sim 10^8$ yr. Several of them continue to form stars, the most active being UGC 5272 A (see Hopp & Schulte-Ladbeck, 1991, A&A 248, 1) while others, like DDO 210 (Hopp & Schulte-Ladbeck, 1994, ESO Conf. Workshop Proc. 49, 511), are pausing in their star formation activity. The CMDs enable us to select the brightest blue supergiants in these galaxies and to estimate their distances, D. Our values agree with the estimates based on the Virgo inflow model of Kran-Korteweg (1986, A&AS 66, 255) at the 30%-level. Prelimanary values are given in the table below.

TABLE 1.   Distances to the studied galaxies, for DDO 210 see the discussion in Hopp&Schulte-Ladbeck, 1994

UGC	D_{here}	D_{KKT}	other names	UGC	D_{here}	D_{KKT}	other names
4459	4.7	3.5	DDO 53	7559	3.9	5.0	DDO 126
5272A	6.2	9.8	DDO 64	8024	4.3	4.5	DDO 154
5272B	6.2	9.8		8091	1.4	1.0:	GR8, DDO 155
5340	7.6	9.1	DDO 68	8320	5.4	5.3	DDO 168
6456	2.1	2.2		8760	6.1	4.6	DDO 183

THE GALAXY MASS DISTRIBUTION FROM MERGERS IN A COLLAPSING SPHERICAL CLUSTER

NIR J. SHAVIV AND GIORA SHAVIV
Department of Physics, Israel Institute of Technology, Haifa, Israel, 32000

We examine the role of mergers in the formation and statistics of galaxies and clusters of galaxies. First steps in this direction were carried out by Shaviv & Shaviv (1993, ApJ.Lett., 412, L25) where the 0D case was treated. The work is extended to examine the effect of self gravity. Thus, we essentially explore the combined effect of violent relaxation and mergers on the resulting structure and mass distribution of galaxies in clusters of galaxies. We developed a numerical method to treat the spherical 2D problem (radial position and velocity) of the evolution of a collapsing cluster of gas clouds (Shaviv & Shaviv, 1995, ApJ., 448, 514).

We have succeeded to show that: 1. Unlike the 0D case, which does not depend on the initial conditions, the 1D case depends on the initial density distribution. This is so because the initial density distribution determines the rate at which various layers fall in and consequently they control the merger rate. 2. The upper part of the distribution has an asymptotic shape of the following form: $N(m) \sim (m/\overline{m})^{-\lambda} \exp(-\beta m/\overline{m})$, where both λ and β depend on the parameter α of the cross section merging, and unlike the zero dimensional case, they depend on the initial geometry. $\overline{m}$ is the mean mass. For evolved systems, the time dependence is only through $\overline{m}$. 3. The radial distribution of the mean galaxy mass is found and compared with the observations. We find that for those parameters of the galaxy-galaxy merger for which the observed Shechter luminosity function is reproduced, also the mean mass distribution agrees with the observations. 4. The total number of galaxies as a function of radial distance, or in other words, the total mass density as a function of radial distance, agrees with the observations as well.

The present approach relies on the dynamics of two body interactions and should be contrasted to the hierarchical model where many body interact simultaneously.

CHEMICAL EVOLUTION OF
STAR-FORMING VISCOUS DISKS

Y. SHIOYA AND M. CHIBA
Astronomical Institute, Tohoku University
Aoba-ku, Sendai 980-77, JAPAN

Star-forming viscous disk model was proposed by Lin & Pringle (1987) as an ubiquitous mechanism to form the exponential stellar disk of spiral galaxies. To clarify whether the viscous evolution models can explain the chemical properties of general spiral galaxies, we have studied the chemical evolution of star-forming viscous disks under the several physical conditions in relation to the possible processes of the disk formation, and compared with the observed gas and metallicity distributions of spiral galaxy disks.

First, we compared the abundance gradient predicted our models with those observed for spiral galaxies (Zaritsky et al. 1994). In most cases, the model abundance gradient is consistent with the observed gradient. Second, we studied the relation between the gas mass fraction and metallicity (f_g - Z relation). We found that there are two distinct classes in the chemical properties of spiral galaxy disks: for the galaxies classified as the class A, their f_g - Z relations are similar to the curve given by the Simple model with fixed y as well as by viscous disk models in all radii, whereas for the class B their gas mass fractions are larger than the model prediction for the corresponding metallicity at the molecular gas dominant region. This fact implies that the time scale of star formation is larger than the viscous time scale in molecular gas dominant regions. If so, it is suggested that the value of the viscosity changes with physical condition of gaseous phase: diffuse gas or dense clouds.

We conclude that the chemical evolution of the star-forming viscous disk, which holds the physical basis from the theory of disk formation, is generally consistent with the chemical properties of the Milky Way and external disk galaxies – except the apparent discrepancy for molecular-gas dominated galaxies. For further understanding, it is thus necessary to consider the role of molecule formation and/or destruction in evolution of spiral galaxies.

LOW SURFACE BRIGHTNESS GALAXIES AND THE FIELD GALAXY LUMINOSITY FUNCTION

D. SPRAYBERRY

Kapteyn Laboratorium, Postbus 800, 9700 AV Groningen, NL

Co-authors: C. D. Impey, Steward Observatory; G. D. Bothun, Dept. of Physics, Univ. of Oregon; M. J. Irwin, RGO.

We have developed a catalog of local low surface brightness galaxies (LSBGs) which is selected by objective criteria. We present here a luminosity function (LF) for LSBGs based on that catalog. This LF includes the effects of the completeness corrections to the LSBG catalog, and includes only galaxies with surface brightnesses ($22.25 \leq \mu_B(0) \leq 24.5$) fainter than those included in the CfA Redshift Survey (see Marzke et al. 1994, AJ 108, 437). The best-fitting Schechter function has parameters $\alpha = -1.42$, $M_B^* = -18.34$, and $\phi^* = 0.0036\, h^3\, \mathrm{Mpc}^{-3}\, \mathrm{mag}^{-1}$. Thus, surveys which do not take account of the observational selection bias imposed by surface brightness are missing a substantial fraction of the local galaxies, but, this missed fraction is not large enough to explain the counts of faint blue galaxies observed at moderate redshift.

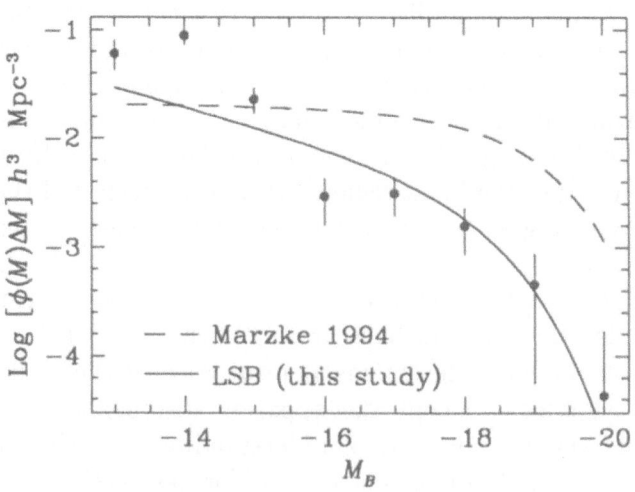

THE REDSHIFT EVOLUTION OF CIV IN GALAXY HALOS

E. A. STENGLER LARREA
Royal Greenwich Observatory
Madingley Road, Cambridge CB3 0EZ, UK

Improved and reliable profile fitting codes have made it recently possible to calculate reliable column densities of gas producing QSO absorption lines. Even using intermediate to low resolution spectra, the obtained column densities are very accurate, as has been extensively proved by Jenkins (1986). We applied this technique to the 131 CIV absorption systems in the survey by Sargent Boksenberg and Steidel (1988) and Steidel (1990). Since several lines were saturated, many CIV column densities, $N(\text{CIV})$, are only lower limits, and a survival analysis (Avni et al. 1980) is required to obtain the underlying redshift distribution of CIV column densities, which we present in Figure 1. Together with the number density distribution of strong ($\log N(\text{CIV}) \geq 14.2$) and weak systems shown in Figure 2, we have found that I. Both strong and weak systems become generally weaker below $z \sim 2$, and II. Weak systems become less numerous below $z \sim 2$, whereas the stronger systems grow steadily in number density down the lowest redshifts sampled. For details of the analysis, and on how this fits into a model for the CIV absorption systems see Stengler-Larrea et al. (1995, *in preparation*).

References Avni, Y., Soltan A., Tananbaum H., and Zamorani, G. 1980, Ap.J. 238, 800. Jenkins, E.B. 1986, Ap.J. 304, 739. Sargent, W.L.W., Boksenberg, A., and Steidel, C.C. 1988, Ap.J.S. 68, 539. Steidel, C.C. 1990, Ap.J.S. 72, 1.

Figure 1 (left) Redshift distribution of the CIV column densities obtained applying a survival analysis to the results of profile fitting 131 CIV systems.
Figure 2 (right) Redshift distribution of the number density of the CIV systems in our sample, for the weak (solid) and strong (dashed) subsamples.

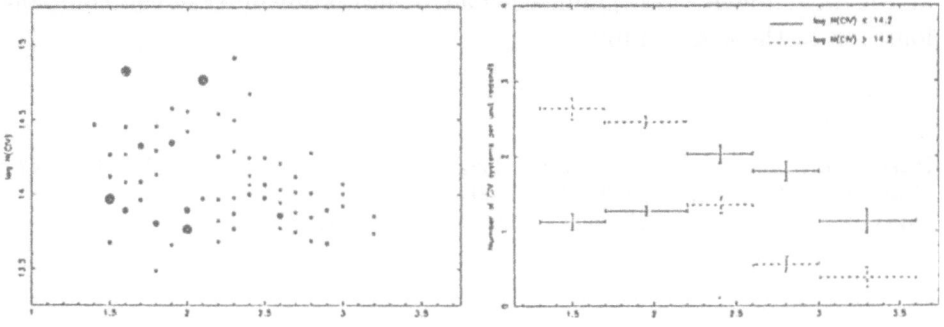

SPECTROSCOPIC STUDY OF NGC 6119 - NGC 6120 SYSTEM

C. SURACE
Observatoire de Marseille
2, place Le Verrier, 13248 Marseille cedex 4 (France)

AND

J. HECQUET AND M. AURIERE
Observatoire Midi Pyrennées
14, Avenue E. Belin 31400 Toulouse (France)

NGC6119 and NGC6120 have been listed in Kiso catalog (Takase *et al.*, 1987) as Sc and Irr galaxies. They have been selected by their clumpy features in order to investigate their nature and star formation history using slit spectroscopy. They have similar systemic velocities (respectively 9215 and 9200 km $\cdot$s^{-1}) and seem to be members of a low gravitational interacting system of 4 galaxies (with MCG-6-36-30 and NGC6122).

We have not got any photometry of NGC6119. However three emitting line regions have been spectroscopicaly resolved and present same characteristics as typical high ionized HII regions with solar metallicity.

NGC6120 have been classified as spiral galaxy. The three emission line regions spectra are characterized by low excitation Hydrogen Balmer emission lines superimposed to an old stellar component. Red color, old stellar component of central clump are consistent with Maehara observations (Maehara *et al.*, 1988), and his definition of this clump as center of NGC6120. The disturbed spiral appearance of NGC6120 and the analysis of the spectrum at 33 Åmm^{-1} seem to confirm the hypothesis of a gas rich companion falling into the disk. The ages of the clumps were derived from Mas Hesse and Kunth models (1991) indicating a star formation older in this companion than in the other clumps.

References

Maehara H., Hamabe H., Bottinelli L., Gouguenheim L.,Takase B., 1988, PASJ, 40, 47
Mass Hess J.M., Kunth D., 1991, A&AS, 88, 399
Takase B., Miyauchi-Isobe N., 1987, ATAO 2nd series, 21, 251

A BVRIJKHα SURVEY OF 355 NEARBY GALAXIES

PETER SURMA
Institute of Astronomy, Cambridge CB3 0HA, UK
surma@ast.cam.ac.uk

A multiband CCD-survey of 355 UGC galaxies in our local cosmological neighbourhood is undertaken in collaboration with S.D.M. White (MPIfAstrophysik Garching), S. McGaugh (IoA, Cambridge), M. Dennefeld (IAP Paris), H. Ferguson (STScI), M. Rieke, A. Grauer (Steward Obs. Arizona). Optical observations are obtained on Calar Alto (MPIA Heidelberg) and at La Palma (RGO) – with data 40% complete at the moment. The selection criterion is diameter $1.5' < D_{25} < 2.5'$. A database of local galaxy properties is being established (including total luminosities, mean SB, diameters, colours, colour gradients, D/B-ratios, present SFRs). Using theoretical evolution models we can predict the bona fide appearance of the galaxy population at any given redshift and thus provide a secure reference point to interpreting galaxies observed at intermediate and high z (e.g. faint galaxy counts).

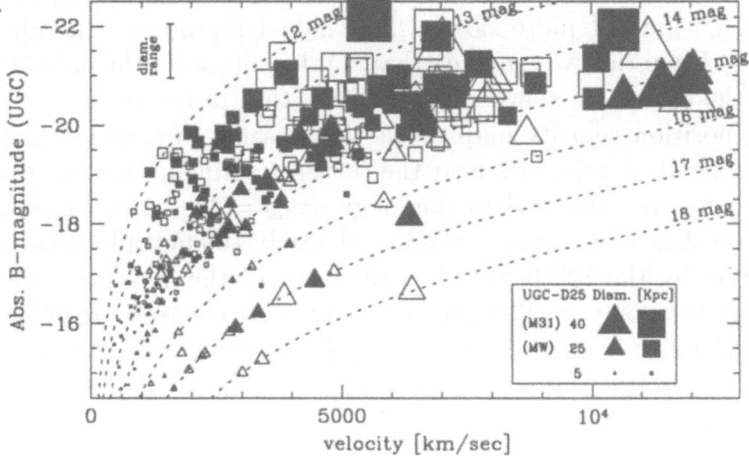

Figure 1. Absolute Magnitude versus redshift for the survey sample (filled symbols: data available, open: to be observed, symbol size shows physical diameter D_{25}), with lines of constant apparent magnitude overplotted. HSB (squares) and LSB galaxies (triangles) are surveyed in roughly equal numbers, a range of ≈ 5 mag is mapped in mean SB, D_{25} size ranges from 1 to 70 kpc, M_B from -22 down to -13 mag (all data from UGC).

THE HI HALO OF NGC 891

R.A. SWATERS, R. SANCISI, J.M. VAN DER HULST

Kapteyn Astronomical Institute, Groningen, The Netherlands

NGC 891 is a spiral galaxy very similar to our Galaxy, almost perfectly edge-on ($i \simeq 89°$) and nearby (D = 9.5 Mpc). For these reasons it has become a key system for the study of the vertical distribution of the interstellar medium in a spiral disk.

Here we discuss two results from new, very sensitive WSRT observations of NGC 891. The HI image shows faint HI emission extending up to several kpc on both sides of the plane. The data have been analysed using three-dimensional modeling to test the effects of inclination, outer flaring, warping or combinations of these. Initially we have assumed that the gas at large z heights rotates with the same velocity as the gas in the plane. But none of these models has given a satisfactory description of the observations. Subsequently, a model has been tried with a two component structure: a thin disk rotating according to the derived rotation curve, and a thick disk rotating 25 km s^{-1} more slowly. This model reproduces the observations very well. From this we conclude that: 1) there is gas in the halo of NGC 891 and 2) this gas has a slower rotation than that in the plane.

The position velocity map shows the presence of emission at high positive and negative velocities near the center. Its kinematics and distribution suggest a fast rotating disk or ring of gas ($v_{rot} = 230$ km s^{-1}) coplanar with the galaxy disk with a radius of approximately 700 pc and unresolved in the z direction by the 12″ beam. The thickness of the disk must therefore be much less than 500 pc. This seems to be the HI counterpart of the rotating nuclear CO disk already discovered by Sofue (1987). Its estimated HI mass is $1 \cdot 10^7$ M$_\odot$, with a factor of two uncertainty. This is significantly less than the amount of molecular gas ($2 \cdot 10^8$ M$_\odot$, García-Burillo et al. 1992).

References

Sofue, Y., Nakai, N., Handa, T., 1987, PASJ **39**, 45
García-Burillo, S., Guélin, M., Cernicharo, J., Dahlem, M., 1992, A&A **266**, 21

FORMATION OF ANISOTROPY IN GALAXIES

CHRISTIAN THEIS
Inst. of Astronomy and Astrophysics,
Olshausenstr. 40, 24098 Kiel, Germany

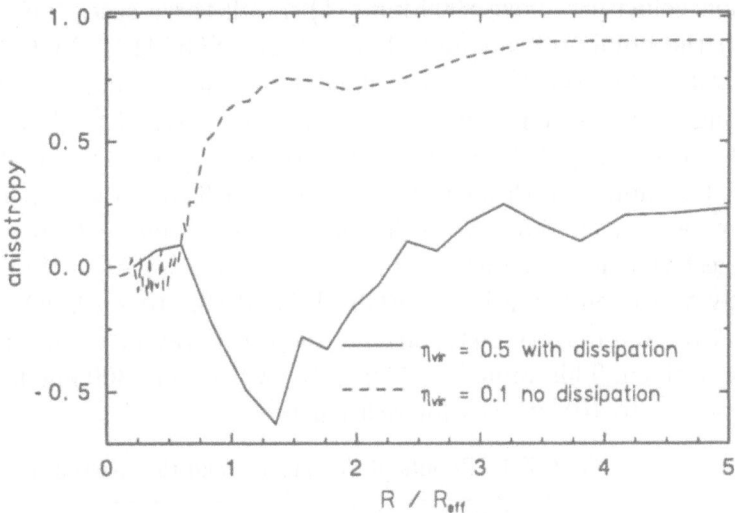

Since observations of elliptical galaxies show only a small fraction of gaseous matter, *dissipationless* N-body models are mainly used to follow the dynamical evolution of these systems. In order to end up with the observed de Vaucouleurs profile these models must be initially very cold which leads finally to a velocity distribution that is isotropic in the core and has a *positive* anisotropy $A \equiv 1 - (\sigma_\theta/\sigma_r)^2$ in the outer region (σ_θ, σ_r are the velocity dispersions in meridional and radial direction.). Contrary to these models, *dissipative* N-body simulations based on the idea of protogalaxies consisting of inelastically colliding clouds give *negative* anisotropies outside the core (see Figure). This dominance of circular orbits is caused by a combination of an orbit-dependent collision rate and a conservation of a negative anisotropy formed during the early collapse phase. Therefore, measurements of the anisotropy in galaxies can provide a tool for the study of the dissipative processes in the early galactic evolution.

MAPPING THE TOTAL GRAVITATING MASS IN THREE GROUPS OF GALAXIES

THIERING I.[1], DAHLEM M.[2]

[1] *Landessternwarte, Heidelberg, Germany*

[2] *STSCI, Baltimore, U.S.A.*

We report on the detection and analysis of hot ionized gas in three group of galaxies (MKW 4s, AWM 5, NGC 6329 group) at distances above 100 M (recession velocities above 8000 km s^{-1}). In all three cases their emission is ce tered on the dominant member of the group (NGC 4104, NGC 6269, NGC 632 respectively). The central electron densities of the gas $n_{e,0}$ are about 0.02 cm$^-$ decreasing to the outside as: $n_e = n_{e,0} [1 + (r/r_c)^2]^{-3\beta/2}$, with an exponen β=0.40 .. 0.77, (see Tab. 1). In each case, the spectra are well-represented thermal Raymond-Smith spectra with $kT \simeq 1.0$ to 1.3 keV in the center and weak increase by ca. 0.5 keV to the outside, indicating a slight cooling flow. Th fits suggest that none of the central galaxies hosts an AGN that contributes si nificantly to the emission in the ROSAT band (0.1 to 2.4 keV). Using the radi. temperature and density distributions, the total gravitating mass within a radi r can be derived. This value is 1.7±0.4 10^{13} M$_\odot$ within 300 kpc for all three case as opposed to to 10$^{14..15}$ M$_\odot$ for rich clusters.

TABLE 1. Results of the imaging and the spectral fits

group:	MKW4s	AWM5	NGC 6329 gr.
cent.surf.br. Σ_0 [$\frac{cts}{\Box'' s}$]	0.072±0.005	0.047±0.007	0.068±0.005
core radius r_c [kpc]	11.0 ± 1.1	9.4±1.4	18.7 ± 2.2
exponent β	0.47±0.002	0.40±0.002	0.77±0.09
$L_{x,tot}$ [10^{42} erg/s]	4.72	7.85	1.90
spectral fit: χ^2/ν	0.85	1.00	1.16
$N_{Hgal}^{\#}$ [10^{20} cm^{-2}]	1.81	5.32	1.24
$z_{inner}/z_\odot$	0.78±0.34	0.58±0.43	1.49±0.55
$z_{outer}/z_\odot$	0.05±0.06	(0.26±0.31)	0.06±0.02
kT_{center} [keV]	1.28±0.12	1.19±0.14	1.02±0.07
kT_{outer} [keV]	1.80±0.48	1.6±0.3	1.25±0.20
$n_{e,0}$[cm^{-3}]	0.020	0.013	0.022
cooling time t_{br} [10^{9}yr]	1.4	2.1	1.1
cooling radius r_{cool} [kpc]	51	42	59
accreting mass $\dot{M}$[M$_\odot$/yr]	6.1	6.5	5.4
M_{grav}(<r=300kpc) [M$_\odot$]	2 10^{13}	1.5 10^{13}	2 10^{13}

values from Hartmann & Burton, 1995, Cambridge Univ. press

THE X-RAY EMITTING GALAXY GROUP SHKH 360

H. TIERSCH[1,6], H. OLEAK[1], D. STOLL[1,6], A.D. SCHWOPE[2],
S. NEIZVESTNY[3], H. BÖHRINGER[4] AND L. CORDIS[5]
[1] *Universität Potsdam, An der Sternwarte 16, Germany*
[2] *Astrophysikalisches Institut Potsdam, Potsdam, Germany*
[3] *Special Astrophysical Observatory, Nizhny Arkhyz, Russia*
[4] *MPI Extraterrestrische Physik, Garching, Germany*
[5] *Hamburger Sternwarte, Hamburg, Germany*
[6] *Visiting astronomer at the DSAZ, Spain*

Shkh 360 has the characteristic signature of a strongly interacting group. Seven galaxies are embedded in a common extended halo and the isophotes indicate clear signs of alignment in B,V, and R. The parameters of the group as the redshift, z, the distance, d, the projected diameter, D, (basing on $H = 55$ km/s/Mpc), the virial radius, R_{vir}, the velocity dispersion, σ_v, the virial mass, $\mathcal{M}_{vir}$, the crossing time, τ, and the space density of galaxies, n, are given in the Table.

z	d [Mpc]	σ_v [km/s]	D [kpc]	R_{vir} [kpc]	$\mathcal{M}_{vir}$ [$10^{12}\mathcal{M}_\odot$]	$\mathcal{M}/\mathcal{L}$ [$\mathcal{M}_\odot/\mathcal{L}_\odot$]	τ [Gyr]
0.1082	590	258	250	77	5.6	10	0.122

The space density of the galaxies in Shkh 360 is $2\ 10^3$ galaxies/Mpc^3, much higher than in galaxy clusters. The interaction between the galaxies results in a hot X-ray emitting intracluster medium which was investigated from the ROSAT PSPC all sky survey. The gas distribution is roughly symmetric. The center of the X-ray emitting region is located about 15 arcsec north-east of the most luminous galaxy. The X-ray luminosity, L_x, of Shkh 360 amounts to $8.3\ 10^{43}$erg/s. The values of L_x and σ_v, found for this and other Shakhbazian groups (Tiersch et al. 1994 in: H.T. MacGillivray et al.: Astronomy from Wide-Field Imaging, Kluwer, p. 623), confirm the finding that the correlation $L_x \sim \sigma_v{}^4$, established for galaxy clusters (Quintana & Melnick 1982, Astron.J. 87, 972), is also valid for galaxy groups, representing the lower end of the scale.

NUMBER DENSITY PREDICTIONS FOR PRIMEVAL GALAXIES

E.THOMMES & K.MEISENHEIMER
Max-Planck-Institut für Astronomie,
Königstuhl 17, D-69117 Heidelberg, Germany

In the last two decades several methods have been employed to search for the ancestors of present day galaxies undergoing their first starburst at high redshift, the so-called „primeval galaxies" (PGs). According to predictions of the Ly-α flux from the PG phase current surveys for Ly-α bright PGs should have detected $10^1 - 10^3$ PGs. Yet, no really good candidate has been found. One possible explanation for this discrepancy would be the presence of dust which could strongly suppress Ly-α. In fact, models of early galaxy formation are able to produce gas to dust ratios of about 1/10 solar after some 10^8 years. This would suppress the emerging Ly-α flux below current detection limits. Thus, to detect a PG in its Ly-α bright phase, one has to hit exactly the redshift of their first star formation. On the other hand, the epoch of galaxy formation is likely to extend over much longer time. So when searching a finite volume of the universe (given by the depth Δz and the field $\Delta\Omega$), the protogalaxies are not simultaniously in the Ly-α bright PG phase and only a fraction of them exceed the detection limit. Our new attempt to predict the expected number of PGs takes this reduced propability into account. Further contraints of the Ly-α luminosity were derived from recent observations and theoretical work (for details see E.Thommes & K.Meisenheimer in *Galaxies in the Young Universe*, Eds. H.Hippelein *et al.*, Springer-Verlag, 1995, p.242). As expected, we find that the number density of Ly-α emitting PGs is drastically reduced in comparison to previous predictions. Thus previous surveys had no realistic chance to find them. Nevertheless, our calculations also show, that the detection of Ly-α bright PGs is in the scope of present day techniques. Specifically, we predict that the Calar-Alto-Deep-Imaging-Survey (see contribution of H. Hippelein et. al., these proc.) which will search for Ly-α in an area of $\approx 0.3\square°$ down to a limiting line flux of $3 \times 10^{-20} W/m^2$ should detect 10 ...100 PGs in three redshift intervals $\Delta z = 0.1$. The uncertainty accounts for our ignorance of q_0 and the unknown epoch of galaxy formation.

THE GALAXY POPULATIONS OF CLUSTERS AT HIGH REDSHIFTS: A "HUBBLE ATLAS"

S.C. TRAGER AND S.M. FABER

UCO/Lick Observatory, Santa Cruz, CA 95064 USA

A. DRESSLER

Carnegie Observatories, Pasadena, CA 91101 USA

AND

THE WF/PC-I INSTRUMENT DEFINITION TEAM

We present first results of a *Hubble Space Telescope* imaging and a Palomar and Keck Observatories spectroscopy program of distant, rich galaxy clusters in the form of a "Hubble Atlas" of morphological types at $z \geq 0.75$. Two clusters from the compilation of Gunn, Hoessel & Oke (1985) have been studied to date, Cl1322+3027 at $z \approx 0.76$ and Cl1603+4313 at $z \approx 0.90$.

(1) If the two clusters are combined, the 26 spectroscopically-confirmed cluster galaxies appear to be slightly distorted but recognizable, with the morphological content split approximately evenly between early and late types. This morphological ratio is much more similar to clusters at $z \sim 0.4$ than to nearby rich clusters like Coma (Dressler et al. 1994). (2) The red envelopes in both clusters are populated by early-type galaxies, primarily by ellipticals. (3) In both clusters, the brightest cluster members are early-type spirals (Sa/b), with luminosities approaching those of nearby cD galaxies. (4) Four "E+A" galaxies are found among the WFPC2-imaged members of the two clusters. Three are spiral galaxies (Sa/b–Sc/d), one of which is undergoing an interaction, and one is an early-type galaxy (E/S0). This morphological spread of E+A's is similar to the wide variety of types seen in clusters at $z \sim 0.3$–0.4 (Couch et al. 1994; Dressler et al. 1994; Wirth, Koo & Kron 1994).

References

Couch, W.J., Ellis, R.S., Sharples, R.M., & Smail, I. (1994) *Ap.J.*, **430**, 121
Dressler, A., Oemler, A., Butcher, H.R., & Gunn, J.E. (1994) *Ap.J.*, **430**, 107
Gunn, J.E., Hoessel, J.G., & Oke, J.B. (1985) *Ap.J.*, **306**, 30
Wirth, G.D., Koo, D.C., & Kron, R.G. (1994), *Ap.J. (Lett.)*, **435**, L105

MAGNETIC FIELDS AND THE INTERSTELLAR MEDIUM
IN SPIRAL ARMS

M. URBANIK AND M. SOIDA
Astronomical Observatory, Jagiellonian University
ul. Orla 171, Krakow, Poland

AND

R. BECK
Max-Planck Institut für Radiastronomie
Auf dem Hügel 69, Bonn, Germany

We performed the high frequency radio studies of spiral galaxies using the 100 m MPIfR radio telescope at 10.55 GHz. Two objects: NGC 4254 and NGC 3627 possess perturbed spiral structures while two others, NGC 3521 and NGC 5055 are flocculent objects, lacking organized spiral patterns. NGC 3521 possesses also a peculiar dust lane. For NGC 4254, NGC 3627 and NGC 5055 deep polarization maps were made, for NGC 3521 the total power data only were analyzed (see Urbanik et al. 1989).

We found that the uniform magnetic field component yielding a symmetric spiral pattern of observed polarization B-vectors may exist away from spiral arms, in regions of chaotic spiral structures or in the absence of "grand-design" spiral arms (details can be found in Soida et al. 1995). Axisymmetric fields generated by e.g. a dynamo action (Elstner et al. 1992) constitute a good candidate for this component. This axisymmetric field may be locally strongly modified by compressional phenomena like the formation of dust lanes or external interactions with the ambient gas. The uniform fields tend to be enhanced in dust lanes and to run along them. However, this process may be counteracted by the uniform field destruction if the dust lane is accompanied by regions of an intense star formation.

References

Elstner D., Meinel R., Beck R.: 1992, Astron. Astrophys. Suppl. **94**, 587
Soida, M., Urbanik, M. Beck, R.: 1995, Astron. Astrophys. submitted
Urbanik, M., Beck, R., Klein, U., Gräve, R.: 1989 Astrophys. Space. Sci. **156**, 195

ON THE GROWTH RATE OF THE CORRELATION FUNCTION
OF FAINT GALAXIES

DAVID VALLS–GABAUD[1] AND BOUD ROUKEMA[2]
[1] *Observatoire de Strasbourg,*
11 Rue de l'Université, 67000 Strasbourg, France.
[2]*Astronomy Centre, University of Sussex, Brighton BN1 9QH,*
UK.

We constrain the growth rate of structure, as represented by the spatial two–point galaxy auto–correlation function, at redshifts where this has not yet been measured directly by combining recent measurements of the amplitude of the angular two–point galaxy auto–correlation, at magnitudes as faint as $V_{median} \leq 25$, with new observations of the redshift distribution of very faint galaxies. We show that ξ for the overall galaxy population (at a fixed proper separation r) grows $(1+z)^{4\pm1}$ times as fast as clustering which is fixed in proper coordinates. Even extreme models where "blue" galaxies have a smaller, IRAS-like, correlation function do not reduce the growth rate below $(1 + z)^{2.5\pm1}$ times the clustering fixed in proper coordinates (Roukema, B.F. and Valls–Gabaud, D. (1995) *A&A*, submitted).

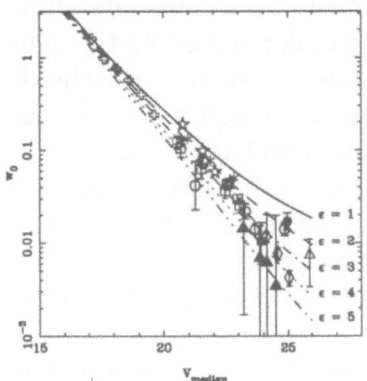

Figure 1. Amplitude of the galaxy two–point angular auto–correlation function $w_0 = w(1')$ against median V magnitude.

NUCLEAR DISKS EMBEDDED IN ELLIPTICAL GALAXIES

FRANK C. VAN DEN BOSCH AND TIM DE ZEEUW
Leiden Observatory
P.O. Box 9513, 2300 RA Leiden, The Netherlands

The Hubble Space Telescope has discovered a number of ellipticals and S0's with small nuclear, *stellar* disks (eg. van den Bosch et al. 1994, AJ, 108, 1579). The kinematics of these disks may allow a derivation of the central mass density of the host galaxy, in much the same way as is possible with ionized gas disks (e.g. M87; Harms et al. 1994, ApJ, 435, L35). In order to understand the kinematic signatures of these disks, we have constructed two-integral, axisymmetric models of ellipticals with nuclear stellar disks. We use the method developed by Hunter & Qian (1993, MNRAS, 262, 401) to calculate $f(E, L_z)$, from which we derive the velocity profiles (VPs). Depending on the choice of the odd part of the DF, one can construct a large variety of models including some with counter-rotating cores.

The main kinematic signature of a nuclear disk is a strong central decrease of the velocity dispersion (a disk is dynamically cold). The VPs clearly reveal a broad, mildly rotating component, and a narrow, rapidly rotating component. Seeing has important effects on the observables of the nuclear disks. When the seeing FWHM exceeds $2 - 3$ horizontal disk-scalelengths, the measured rotation curve becomes dominated by the light of the elliptical. The only signature of the nuclear disk that remains is a central *increase* in velocity dispersion, due to seeing convolution of the disk's rotation curve. Although such an increase could be interpreted as evidence for a nuclear black hole, the increase of σ_0 is rather small, not exceeding 10%. Counter-rotating, nuclear disks can explain the observed counter-rotation in a number of galaxies that have central disky isophotes. However, the counter-rotation is only visible when the disk light contributes significantly to the central VP's. In those cases a central *decrease* in velocity dispersion will be observed. Surprisingly, in most cases where counter-rotation is detected, one finds an additional strong, central *increase* in velocity dispersion. Although this might indicate the presence of a central black hole, further dynamical modelling is required to confirm this.

THE FUNDAMENTAL PLANE AT $Z = 0.4$

PETER VAN DOKKUM, MARIJN FRANX
Kapteyn Astronomical Institute
University of Groningen
The Netherlands

Measurements of the mass-to-light ratios of galaxies are crucial to a proper understanding of galaxy dynamics, dark matter distributions, and the general matter content of the universe. We report on results of a program to measure the change of the M/L ratio with redshift. Since the evolution of the M/L can be parameterized as

$$\ln M/L\,(z) - \ln M/L\,(0) = -0.83z(1 + q_0 + 1/z_{\text{form}}), \qquad (1)$$

measurements of the evolution of the M/L ratio give valuable information about the epoch of galaxy formation.

Studies of nearby galaxies have shown that their central M/L ratios, as computed from the structural parameters of the galaxies, are very regular. Here, we present first results on the determination of the structural parameters r_e, μ_e, and σ, for galaxies in the rich and concentrated cluster CL0024 + 1654 at $z = 0.39$.

Hubble Space Telescope imaging data were combined with 1991 spectroscopic observations made at the *Multiple Mirror Telescope*. In a 19.5 hour integration through a multi-aperture mask sufficient signal to noise ratios were obtained to measure velocity dispersions reliably. With the combination of HST imaging and high S/N spectroscopy it is possible to determine the required parameters accurately out to $z = 0.4$. These measurements will be used to determine the change in M/L with redshift. A preliminary analysis shows that the slope and offset of the Fundamental Plane are very similar to the FP at low z, indicating little evolution since $z = 0.4$.

SPECTROPHOTOMETRIC POPULATION SYNTHESIS OF EARLY TYPE GALAXIES

A. VAZDEKIS

Instituto de Astrofísica de Canarias, E-38200 La Laguna, Tenerife, Spain

R. PELETIER

Kapteyn Institute, Groningen, the Netherlands

AND

E. CASUSO AND J. BECKMAN

Instituto de Astrofísica de Canarias, E-38200 La Laguna, Tenerife, Spain

We have developed a new stellar population synthesis model for calculating colours and absorption line indices in early type galaxies. This model can work either for single-age stellar populations or in an evolutionary scheme following the chemical evolution. The model is based on the isochrones of the Padova group and we have developed our own method of conversion to colours. Details can be found in Vazdekis A., Casuso E., Peletier R. & Beckman J. (submitted, 1995). To test the model we have obtained accurate observations in many colours and line indices of the three standard galaxies: NGC 3379, NGC 4472 and NGC 4594.

In general we can find reasonably solutions for the galaxies, with $Z > Z_\odot$ (Casuso *et al.*, ApJ. 1995, In Press.) Fits are good for most colours and absorption line indices except for those of Fe and Ca. Including α-enhancement improves the fit for Fe, but worsens the fit for the NaD index. For an evolutionary scheme with a single constant IMF slope, in a closed-box approximation, the Mg_2 index from the models always falls short of the observed values (Vazdekis *et al.* (1995). Finally, we obtain much better fits if we introduce a significant change in the IMF slope, favouring massive stars in the early stages of galactic evolution, and low-mass stars for the remaining time.

B AND R SURFACE PHOTOMETRY OF FAINT GALAXIES IN THE AREA OF THREE COSMIC VOIDS

J. VENNIK
Tartu Observatory, EE2444 Tartu, Estonia

B. KOVACHEV
Astronomical Institute, 1784 Sofia, Bulgaria

U. HOPP
Universitätssternwarte München, D 81679 München, FRG

AND

B. KUHN
MPI für Astronomie, D 69117 Heidelberg, FRG

We performed B and R surface photometry of 92 faint mostly late type galaxies, selected towards three nearby voids (Hopp et al., 1995 A&AS **109**, 537). CCD frames were taken with the Calar Alto 3.5m telescope and with the MPIA/ESO 2.2m telescope. We calculated the azimuthally averaged equivalent profiles, asymtotic total magnitudes and colours, and performed ellipse-fitting of isophotes. Observed light profiles were fitted with a power law $\mu = \mu_0 + 1.086\alpha r^{1/n}$. About 40% of profiles are pure exponentials ($n = 1$); 38% of the profiles reveal a bulge/disk or disk/disk composition; 20% of them show central light depression. We classified the nearby galaxies ($z \leq 0.04$) as being related either to clusters (11 galaxies), to sheets (23) or to voids (9). The obtained parameters are compared to those of HSB and of LSB field galaxies.

The main results are as follows: – most of the studied luminous galaxies ($M_B < -19$) are typical Freeman's disks residing behind of the nearby voids; – the faint galaxies ($M_B \geq -19$) are similar to the brightest LSB dwarfs: – the profile type frequencies resemble those of LSB field galaxies; – the isolated galaxies are not necessarily faint, but they have blue colours ($B - R \simeq 0.8$) ; – one needs a larger sample of isolated galaxies in order to reveal their possible specific properties.

AN HI STUDY OF URSA MAJOR SPIRALS.

Dark Matter in spirals and the TF-relations

M.A.W. VERHEIJEN
Kapteyn Institute
Postbus 800, 9700 AV Groningen, The Netherlands

We investigate the scatter in the TF-relations and study the Dark Matter component of spiral galaxies as a function of luminosity, morphology, scale length etc. by means of a detailed kinematic and photometric study of individual galaxies in a cluster. Because all the galaxies are at the same distance there is no doubt about their relative masses and luminosities.

A galaxy is considered to be a member of the Ursa Major cluster if its position on the sky < 7.5 degrees from $\alpha = 11^h 56^m.9$ $\delta = 49^d 22'$ and $700 < V_{sys} < 1210$ kms^{-1}. So far, 79 galaxies are identified as cluster members. The currently forming cluster is rich in spirals and has a low velocity dispersion of ≈ 150 kms^{-1}. The estimated depth of the cluster contributes $\approx 0^m.17$ to the scatter in the TF-relations.

B, R, I and K' surface photometry obtained with the UH 24" and 88" telescopes on Mauna Kea is available for 78 galaxies. Analysis of the luminosity profiles is currently in progress. HI 21cm-line synthesis observations done with the WSRT provide HI line widths and rotation curves for 62 observed galaxies of which 32 galaxies are reduced and analyzed.

The cluster environment does not seem to influence the HI properties of the spirals as is the case in Virgo (Cayatte *et al*, 1994). Of the 32 galaxies analyzed so far, 19 have published H-band magnitudes (Tormen and Burstein, 1995). Using the maximum rotational velocities from the 19 rotation curves, the biweight (Beers *et al*, 1990) scatter in the TF-relation after an inverse fit is $0^m.26$ while using the amplitude of the last measured point on the rotation curve results in a biweight scatter of $0^m.40$. A quadrature subtraction of $0^m.17$ from $0^m.26$ results in a observational plus intrinsic scatter of $0^m.20$ or a 9% uncertainty in distance.

References

Beers, T.C., Flynn, K. and Gebhardt, K. (1990), *A.J.*, **100**, p. 32
Cayatte, V., Kotanyi, C., Balkowski, C. and van Gorkom, J.H. (1994), *A.J.*, **107**, p. 1003
Tormen, G. and Burstein, D. (1995), *Ap.J.Suppl.*, **96**, p. 123

THE DYNAMICAL EVOLUTION OF CENTAURUS A

S.G. VINE AND R.C. THOMSON
Institute of Astronomy
Madingley Road, Cambridge, CB3 0HA. UK.

Centaurus A and NGC 5237 — A Spiral-Elliptical Interaction

Centaurus A (NGC 5128) is a nearby Giant Elliptical (3.5Mpc, 10^{12} Solar Masses). It has a prominent dust-lane which has an axis of rotation orientated differently from that of the underlying galaxy. Close to it are two galaxies: NGC 5237 – a dwarf elliptical, and Fourcade-Figueroa – a low surface brightness irregular. We are investigating whether these three objects are the resulting components of a grazing interaction between a Giant Elliptical and a less massive spiral galaxy. We have conducted n-body simulations, modelling all components of the two original interacting galaxies. Preliminary results show that Cen A, its dust-lane, and NGC 5237 are consistent with this simulation model. Centaurus A exhibits many other features which are also indicative of a past interaction.

Thomson (1992, MNRAS, 257, 689) conducted test-particle simulations to investigate the Spiral-Elliptical interaction. Here we report much more sophisticated simulations which mimic both the elliptical and the spiral by a gravitationally self-consistent model. In our model the elliptical is ten times as massive as the spiral galaxy. The components of the spiral: bulge, disc, and dark matter halo are populated by particles according to the observed density and velocity distributions. The spiral and elliptical have initially a parabolic path, with the spiral in a prograde sense grazing the maximum radius of the elliptical.

We have used the HARP n-body integrator. This is a piece of hardware which was developed to calculate the integration algorithm of Makino and Aarseth (1992, PASJ, 44, 141). It can be thought of as a "hardware subroutine" called from simulation code. It's speed enables us to explore a wide range of initial parameters which would otherwise be a severe restriction in this approach to the problem.

HIGH-VELOCITY CLOUDS AND GALACTIC EVOLUTION

H. VAN WOERDEN, U.J. SCHWARZ, R.F. PELETIER
Kapteyn Institute, University of Groningen
Postbus 800, 9700 AV Groningen, The Netherlands

AND

B.P. WAKKER
Dept. of Astronomy, Univ. of Wisconsin, Madison WI, USA

High-velocity clouds (HVCs) are HI clouds deviating strongly from the general Galactic rotation. Their distances, and hence origins, have long remained unknown. However, three major complexes are now known to lie at $z > 1$ kpc, while two smaller clouds are closer to the plane (Wakker et al. 1995). These incomplete results still allow a variety of origins.

Ca^+ and other metal ions have been found in at least seven HVCs (see reviews by Wakker et al. 1995; Schwarz et al. 1995). Abundances vary from cloud to cloud and from ion to ion, with values relative to solar ranging from 0.002 to about 1. These variations may be due to different origins, ionization conditions, and depletion of metals onto dust. Clearly, these HVCs are not primordial; they may consist of enriched intergalactic gas, or depleted Disk gas circulated into the Halo, or a mixture of both.

We have detected CaII absorption by the largest HVC,Complex C, in the quasars PG1351+640 and Mark 290 (Wakker et al. 1996), giving Ca^+/HI ratios of 0.007 and 0.012 times the solar total calcium abundance, similar to those found in the Magellanic Stream and in Complex A.

The detection of Ca^+ in Complex C opens the road to a determination of its distance and nature. Further work will clarify the origin of HVCs, and their role in the evolution of the Galaxy and its interstellar medium.

References

Schwarz, U.J., Wakker, B.P. van Woerden, H. 1995, A&A 302, 364
Wakker, B.P., van Woerden, H., Schwarz, U.J., Peletier, R.F., Douglas, N.G., Danly, L., de Boer, K.S. 1995, IAU Symp. 169, in press.
Wakker, B.P., van Woerden, H., Schwarz, U.J., Peletier, R.F., Douglas, N.G. 1996, A&A (Letters), submitted.

THE K-BAND LUMINOSITY FUNCTION

S.J. WAGNER, M.W. KÜMMEL

Landessternwarte, Königsstuhl, 69117 Heidelberg, Germany

Investigations of broad band energy distributions of specific classes of sources requires homogeneous samples of a sufficiently large number of objects. Deep and homogeneous surveys in those energy ranges which are accessible to satellites only are rare. One such a field in the north ecliptic pole (NEP). The ROSAT and IRAS whole sky surveys and deep additional observations by other satellites make the NEP the region of the deepest mid-IR and X-ray observations. We performed complete surveys in three optical/IR colors at 460 nm, 700 nm, and 2.1 μm (B, R, and K') within a one square degree field around the NEP. Limiting magnitudes in the three bands are 23, 24, and 19, respectively. The optical bands are observed with sufficient spatial sampling to classify extended and point sources. Down to levels which still correspond to high completeness limits we detect 80.000, 240.000 and 25.000 sources in B, R, and K', respectively.

From these surveys we derive the luminosity function and color luminosity relation in the near-IR regime from a sample of sources which is much larger than those used previously. In addition we also study the spatial correlation function of near-IR bright sources. Image classification is carried out with the optical frames down to 22nd magnitude.

We compare our K' band LF to that of Gardner et al. (ApJ 415, L9, 1993), who derived a LF over 12 magnitudes from several surveys of different depths and field sizes, which were, however, taken at different galactic latitudes. They clearly demonstrate that the LF displays a break at K'=16.8 where the average B-K colors are largest. From our survey we find that the break of the LF and its slope at the bright side depends crucially on the separation of point sources. In the brighter part the point sources are exclusively stars. The fraction of point sources increases for redder objects (B-K>5). This indicates, that the shape of the K' LF should not be derived from ensembles of data taken at different galactic latitude unless a clear separation of stars and galaxies is possible.

DYNAMICS OF MASSIVE BLACK HOLES IN THE CENTER OF INTERACTING GALAXIES

N. WASSMER, R.BIEN, R. WIELEN
Astronomisches Rechen-Institut Heidelberg

Our numerical tool is SUPERBOX (Bien et al., 1991) which can treat an arbitrary number of galaxies. Simulations including Black Holes are important to understand recent observations, see e. g. Mediavilla and Arribas (1993).

In our opinion, Black Holes should represent regions of highest activity before and after an interaction or merging. We expect that the position of the Black Hole does not neccessarily coincide with regions of highest stellar density after an interaction. Possibly this allows us to estimate the efficiency and duration of interactions.

¿From our whole material we propose the following interpretation. A large impact parameter causes a small offset, i. e. the distance between a Black Hole and the central density. After about 20 Myr the Black Hole falls back to the center of the parent galaxy. When the impact parameter is small the offset is 200 – 500 pc and the time-scale is 30 – 40 Myr. In other words: when the interacting partners are well seperated, the offset has already disappeared. We expect a long-lasting offset when the two galaxies are merging. The scattering of a Black Hole is mainly caused by the the core of the disturbing partner. A massive Black Hole falls into the center quite rapidly. If the collision is head-on both Black Holes can be expelled.

References

Bien, R., Fuchs, B., Wielen, R.: 1991,Proc. CP90 Europhys. Conf. Phys.
Mediavilla, R., Arribas, S.: 1993, Nature, vol. 365

INTRODUCTION TO THE JCMT-CSO INTERFEROMETER

M.C. WIEDNER AND R.E. HILLS
Mullard Radio Astronomy Observatory (MRAO)
Madingley Road, Cambridge, CB3 0HE, Great Britain

AND

J.E. CARLSTROM AND O.P. LAY
Devision of Physics, Mathematics and Astronomy, Caltech,
Pasadena, CA 91125, USA

The James Clerk Maxwell Telescope (JCMT) and the Caltech Sub-millimeter Observatory (CSO) are located 160m apart on Mauna Kea, Hawaii. Recently they have been linked together to allow the first submillimetre interferometric observations (between 300GHz and 490GHz, soon to 690GHz). Submillimetre observations can be used to probe the physical conditions of molecular clouds and dust in starforming regions. Lines from rotational transitions of molecules such as CO, HCN and HCO^+ trace densities, temperatures and masses of molecular gas. The thermal emission of cold dust, typically found in dense clouds, also peaks in the submm regime.

The interferometer has a resolution of 0.5 arcsec at 345GHz with which we would be able to observe the central 100pc of a galaxy at z=0.01. The sensitivity of the instrument is 16mJy at a bandwidth of 1GHz for continuum observations and 50mJy for a linewidth of 100km/s for line observations when integrating over 1000 seconds at 345GHz (Carlstrom et al., 1994). Continuum and line emission of starburst galaxies are greater at submillimetre than at millimetre wavelengths: continuum emission of optically thin dust is proportional to $\sim \nu^3$, the intensity of line emission is proportional to ν^2 to ν^4. The combination of the resolution and the sensitivity obtainable with the JCMT-CSO Interferometer is clearly very valuable for investigating the nature of starburst galaxies and of their energy sources.

References

Carlstrom, J.E., Hills, R.E., Lay, O.P., Force, B., Hall, C.G., Phillips, T.G. and Schinckel, A.E. 1994, in *Astronomy with Millimeter and Submillimeter Wave interferometry*, eds. Ishiguro, M. and Welch, W.J., ASP Conference Series, p.135

MOLECULAR HYDROGEN AT $Z = 2.8108$

G.M. WILLIGER[1,2], K.M. LANZETTA[3]. R.F. CARSWELL[4] AND
J.A. BALDWIN[1]
[1] *Cerro Tololo Inter-American Observatory. La Serena, Chile*
[2] *MPI für Astronomie, Heidelberg, Germany*
[3] *State University of New York, Stony Brook. New York, USA*
[4] *Institute of Astronomy, Cambridge, England*

The Lyman and Werner bands of H_2 in interstellar gas provide information about gas temperature, density and the ultraviolet radiation field, and possibly about dust content. This is especially useful for high redshift QSO absorption systems, where usually the only data available arise from absorption lines. We present $25 - 50$ km s^{-1} resolution data taken with the CTIO 4m telescope plus echelle spectrograph of the Lyα forest region of 0528−250, which has a damped Lyα absorption system at $z = 2.81$. Using a χ^2 profile fitting routine (Lanzetta & Bowen 1992, ApJ, 391, 48), we find an H_2 fraction of $\sim 10^{-2}$, an order of magnitude below that of Galactic diffuse interstellar clouds. This may be caused by some combination of a less efficient H_2 formation rate or an increased H_2 dissociation rate. Using the relative populations of the $J'' = 0, 1$ rotational levels, we derive a kinetic temperature of $T_K = 136 \pm 16$ K. The total velocity spread as traced by sensitive metal transitions is 250 km s^{-1}, consistent with a highly inclined, rotating ensemble of clouds associated with a luminous spiral galaxy. A representative section of the spectrum is shown below, binned at roughly the Nyquist rate with the χ^2 fit to H_2 of the $z = 2.8108$ absorption system.

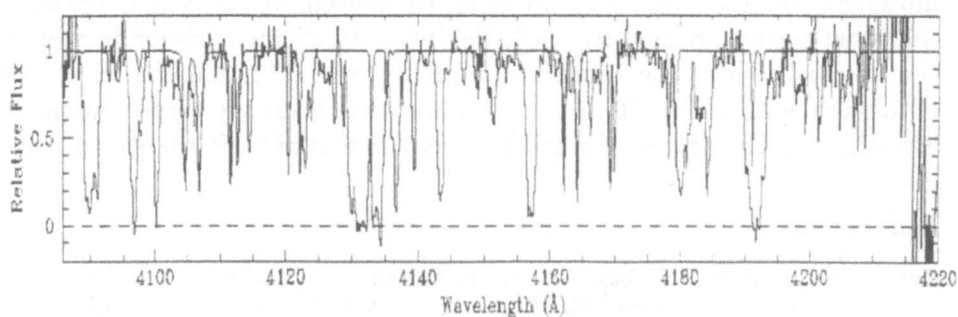

MORPHOLOGY OF E+A GALAXIES IN CL0016+16 (Z=0.54)

GREGORY WIRTH

Lick Observatory, Santa Cruz, CA 95064

AND

PAOLA BELLONI

Universitätssternwarte, Scheinerstraße 1, 81698 München

We present new results on the morphology of member galaxies in the distant cluster Cl0016+16 from HST images (WFC1). Based on narrow multiband ground-based photometry and spectra obtained with the Keck 10 m telescope we identify 7 new cluster members which appear to have strong Balmer absorption features but no detectable emission lines, doubling the number of such galaxies previously observed with HST in this cluster. These candidate E+A galaxies have been identified in other distant clusters, but the morphology of this population has appeared bulge-like in AC114 (Couch et al., 1994, ApJ 430, 107) and disk-like or irregular and interacting in Cl0939+47 and Cl0016+16 (Wirth et al., 1994 ApJ 435, L105). By means of the image concentration index as a quantitative measure of morphology we show that our enlarged sample of E+A objects in Cl0016+16 now contains some galaxies resembling bulge systems as well as the previously-identified disk-like objects. The observed heterogeneity suggests that both galaxy mergers (rapidly resulting in an $r^{1/4}$ profile) and ram-pressure stripping of isolated late-type systems may originate E+A objects.

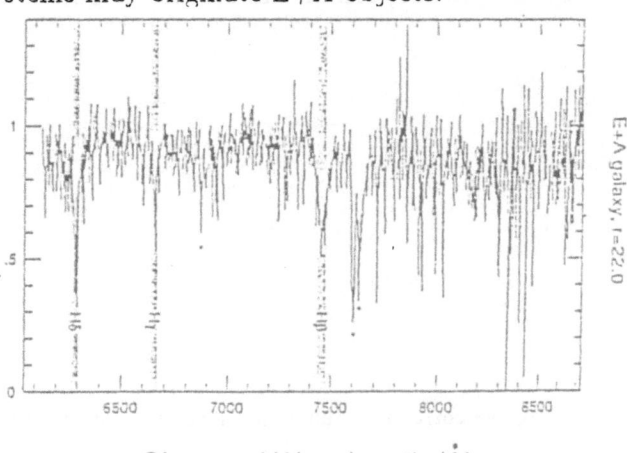

Observed Wavelength (Å)

Figure 1: Keck spectrum of a moderately-bright new identified E+A galaxy showing the strong Balmer lines which define this class of objects.

MEASURING THE ANGULAR CORRELATION FUNCTION FOR FAINT GALAXIES AT HIGH GALACTIC LATITUDES

D. WOODS, G.G. FAHLMAN AND H.B. RICHER

Dept. of Geophysics and Astronomy, U.B.C.
2219 Main Mall, Vancouver, B.C., Canada.

The angular correlation function ($\omega(\theta)$) for faint, magnitude-limited samples is an important test of large scale structure formation scenarios, as well as being a valuable probe of galaxy evolution. By measuring $\omega(\theta)$ one hopes to better understand the mechanism or galaxy species responsible for the number counts excess (relative to "no-evolution" models of galaxies) typically observed at blue wavelengths. Images of three 'blank' fields were obtained at the prime focus of the CFHT with sub-arcsecond seeing in V, R and I, to magnitude limits of 25, 24.5 and 24, respectively. The angular correlation functions calculated for one field, NF1, in V is shown in Figure 1. Clearly the amplitude of $\omega(\theta)$ is decreasing at fainter magnitude limits. Note the number of objects detected are not sufficient to accurately measure $\omega(\theta)$ for an individual field, in a given colour, to significantly small angular separations. In order to do this we must combine the data from our three fields and perform a multi-field fit. A full summary of this analysis will be presented in Woods *et al.* (1995, in prep.) including determinations of $\omega(\theta)$ using galaxy samples culled from all our fields and selected by magnitude, colour and surface brightness.

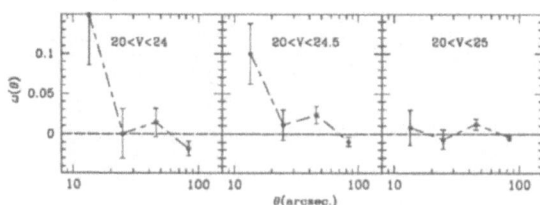

Figure 1. Angular correlation functions calculated for listed magnitude limits for NF1.

STAR FORMATION HISTORIES OF SPIRAL GALAXIES FROM A MULTI-WAVELENGTH STUDY

C. XU

Max-Planck-Institut für Kernphysik
Postfach 103980, D-69117 Heidelberg, Germany

We study the star formation histories of a sample of 113 nearby spiral galaxies using their radio continuum (20cm), FIR (40–$120\mu m$), H ($1.65\mu m$) and B (4400Å) luminosities. The first two are used as indicators of star formation rate over the past $\sim 10^8$ years, as suggested by the tight and nearly universal FIR/radio correlation (Xu et al. 1994). Compared to other indicators of recent star formation rate such as $H\alpha$ and UV, FIR and radio continuum have the advantage of being insensitive to extinction. The B luminosity is taken as star formation indicator for the time scale of $3\ 10^9$ years, and the H luminosity for the time scale of 10^{10} years. We find:

1. The long-term star-formation history (from a few billion years to the entire Hubble time), as indicated by the B-to-H luminosity ratio, depends strongly on the Hubble type.
2. The recent star-formation history in the last a few billion years, as indicated by the radio-to-B luminosity ratio and the FIR-to-B luminosity ratio, does not depend on the Hubble type.
3. Galaxies of a given Hubble type have similar long-term star-formation histories. On the other hand, their recent star-formation histories in the last a few billion years can be much different: the ratio between the star-formation rate averaged over the past 10^8 years to that over the past $3\ 10^9$ years can be different by two orders of magnitude, as indicated by the scattering of the radio-to-B luminosity ratio and that of the FIR-to-B luminosity ratio. This is not likely to be due to the extinction on the B luminosity, because the scattering of the B-to-H luminosity ratio for a given type is much smaller.

References

Xu, C., Lisenfeld, U., Völk, H.J., Wunderlich, E. 1994, A&A 282, 19

OBSERVABLE PROPERTIES OF PASSIVELY-EVOLVING GALAXIES AT HIGH REDSHIFT

TORU YAMADA
Cosmic Radiation Laboratory
The Institute of Physical and Chemical Research

AND

NOBUO ARIMOTO
Institute of Astronomy
University of Tokyo

1. Basic Idea

There are evidenves which suggest that many of the early-type galaxies in rich cluster environment formed at fairly high redshift, z > 2. If the galaxies formed at such early epoch have experienced no intensive star-formation events, their photometric properties can be traced with less ambiguity by using the stellar evolutionary sinthesis models. Here we demonstrate what conspicuous feature can be observed for those passively-evolving galaxies at high redshift and how we can constrain the epoch and period of their formation.

2. Conspicuous Turn-Off in the Optical-NIR Colours

If the model galaxies are observed at higher redshift, their opt-NIR colours become redder by the shift of the 4000 Å break, while they becomes bluer by the "revival" of more massive and bluer main-sequence stars. As a result, conspicuous colour-turn-off points, which depend on the galaxy-formation epoch very strongly, are seen in their colour evolution. We also simulated the expected $I - K$ distributions of K-band magnitude-limited sample in order to make more realistic predictions. Unfortunately, present-day NIR detector is so small that we can obtain any statistical data set to compere with models. However, next-generation wide-field NIR cameras will bring us the opportunities to search and investigate properties and distributions of those passively-evolving galaxies at high-redshift.

THE AGE OF ELLIPTICAL GALAXIES IN CLUSTERS

BODO ZIEGLER & RALF BENDER
Universitätssternwarte, Scheinerstraße 1, D-81698 München

Nearby cluster ellipticals follow a very tight relation between velocity dispersion σ and Mg absorption (e.g. Bender *et al.* 1993, *ApJ* **411**, 153). The small scatter in Mg implies that the age and metallicity spread at a given σ in ellipticals is smaller than 15% (applying Worthey's population synthesis models 1994, *ApJS* **95**, 107). This means that ellipticals cannot have formed continuously over the Hubble time and ongoing merger processes represent only a tail of latecomers.

More reliable information about the major star formation epoch in ellipticals can be obtained from the redshift evolution of the Mg$-\sigma$ relation. We have observed several clusters at $z \sim 0.37$ with the Calar Alto and La Silla 3.5m telescopes and measured the weakening of Mg in member ellipticals (see the figure). Translating the observed Mg difference between today and $z \sim 0.37$ with Worthey's models into a relative change in age indicates that the bulk of stars in these cluster ellipticals formed at redshifts above 3. (Bender & Ziegler 1995, MPA Report, Garching).

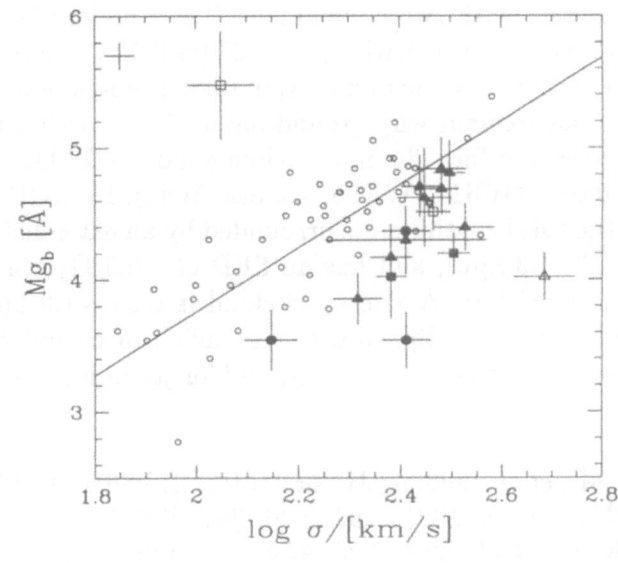

DEEP HST IMAGING OF A GALAXY CLUSTER AT Z=2.40

ROGIER A. WINDHORST AND SAM M. PASCARELLE
Arizona State Univ., Dept. of Physics, Tempe, AZ 85287-1504

AND

WILLIAM C. KEEL
Univ. of Alabama, Dept. of Physics, Tuscaloosa, AL 35487

We present a 67-orbit HST/WFPC2 exposure on the weak radio galaxy 53W002 at z=2.390 and its surrounding cluster. Color Plate 1 shows 12 orbits in I_{F814W} & V_{F606W}, and 24 in B_{F450W}. Potential cluster members were identified through 15 orbits in $F410M$, optimized for narrow-band searches for compact $Ly\alpha$ objects at z≃2.4 (P96), and confirmed through spectroscopy (W91, P96); 16 candidates were found with significant narrow-band emission in $F410M$: 4 out of 5 had a confirming MMT spectroscopic redshift at z≃2.40 (P96). All are located within 60" from 53W002, or $\sim 0.24h_{100}^{-1}$ Mpc (q_o=0.5) at $z \simeq 2.4$, the physical scale of a group or small galaxy cluster. One object contains a weak (variable) AGN, another is a merger with two companions. Their underlying young stellar population is very compact, with $r_{h.l.} \simeq 0.2$" ($\simeq 0.8h_{100}^{-1}$ kpc), and considerably fainter than the L^*-value at z~2.4, implying sub-galactic sized objects. These results may explain why ground-based $Ly\alpha$ searches for PG's have been largely unsuccessful. The narrow-line galaxy 53W002 was imaged in the PC at ~0.07" FWHM (WK95, see also W94). Its AGN component is $\leq 20\pm4\%$ of the total continuum, surrounded by an extended $r^{1/4}$-envelope with $r_{h.l.} \simeq 1.1$" (4.3 kpc), and has an SED of ~0.3 Gyr in the center to ~0.5–1.0 Gyr at ~4 kpc. A *one-sided* cloud is seen ~1.8 kpc West, ~0.3 mag bluer than the SED, aligned with the radio source and its Ly-α cloud, presumably weak scattered AGN light, and/or jet-induced star-formation.

References
Pascarelle, S. M., *et al.* 1996, ApJL, **456**, (Jan. 1; P96); Also: Nature, subm.
Windhorst, R. A., *et al.* 1991, ApJ, **380**, 362 (W91)
Windhorst, R. A., *et al.* 1994, ApJ, **435**, 577 (W94); Also: ApJL, **400**, L1
Windhorst, R. A., Keel, W. C., 1995, ApJL, submitted (WK95)

Plate 1. True color image of our 48-orbit Cycle 4-5 HST/WFPC2 exposure in *both* B_{F450W} (16 hours; blue gun), V_{F606W} (5.7 hours; green), and I_{F814W} (5.7 hours; red). *V* was rotated by -6.721° to match the *B*- and *I*-exposures, resulting in slanted borders. North is 39.7° counterclockwise from vertical. This WFPC2 image is 64 x 64", has 0.07" resolution (FWHM) and a 3-σ point source sensitivity of $R \cong 29.3$ mag. Objects labelled A, B, C, etc., are canditate cluster galaxies at z≅2.40 with significant excess in the WFPC2 *Lyα* filter (Pascarelle *et al.* 1996). Object 53W002 (not visible here) plus object A, B, and E are spectroscopically confirmed at z≅2.40.

SUBJECT INDEX